Deterministic Scheduling Theory

R. Gary Parker
School of Industrial and Systems Engineering
The Georgia Institute of Technology

CHAPMAN & HALL
London • Glasgow • Weinheim • New York • Tokyo • Melbourne • Madras

Published by
Chapman & Hall, 2–6 Boundary Row, London SE1 8HN, UK

Chapman & Hall, 2–6 Boundary Row, London SE1 8HN, UK

Blackie Academic & Professional, Wester Cleddens Road, Bishopbriggs, Glasgow, UK

Chapman & Hall GmbH, Pappelallee 3, 69469 Weinheim, Germany

Chapman & Hall USA, 115 Fifth Avenue, New York, NY 10003, USA

Chapman & Hall Japan, ITP-Japan, Kyowa Building, 3F, 2-2-1 Hirakawacho, Chiyoda-ku, Tokyo 102, Japan

Chapman & Hall Australia, 102 Dodds Street, South Melbourne, Victoria 3205, Australia

Chapman & Hall India, R. Seshadri, 32 Second Main Road, CIT East, Madras 600 035, India

First edition 1995

ISBN (HB) 0-412-99681-2

A catalogue record for this book is available from the British Library

Library of Congress Cataloging-in-Publication Data

Parker, R. Gary.
Deterministic scheduling theory / by R. Gary Parker.
p. cm.
Includes bibliographical reference and index.
ISBN 0-412-99681-2
1. Scheduling (Management) I. Title.
TS157.5.P37 1995
658.5'3—dc20 95-15181
CIP

∞*

Printed on permanent acid-free text paper, manufactured in accordance with ANSI/NISO Z39.48-1992 and ANSI/NISO Z39.48-1984 (Permanence of Paper).
[Paper = Magnum, 70 gsm]*

Printed on acid-free text paper, manufactured in accordance with ANSI/NISO Z39.48-1992 (Permanence of Paper). [Paper = Fineblade, 100 gsm]*

I dedicate this book to my parents.

Contents

List of Figures xi

List of Tables xvii

Preface xix

1 INTRODUCTION **1**
- 1.1 Some Scheduling Problems 2
  - 1.1.1 Bicycle Assembly 2
  - 1.1.2 Classroom Assignment 5
  - 1.1.3 Scheduling Athletic Events 6
  - 1.1.4 Soft Drink Bottling 8
- 1.2 Classification of Scheduling Problems 9
  - 1.2.1 Machine Environment 11
  - 1.2.2 Job Characteristics 11
  - 1.2.3 Optimality Criteria 12
- 1.3 Outline of the Book 13

2 MATHEMATICAL PRELIMINARIES **15**
- 2.1 Computational Complexity 15
  - 2.1.1 Problems 17
  - 2.1.2 Computational Orders 17
  - 2.1.3 Problem Size and Encoding 19
  - 2.1.4 Problem Reductions 20
  - 2.1.5 Fundamental Problem Classes 22
  - 2.1.6 What to Do with $\mathcal{NP}$-Hard ($\mathcal{NP}$-Complete) Problems 26
- 2.2 Graph Theory 28
  - 2.2.1 Fundamental Definitions 28
  - 2.2.2 Trees, Forests, Arborescences, and Branchings 31
  - 2.2.3 Covering, Matching, and Independence 31
  - 2.2.4 Vertex Colorings, Edge Colorings, and Perfect Graphs 34
- 2.3 Partial Enumeration/Branch-and-Bound 36

3 SINGLE-PROCESSOR PROBLEMS 43

3.1 Independent Jobs 43

3.1.1 Completion Time Models 43

3.1.2 Lateness Models 45

3.1.3 Tardiness Models 46

3.2 Dependent Jobs 58

3.2.1 Special Precedence Structures 58

3.2.2 Total Completion Time Models 68

3.2.3 Due-Date Problems 71

3.2.4 Sequence-Dependent Setup Problems 73

3.3 Exercises 75

4 PARALLEL-PROCESSOR PROBLEMS 83

4.1 Independent Jobs 83

4.1.1 Makespan Case 83

4.1.2 Total Completion Time Problems 99

4.2 Dependent Jobs 102

4.2.1 General Precedence 102

4.2.2 Special Precedence Structures 107

4.2.3 The Two-Processor Case 109

4.2.4 Approximations 119

4.3 Exercises 129

5 FLOW SHOPS, JOB SHOPS, AND OPEN SHOPS 135

5.1 Flow Shops 135

5.1.1 Permutation Schedules 136

5.1.2 Solvable Flow Shop Cases 141

5.1.3 General Flow Shops 149

5.2 Job Shops 153

5.2.1 Intra-Job Precedence 154

5.2.2 Inter-Job Precedence 157

5.3 Open Shops 165

5.4 Exercises 171

6 NONSTANDARD SCHEDULING PROBLEMS 177

6.1 The Classroom Assignment Problem 177

6.1.1 Vertex Coloring and the Fundamental Problem 178

6.1.2 Modifications 182

6.2 Staffing Problems 184

6.3 Timetabling Problems 190

6.3.1 Edge-Coloring and Class-1 Graphs 192

6.3.2 Modifications 194

6.4 Exercises 198

7 PROJECT SCHEDULING 203
7.1 Project Network Construction 204
7.2 Basic Scheduling Calculations 208
7.2.1 Critical Paths 208
7.2.2 Late Start and Slack Times 210
7.3 Time–Cost Optimization 212
7.3.1 Linear Time–Cost Data 212
7.3.2 Nonstandard Time–Cost Data 220
7.4 Exercises 223

8 CHINESE POSTMEN AND TRAVELING SALESMEN 231
8.1 Eulerian Traversals and the Chinese Postman Problem 231
8.1.1 Eulerian Graphs 232
8.1.2 Postman Problems 235
8.2 Hamiltonian Cycles and the Traveling Salesman Problem 244
8.2.1 Hamiltonian Cycles 245
8.2.2 Heuristics for the Traveling Salesman Problem 250
8.3 Exercises 254

References 259

Index 285

List of Figures

1.1 Bicycle assembly representation 3
1.2 Timing diagram of a task assignment 4
1.3 An improved assembly schedule 4
1.4 An optimal schedule 5
1.5 Representation of event orders for boys and girls 7
1.6 Schedule of girls (resp. boys) first, boys (resp. girls) next 7
1.7 Alternating schedule for girls and boys 8
1.8 An improved schedule 8

2.1 Schedule construction from knapsack solution 22
2.2 Graphs (right) and multigraphs (left) 29
2.3 A sample digraph 29
2.4 Walks, paths, chains, cycles, and circuits 30
2.5 Subgraphs 32
2.6 Regular and complete graphs 33
2.7 Isomorphism and homeomorphism 34
2.8 Trees and arborescences 35
2.9 Covers, independent sets, matchings, and cliques 35
2.10 Colorings in graphs 37
2.11 Branch-and-bound tree for Example 2.2 41

3.1 Bipartite graph of Example 3.4 49
3.2 Sense of ordering from Theorem 3.12 51
3.3 EDD ordering in Example 3.6 53
3.4 Outcome of iteration 2 in Example 3.6 54
3.5 Final schedule for Example 3.7 57
3.6 Precedence chains 59
3.7 In/out forests/trees 60
3.8 Construction of a vertex transitive series-parallel graph 61
3.9 A vertex series-parallel digraph 62
3.10 A digraph that is not vertex series-parallel 62

3.11 Forbidden subgraph recognition 63
3.12 Decomposition tree for graph of Figure 3.9 64
3.13 Two decomposition trees of the same graph 64
3.14 Edge series-parallel decomposition 66
3.15 Forbidden subgraph of 2-terminal, edge series-parallel digraphs 66
3.16 Line graph construction 67
3.17 Relationship between forbidden graphs 68
3.18 Vertex series-parallel graph and its decomposition tree 70
3.19 Final sequence 71
3.20 A suitable schedule from a clique of size k 72
3.21 Graph of Example 3.11 73
3.22 Timing diagram relative to an instance of $1|seq.dep.|C_{\max}$ 74
3.23 Exercise 3-13 77
3.24 Exercise 3-19 78
3.25 Exercise 3-20 78
3.26 Exercise 3-25 79
3.27 Exercise 3-26 80
3.28 Exercise 3-28 80

4.1 Timing diagram of Example 4.1 84
4.2 Timing diagram of Example 4.2 85
4.3 Bad instance for A_{KK} with $k = 0$ 87
4.4 Schedule generated by $L(k)$ 87
4.5 An optimal schedule 88
4.6 An LPT schedule 89
4.7 LPT for $k = 2m - n$ 90
4.8 Suboptimality of LPT with $n = 2m$ 91
4.9 Parallel-processor scheduling $\leftrightarrow$ bin-packing 92
4.10 A_{FFD} application 93
4.11 Schedule generated for Example 4.7 94
4.12 A_{FFD} anomaly 95
4.13 Anomaly-proof target capacity $\bar{C}$ 95
4.14 Bound on r_2 96
4.15 Bound on r_3 97
4.16 Bound in r_m, $m \geq 4$ 97
4.17 Schedule for instance of $R||\Sigma c_j$ 100
4.18 Graph of Example 4.9 101
4.19 Construction of Theorem 4.10 104
4.20 A suitable schedule with $C_{\max} = 3$ 105
4.21 Sample reduction from Theorem 4.10 106
4.22 Digraph for Example 4.11 108
4.23 Digraph on which A_H fails 109

4.24 Graph of Example 4.13 111
4.25 Instance of Example 4.14 113
4.26 Non-transitively reduced graph 114
4.27 A_{CG} failure when $m = 3$ 115
4.28 Application of A_{GJ} 119
4.29 Bicycle assembly graph 120
4.30 Effect of increasing m from 3 to 2 120
4.31 List-processing generated schedule 121
4.32 Worst-case sense of Theorem 4.20 bound 123
4.33 Worst-case arborescences 124
4.34 Insufficiency of optimality and robust schedules 125
4.35 Schedules generated from instance of Figure 4.34 126
4.36 Optimal/suboptimal $m+1\backslash m$ processor schedules 126
4.37 Schedules of Example 4.20 127
4.38 Tovey's construction 127
4.39 Four to three-processor schedules for Example 4.21 128
4.40 Three to four-processor schedules for Example 4.21 128
4.41 Exercise 4-9 130
4.42 Exercise 4-12 130
4.43 Exercise 4-24 132
4.44 Exercise 4-35 133
4.45 Exercise 4-36 133
4.46 Exercise 4-37 134

5.1 Flow shop structure 136
5.2 Flow shop solution/schedule 137
5.3 Properties P_1 and P_2 138
5.4 P_2 does not hold for Σc_i 139
5.5 Permutation solutions are not sufficient with $m \geq 4$ 140
5.6 Jobs i and j are initial jobs 141
5.7 Jobs i and j are not initial jobs 142
5.8 Example 5.1 143
5.9 Demonstration of property P_4 145
5.10 Example 5.4 148
5.11 Primative branch-and-bound tree for flow shops 150
5.12 Concept of $B_k^L(S_p)$ 151
5.13 Tree for Example 5.7 152
5.14 Job shop structure 154
5.15 $J2||C_{\max}$ structure for proof of Theorem 5.7 155
5.16 Relationship between a suitable schedule and a knapsack solution 156
5.17 Example 5.9 157

5.18 Nonactive and active schedules 158
5.19 General job shop structure 159
5.20 Example 5.10 161
5.21 Tree of active schedules for Example 5.10 161
5.22 Example 5.11 163
5.23 Reference Example 5.12 164
5.24 Open shop illustration 166
5.25 Concept of proof of Theorem 5.12 167
5.26 Final schedule for Example 5.13 168
5.27 Example 5.14 171
5.28 Matchings of Example 5.14 172
5.29 Schedule construction of Example 5.14 173
5.30 Exercise 5-4 174
5.31 Exercise 5-15 175
5.32 Exercise 5-16 175

6.1 Graph of Earth Day illustration 179
6.2 Sample construction of Theorem 6.1 180
6.3 Schedule generated from the coloring of Example 6.1 181
6.4 Staffing of Example 6.2 186
6.5 Staffing of Example 6.3 187
6.6 Constraint matrices of Example 6.4 189
6.7 Row circularity determination 191
6.8 Timetable for Example 6.5 192
6.9 Sample graph formed from R 193
6.10 Disjoint matchings of graph in Figure 6.9 194
6.11 Construction of Lemma 6.9 196

7.1 AOA depiction of the bicycle assembly requirements 205
7.2 Alterative dummy activity use 206
7.3 Illustration of the construction of Theorem 7.1 207
7.4 AOA network of example 210
7.5 Time–cost illustration 214
7.6 Compressions for Example 7.6 214
7.7 A_{TC} computations for Example 7.7 218
7.8 Modified network structure for the case $0 < f_{ij} < s_{ij}$ 219
7.9 Pseudo-activity construction 221
7.10 Nonconvex time–cost functions 222
7.11 Discrete time–cost data 223
7.12 Exercise 7-2 224
7.13 Exercise 7-4 224
7.14 Exercise 7-5 225

7.15 Exercise 7-7 225
7.16 Exercise 7-12 226
7.17 Exercise 7-13 227
7.18 Exercise 7-18 228
7.19 Exercise 7-20 228
7.20 Exercise 7-21 229

8.1 Eulerian and non-Eulerian graphs 232
8.2 Directed Eulerian traversal 234
8.3 Eulerian traversal in a mixed graph 235
8.4 P_{sym} solution for a mixed graph 236
8.5 Graphs of Example 8.3 237
8.6 Digraph of Example 8.4 239
8.7 Postman construction for mixed graphs 240
8.8 Application of A_{ES} 241
8.9 Application of A_{SE} 242
8.10 Behavior of A_{ES} and A_{SE} 243
8.11 Worst-known graph for the composite use of A_{ES} and A_{SE} 244
8.12 A non-Hamiltonian graph 246
8.13 Graphs of Example 8.7 247
8.14 Non-Hamiltonian graph generated by Chvátal's construction procedure 248
8.15 Graph of Example 8.8 249
8.16 Application of algorithm A_C 251
8.17 Relationship between V_0 vertices and a tour 252
8.18 Worst-case instance for A_C 253
8.19 Step 1 construction 253
8.20 Step 2 construction 253
8.21 Exercise 8-1 254
8.22 Exercise 8-2 255
8.23 Exercise 8-4 255
8.24 Exercise 8-5 256
8.25 Exercise 8-9 257

List of Tables

3.1 Computation for Example 3.11 74

7.1 Example 7.4 211
7.2 Modified Network Construction 217
7.3 Capacity specification in the maximum flow network 217

Preface

This is a book about scheduling. In fact, it is really a book about what *I* know about scheduling, and even then, the choice of topics and the manner in which they are presented is very much a function of how my own taste, interest and understanding of the subject has developed over some twenty-five years. As a consequence, this treatise is not nor was it even intended to be encyclopedic in it's coverage and it has not been written with the aim of appealing to everyone who has ever met up with a scheduling problem. My hope is, however, that this book will be seen as a useful addition to the vast literature on the subject of scheduling and also, that it will contribute to the stature of the topic as an important *and* interesting one in applied combinatorics.

My own interest in scheduling theory began when, as a graduate student in the summer of 1968, I took a course of the same title. Indeed, I imagine that if one were to gather a little evidence in the matter, it would be the case that numerous academic programs formalized courses in scheduling during that period, a development most certainly aided and legitimized by the 1967 publication of the seminal work, *SCHEDULING THEORY* by Conway, Maxwell, and Miller. And I am also quite clear in recalling that the appeal for me at that time, and as it has remained throughout my career, was almost exclusively nurtured by the purely combinatorial issues embodied in most scheduling problems–especially ones that seemed particularly "simple."

Of course, the observant reader will quickly note that 1968 was, at least from the perspective of most in the scheduling research community, B.C. (before complexity). Most certainly, this condition greatly influenced the literature on scheduling as well as the way the subject was perceived, indeed the way it was treated in many quarters. In my own case, a sobering reminder of this is no further away than my class notes, prepared the first time I taught a course on the topic at Georgia Tech in 1972. Comparing those old notes with what I do in the same course now, some twenty years later, is for me, and notwithstanding my own scholarly and pedagogical failings, an exercise in tracing the changes in the field itself. Even in settings where the subject matter is not explicitly scheduling, I have consistently found myself turning to the latter as an enormously rich source of clever counterexamples, for surprising anomalies, for nontrivial polynomial-time algorithms, for unexpected complexity results, for illustrations of branch-and-bound approaches, and even for interesting graph-theoretic applications, in order to demonstrate or describe phenomena arising in the broader world of combinatorics and combinatorial optimization. The present treatise is heavily motivated by these attributes.

I believe that a substantial portion of the material here could be covered in a one-quarter, graduate-level course. The aforementioned course taught at Georgia Tech has been composed traditionally of a mixture of Master's and Ph.D. students in fields such as Industrial Engineering, Operations Research, Computer Science, and Mathematics. Depending upon this composition, I have always found the material to be quite amenable to customized offerings, *e.g.*, skipping certain proofs, neglecting any detailed study of various cases, etc. The book has been fashioned with this in mind.

Some introductory material is presented in **Chapter 1**. Included are a few illustrations of scheduling problems that have been deliberately chosen in order to expose the breadth of the field. **Chapter 2** contains background material dealing with complexity, graphs, and partial enumeration. It's inclusion is based on a desire to keep the book reasonably self-contained. Hopefully, users will find it a handy reference when the relevant concepts and issues are raised throughout the text. Our coverage of scheduling results then proceeds in somewhat "standard" fashion in the sense of topical progression starting with single-processor results in **Chapter 3**. In **Chapters 4** and **5**, we present material on parallel-processor problems followed by flow and job shop-like models respectively. The coverage in **Chapter 6** deals with problems that do not tend to fit so easily into the structured framework typically employed for classifying scheduling problems. Included in this chapter are the classroom assignment problem, the workload staffing problem, and the so-called timetabling problem. Finally, the last two chapters deal with subjects that, in my own experience, tend to be expendable as time in the term grows scarce (or if class interests dictate). **Chapter 7** contains basic material in the area popularly known as project scheduling while **Chapter 8** deals with some classic results pertaining to traversals in graphs and that often have value in the broader scheduling context of routing. In the latter, we cover fundamental outcomes involving Eulerian and Hamiltonian traversals including their practical manifestations in the form of the Chinese postman's problem and the celebrated traveling salesman problem.

The general theme within each chapter is to substantiate key results with proofs. There are exceptions to this, occurring mostly when results are fairly obvious and/or when their proofs follow recurring arguments. In some cases, readers are asked to provide details as an exercise. Most algorithms in the book are demonstrated with an example and each chapter (3-8) has a fairly healthy list of exercises ranging from routine algorithm application to rather more strenuous, possibly even open problems.

On the issue of exercises, I would like to point out that in their creation, I have attemtped to span the variety of interests and tastes that are often exhibited by those who I assume would be potential users of the book. In this regard, I do not think that most students are well-served by long lists of problems of only the "brain-teaser" sort. On the other hand, I also believe that it is intellectually dishonest and certainly naive, to fix exclusively on exercises which require little more than what is fairly pedestrian numerical manipulation *or* that invite analyses substantially divorced from an appreciation of a problem's inherent combinatorial characteristics. Also, the reader will see that I have resisted the temptation to specifically denote certain exercises that are to be viewed as particularly challenging. Doing so strikes me as a little pretentious in any event, but more importantly, it has been my experience that truly curious students learn best by addressing this issue on their own and independent of whether or not they are actually able to find solutions.

A final note regarding content. There are two clear omissions in this work: stochastic scheduling results and the explicit treatment of "practical" scheduling models or applications. The title of the book implies the first and a quick perusal of it's contents suggests the second. But while these omissions are not inadvertent they are most certainly not intended to be provocative either. Rather, I have simply chosen to defer coverage of material in these areas to those who are substantially more qualified than me to make the cases they deserve.

The last order of business is a pleasurable one. In this regard, I want to express my gratitude to a host of individuals beginning with all of the students who, over the years, have had to endure my sloppy class notes and then finally, various drafts of this manuscript. Some of the (many) questions that they posed, both inside and outside of class, and to which I often had no intelligent answers, have found their way into the book in the form of exercises, examples and counterexamples. I appreciate the thoughtfulness of my colleagues as well, both at Tech and elsewhere who have reviewed the manuscript and/or used all or at least parts of the book in their own classes. Prominent in this respect are Craig Tovey, Mike Pinedo, Ron Rardin, and Reha Uzsoy. I also want to pay homage to Eugene Lawler. I do this by stating that if every result of his, or at last ones that he has been associated with in the field of scheduling, were omitted, this book would be fundamentally different; it certainly would be much shorter. Finally, I wish to thank Mrs. Barbara Vickery who single-handedly typed the manuscript. The keen wit and enormous patience she exhibited throughout this entire project have been exceeded only by her great skill and unfailing professionalism.

CHAPTER 1

INTRODUCTION

A suitable (and frequently employed) description of the subject of scheduling is that it is a field of study concerned with the *optimal allocation or assignment of resources, over time, to a set of tasks or activities.* In the so-called "real world," these resources (money, labor, machines, etc.) are generally restricted or scarce so this allocation inevitably gives rise to competition among tasks that are vying for their use. In turn, decisions regarding the resolution of these natural scheduling conflicts lead to purely combinatorial questions regarding how lists of tasks are to be arranged or sequenced. As the standard lead-in puts it, this suggests both good news and bad news.

The good news is that scheduling problems exist everywhere. Even a conservative reading of the description above leaves room enough to include important problems in the obvious settings of manufacturing, transportation, and logistics as well as ones not so evident in fields such as communications, media management, and sports. Moreover, many of these settings, and specifically the problems they exhibit, are meaningful in a practical sense. That is, effective scheduling solutions accordingly can produce substantial economic dividends. Now for the bad news.

Many scheduling problems, while mathematically challenging and thus of intellectual interest, are exceptionally difficult to solve. The inherent combinatorial explosiveness implied by the necessary requirement to examine the underlying sequencing problem present in many of these scheduling contexts should at least suggest that arriving at appropriate (*i.e.*, optimal or even near optimal) solutions might be a very tedious undertaking. In fact, this is more than a mere suggestion, for we now possess formal evidence that many scheduling problems are intractable in a particular, well-defined sense, and moreover, that this condition is likely to be a permanent one. Still, such a state of affairs, while depressing, does not render the corresponding problems unworthy of investigation. In addition, their importance is not diluted by simply knowing that they are hard, formality notwithstanding. It may, however, mean that we simply need to alter our approaches to such problems and perhaps be a little (or a lot) less ambitious in our expectations regarding their resolution.

In any event, our use of the good news-bad news theme helps to underscore what must be considered a major influence on the enormous growth of results in modern scheduling theory during the last thirty to forty years. For when combined with the myriad settings in which important scheduling problems arise, the realization that many possess mind-boggling complexity has nurtured an expanse of developments as rich and as varied as any within the so-called applications subject areas of operations research and the mathematical sciences. A key aim of this book is to not only reflect this proliferation, but to create an appreciation for it as well.

1.1 Some Scheduling Problems

In the following examples, we exhibit different types of scheduling problems. Note that our aim in these choices is not only to exhibit variety but also to convey a sense of the sorts of phenomena that accompany many such problems and that contribute directly to the frustration as well as the intellectual challenge in dealing with them.

1.1.1 Bicycle Assembly

This example is borrowed from Graham (1978). A bicycle manufacturer assembles finished parts for each bike by employing assembly teams, *i.e.*, each team consists of three workers who together are responsible for the assembly of a complete unit. There are several teams but the work of each has been standardized and so it is sufficient to examine only the activity at one work station.

The Industrial Engineering department has performed an operations analysis and the assembly of a bicycle has been broken down into 10 distinct operations. These include such tasks as frame preparation, front and rear wheel mounting/alignment, installation of the gear cluster, attachment of pedals, and so forth. Each of the 10 tasks requires a given duration or processing time that has been determined from standard time and motion analysis. In addition, the process of assembly is at least partially governed by certain technological constraints. For example, pedals can be placed on the bicycle crank only after the latter has been mounted on the frame. Overall, these technological *precedence constraints* can be captured by the pictorial representation in Figure 1.1. Tasks are denoted for convenience by $T_1, T_2, \ldots, T_{10}$, and the numbers by each represent the task's respective duration time in minutes.

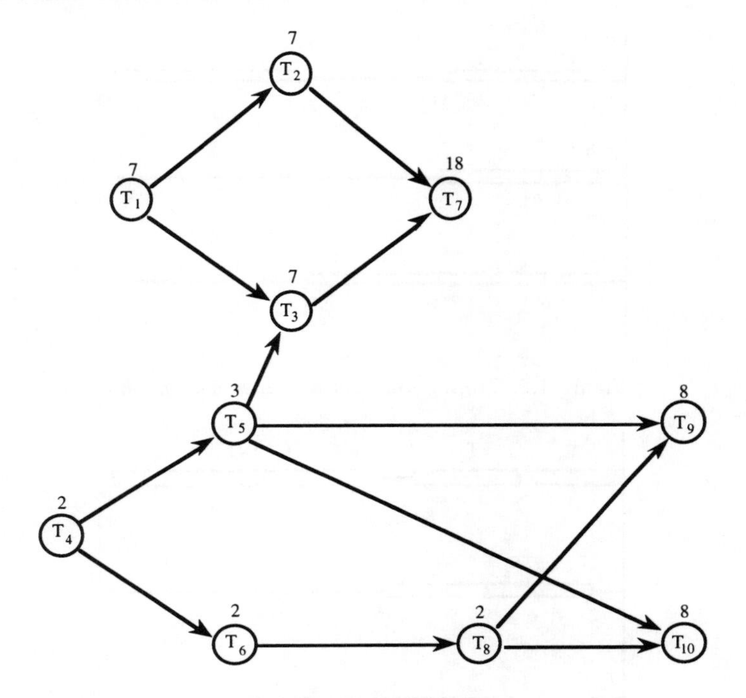

Figure 1.1. *Bicycle assembly representation*

We now come to the problem. Specifically, we seek an assignment of the tasks to the three members of the assembly team that minimizes the overall assembly time for a single unit. The team does not start on another unit until the present one is completed, and each task, once assigned, is performed only by the respective team member. In addition tasks are not preempted, *i.e.*, once started by a team member, a task is performed to completion before another one can be worked on by that member.

First, we determine if the problem possesses interest. For example, if we assign tasks T_1, T_2, and T_7 to the first worker; T_4, T_6, T_8, and T_{10} to the second; and the remainder to the third team member and if everyone performs his assigned tasks as soon as allowed by the aforementioned constraints, then the time required to complete one assembly could be as long as 39 minutes. The corresponding schedule of the tasks is displayed in a timing diagram in Figure 1.2.

On the other hand, the schedule shown in Figure 1.3 indicates that we can do better and still satisfy the precedence constraints. Indeed, the task assignment depicted has a completion time very close to that of an optimal

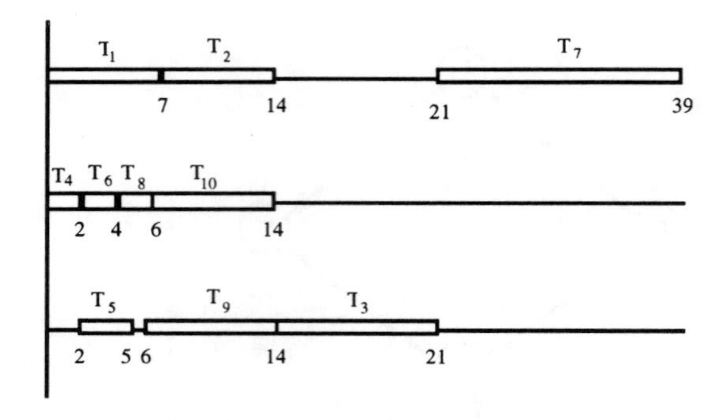

Figure 1.2. *Timing diagram of a task assignment*

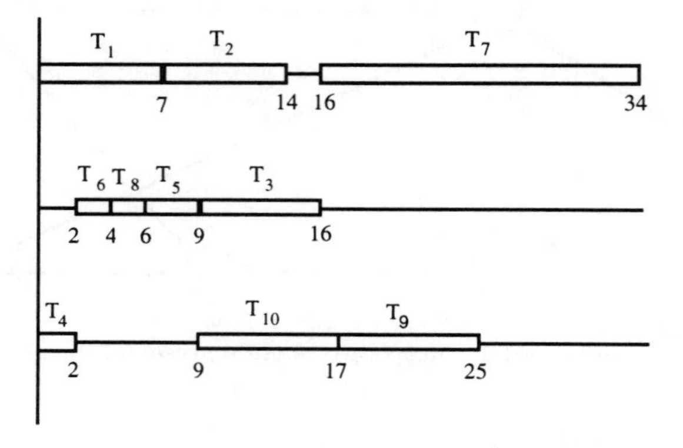

Figure 1.3. *An improved assembly schedule*

schedule. In fact, no (feasible) assignment exists that requires less than 32 minutes since the chain of tasks $T_1 - T_2 - T_7$, which cannot be overlapped, consumes 32 minutes alone.

In any event, the point to be made is that there are many task assignments possible in this case and some are certainly better than others. Moreover, the complication in finding an optimal or at least a "good" schedule for this sort of problem can only be exacerbated by increasing the number of tasks or the number of team members to which they are to be assigned.

But before leaving this illustration, we demonstrate that it is indeed possible to find a schedule of length 32 minutes and, in fact, by even employing assembly teams of reduced size! To see this, suppose we assign the 10 tasks to two team members as (T_1, T_2, T_7) and $(T_4, T_5, T_6, T_3, T_8, T_9, T_{10})$ and let us process them in the order indicated (adhering, of course, to the same technological constraints). The resultant schedule is shown in Figure 1.4.

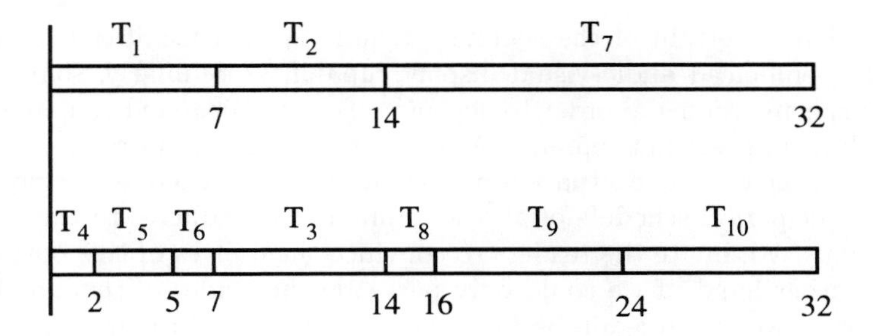

Figure 1.4. *An optimal schedule*

1.1.2 Classroom Assignment

Suppose we are planning an Earth Day program consisting of a series of seminars dealing with topics such as recycling, alternative energy sources, and solid waste management. A local university has agreed to allow the use of classrooms in order to conduct a total of 14 seminars that can begin at 8:00 in the morning but must conclude by 5:00 in the afternoon. No seminars are scheduled to run during the lunch hour (12:00-1:00 p.m.) in order to allow conferees to attend a picnic. The seminar coordinators have requested specific blocks of time during the day for each of their sessions, where the latter are denoted by the letters below. A given session must be scheduled in each of the consecutive, 1-hour periods as shown.

Seminar :	*A*	*B*	*C*	*D*	*E*	*F*	*G*	*H*	*I*	*J*	*K*	*L*	*M*	*N*
Periods :	2	8	4, 5	1, 2	3, 4, 5	6	7, 8	2, 3	1, 2	5	6, 7	3, 4	8	2, 3, 4

Now, the university has offered five classrooms and would prefer that no more than this be used. The issue then is one of finding a schedule (if possible) that accommodates all of the sessions, uses no more than the stated five rooms, and that fits into the eight allowable time periods.

First, we observe that *at least* five rooms will be required since the coordinators of seminars A, D, H, I, and N requested a common period for their sessions, *i.e.*, period 2. On the other hand, this does not mean that five rooms will suffice, although in this particular instance we are fortunate, as the assignment and corresponding schedule below indicates.

Period	1	2	3	4	5	6	7	8
Room 1	*D*	*D*		*C*	*C*	*F*		*B*
Room 2	*I*	*I*	*E*	*E*	*E*		*G*	*G*
Room 3		*H*	*H*		*J*	*K*	*K*	
Room 4		*N*	*N*	*N*				*M*
Room 5		*A*	*L*	*L*				

But what if certain of the seminars require special facilities such as a computer-enhanced audio-visual display capability? Similarly, some may require specific rooms in order to accommodate handicapped conferees. In particular, suppose that seminars A, D, I, and E must be in rooms 2, 3, 1, and 4, respectively. The situation now changes in that there exists no five-room, eight-period schedule of all the seminars that adheres to the added restrictions. We invite the readers to convince themselves of this outcome. On the other hand, if we could only persuade the leader of seminar E to use room 1, we would again be fortunate in that it would now be possible to find a suitable schedule that satisfies the other requirements and that does not compel us to ask the university for another room.

While we also leave the explicit verification of the last claim above as an exercise, we do suggest that the reader be clear on the point that this sort of "classroom assignment" problem, like the bicycle assembly problem before it, possesses serious combinatorial interest. However, doubters are certainly allowed to peek at the contents of Chapter 6.

1.1.3 Scheduling Athletic Events

The final examination in a junior high school coeducational physical fitness class consists of a series of events, *e.g.*, situps, running in place, vertical jumps, etc. The boys are tested together (by event), as are the girls. The events may or may not be the same for each gender and each event is estimated to consume a fixed length of time depending upon such factors as the number of students to be tested, the type of activity required, and equipment setup needed. Only one physical education teacher is available to conduct the testing, and as a consequence only one event (for boys or girls) can be evaluated at a time. In addition to the "event duration" just specified, there is also a rest period following each event. This is a length of time the boys or girls must wait after one event is completed before the next one can begin. The length of these periods depends on the physical demands of the particular event just completed.

Suppose four events are to be evaluated for both the boys and the girls. We shall also assume the events are to proceed for both groups in a pre-specified order. Consider Figure 1.5. Here, the events are labeled relative to gender and by event number, *e.g.*, B_2 is the second event for boys while G_4 is the fourth and last event for the girls. The first number by each event specification is the event duration time and the second is the aforementioned "cooling off" time. After event four in both cases, students are free to leave and so no rest time is specified accordingly. The objective of the teacher is simple: find a schedule of the events that completes all testing as soon as possible.

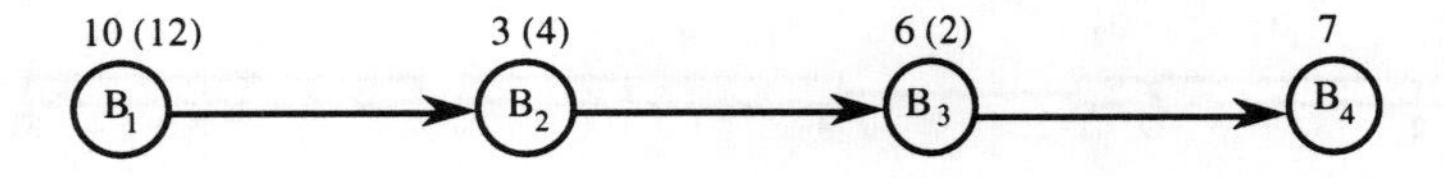

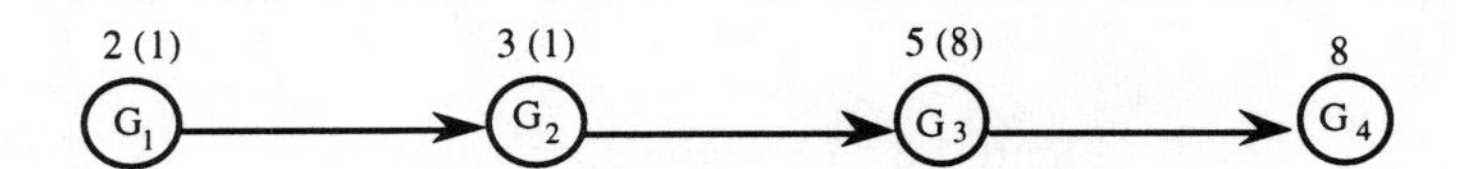

Figure 1.5. *Representation of event orders for boys and girls*

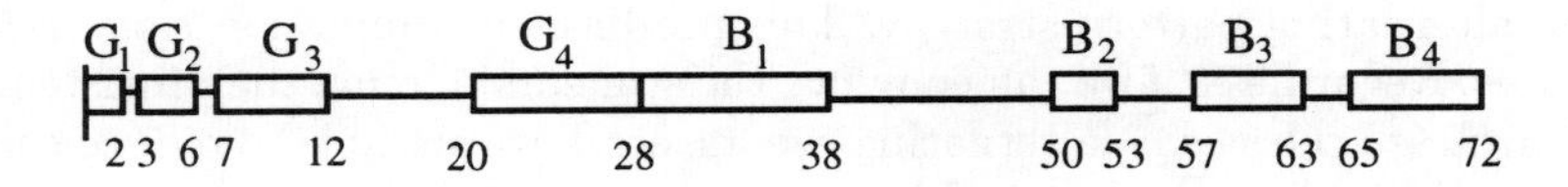

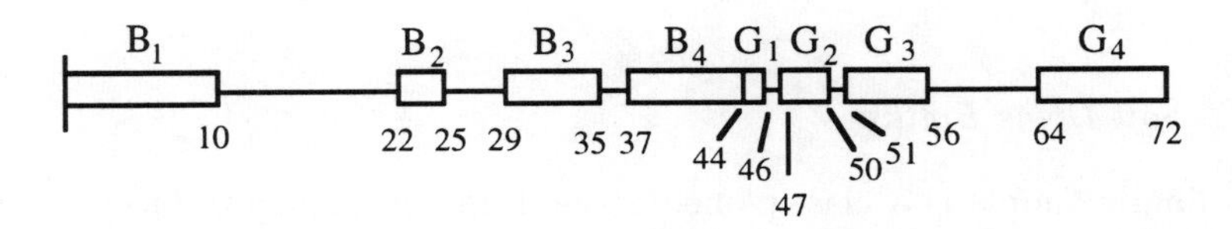

Figure 1.6. *Schedule of girls (resp. boys) first, boys (resp. girls) next*

Again, it is easy to see that the problem possesses combinatorial interest in that alternative orderings of the events give rise to alternative completion times. That is, the stated completion time objective is certainly sequence-dependent. Now, suppose the instructor decides to appeal to chivalry and schedule all the events for the girls first followed by the events for the boys. The timing diagram of the corresponding schedule is shown in Figure 1.6. Chivalry aside, it is also the case (expectedly) that scheduling the boys first followed by the events for the girls produces the same completion time (see Figure 1.6). In any event, neither alternative is particularly good, which is also expected since all rest time for the girls (resp. boys) uses up valuable testing time for the boys (resp. girls).

On the other hand, the instructor may want to invoke a more egalitarian scheduling strategy and alternate events for boys and girls. Letting girls go first and proceeding as $G_1, B_1, G_2, B_2, \ldots$, etc. yields the schedule shown in Figure 1.7, which produces an improvement.

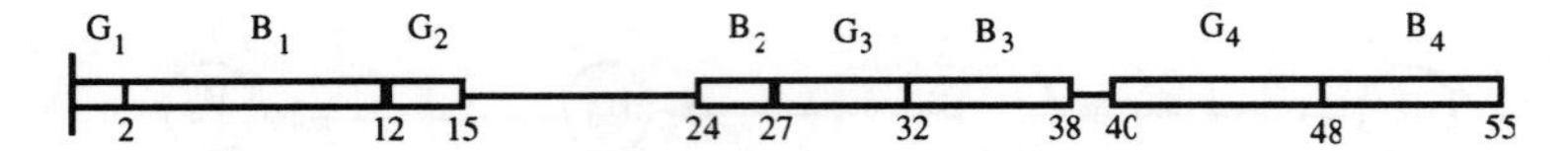

Figure 1.7. *Alternating schedule for girls and boys*

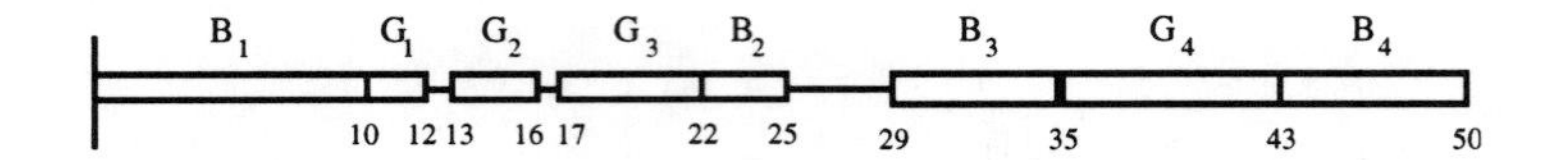

Figure 1.8. *An improved schedule*

We do slightly worse if we begin with the boys and then follow the given alternating pattern strategy. This produces a completion time of 56 minutes. Regardless, if we can convince the students to relax their insistence on "fairness" rules (*e.g.*, alternating events, etc.), we are able to reduce time even further. The schedule in Figure 1.8 illustrates such an improvement. Is this an optimal schedule?

1.1.4 Soft Drink Bottling

Our final example is a classic one. A small, independent producer of soft drinks currently bottles all flavors on a single machine. Further, the company presently serves a relatively local market by offering only four flavors. The bottling operation is repetitive in the sense that all four flavors are bottled in a sequence that repeats but the arrangement of the flavors within the cycle can be arbitrary and this leads to the problem.

While the actual bottling operation takes a fixed amount of time per flavor depending on the number of bottles to be filled of each, the machine must be cleaned after one flavor is completed and prior to the beginning of the next. This cleaning and setup time is not negligible and depends heavily on the flavor that was bottled previously. For example, particularly pungent flavors can require that the machine be cleaned and flushed with a caustic solution while substantially less preparation might be required for other flavors. Hence, two attributes of the bottling operation contribute to the overall cycle time. The first is the filling time for each flavor while the second is the cleanup or changeover time incurred between the bottling of successive flavors. The filling time is a constant, but the changeover time is a function of the sequence of flavors to be bottled. So, if the aim is to minimize cycle time, it is sufficient that we examine the equivalent problem of minimizing only the total changeover time.

Presently, the changeover times for the four flavors (denoted as f_1, f_2, f_3, f_4) can be captured by the matrix shown below. Now, let us assume the flavors are to be bottled in the cyclic ordering given by $f_1 - f_2 - f_3 - f_4$. Note that this ordering is formed by appealing to a sort of greedy mentality that successively selects flavors (after fixing one with which to start) on a smallest, unselected changeover time basis. Obviously, such an ordering strategy might be too shortsighted, resulting ultimately in a very poor overall solution. Nonetheless, the total changeover time for the stated ordering is 57. In fact, if we fix each of the flavors as the first one and repeat the given strategy, we do no better than this, producing total changeover values of 57, 83, and 66, respectively. But this four-flavor instance is small enough to examine explicitly whereupon we find that among the six possible (cyclic) orderings, $f_1 - f_2 - f_4 - f_3$ is the best, having total changeover time of only 20.

$$\begin{array}{c} \\ f_1 \\ f_2 \\ f_3 \\ f_4 \end{array} \begin{array}{c} \begin{array}{cccc} f_1 & f_2 & f_3 & f_4 \end{array} \\ \begin{pmatrix} - & 2 & 70 & 50 \\ 6 & - & 3 & 4 \\ 8 & 3 & - & 2 \\ 50 & 5 & 6 & - \end{pmatrix} \end{array}$$

Suppose, however, that market changes create a need to expand the number of flavors (including a number of diet and sugar-free drinks) to 10. Certainly, the same ad hoc, "next smallest changeover" strategy described above could be used to generate a 10-flavor sequence but whatever was poor about its behavior before is not alleviated now and possibly could even be amplified. Even more obvious is that the fall-back tactic used in the four-flavor setting of listing all possible orderings and choosing the best, while certainly unsophisticated, now begins to become a serious computational question. If minimizing total changeover time is a major economic concern for the bottler, a very hard combinatorial problem now accompanies it.

1.2 Classification of Scheduling Problems

The scenarios described in the previous section provide convenient vehicles for examining the sorts of settings in which scheduling problems arise in practice. There are many from which to choose in order to provide demonstrations, but of these, many are similar, if not identical, when cast (necessarily) in an abstract context; the relevant settings are just disguises. Indeed, these abstract models give rise to a host of structures resulting in something of a "language" of scheduling which, in turn, is so rich that its own codification has been created in order to provide delineation. In this section, we give enough detail in order to employ this language effectively throughout the book.

The bicycle assembly problem is typical of a scheduling environment where a set of jobs, possibly restricted by technological precedence constraints, are to be completed by multiple processors, each of the same capability and to which any of the jobs could be assigned. That is, these processors exist in parallel and may represent machines, or in our case, people, etc. In the refined world of scheduling *theory*, we would refer to our problem as a *precedence-constrained, parallel- (sometimes, multi-) processor problem.* We would most likely need to distinguish further as to whether or not the processors are identical, nonidentical, or related; the number of processors; the value of job duration times; and even the nature of the precedence structure. As we shall observe later, variations in all of these attributes can make an enormous difference in the complexity of the basic problem.

Alternately, the soft drink bottling illustration captures the important aspect in some scheduling problems where changeover or setup times are critical, varying as functions of sequence. Rather than bottling soft drink flavors, this same phenomenon arises in settings such as paint manufacture, rolling mill operations, and scheduling changeovers for general-purpose machine tools. All of these are referred to as so-called single-processor scheduling problems subject to sequenc-dependent changeover times. In the context of our analysis, interest in whether or not soft drinks are being bottled or paint colors are being produced is secondary to the combinatorial problem of finding a correct sequence, *i.e.*, one minimizing total changeover time. Indeed, this problem is easily modeled as the well-known and much studied *traveling salesman problem.* The latter is a celebrated combinatorial optimization problem which has generated a voluminous literature of its own, including a book of the same title (see Lawler *et al.*, 1985).

Throughout this text, we will adhere to what has become the adopted convention for problem classification in the scheduling literature. Following Lawler *et al.* (1989), we employ a three-field classification $\alpha|\beta|\gamma$ where the first field corresponds to the machine or processor setting, the second to job characteristics, and the third to the optimality criterion of interest. We also will employ the following notation relating to various, integer-valued attributes of jobs (when the context is clear, we may substitute terms such as "task" or "activity" in place of "job" as the component of work). First, if a job j possesses more than one *operation*, this number is specified by $m(j)$. The duration or *processing time* of the job is τ_j when $m(j) = 1$ and τ_{jk} otherwise, where the subscript k distinguishes between operations (performed by different machines or possibly different visits to the same machine) in the multiple-operation case. If a job possesses a *due-date* it is denoted by d_j and if the job has an assigned weight we shall use the notation w_j accordingly. In some models, jobs are not assumed available at the same (arbitrary) time, say zero, but rather each is assigned a *release time*. We denote these by r_j. Other job data that might need to be specified will be introduced as the appropriate coverage warrants.

1.2.1 Machine Environment

As indicated, the field α is for specifying the number of machines and, more crucially, for describing the configuration of the processing environment relative to the machines. If a set of single-operation jobs is to be processed on a set of *parallel identical machines*, we denote this by P. The bicycle assembly illustration falls into this category. If the machines are nonidentical but related in some way by, say, uniform machine speeds, a *uniform parallel machine* setting exists and is denoted by Q and if the processors are not related, we say that the setting is one of *unrelated parallel machines*. Here, we denote the setting by the symbol R.

When jobs possess multiple operations and if an ordering is imposed on these operations, two cases arise. If the ordering is the same for each job, a *flow shop* problem results denoted by F and if the ordering is allowed to be different among jobs, we have a *job shop*, specified by J. If there are multiple operations but no ordering on the machines is imposed, an *open shop* results which we denote by O.

When none of these configurations is relevant, we have a *single-machine* problem that is recognized by fixing the parameter for the number of machines at 1. In the cases of multiple machines but when the number is fixed, such specifications are denoted accordingly (*e.g.*, $P2, F3$, etc.) and when no fixed machine number is given, the interpretation is that this attribute is taken to be a variable and is considered as part of the problem *instance*, *i.e.*, it does not play a role in specifying a particular problem *type*.

1.2.2 Job Characteristics

The second field is reserved for attributes of jobs that when explicitly specified also figure prominently in categorizing what can result as varied and numerous outcomes in a problem's solvability status. Generally, four such attributes are recognized: whether or not a job can be interrupted or preempted during processing, denoted by *pmtn*; whether or not a precedence ordering is imposed on the jobs, denoted by *prec* (if this precedence structure is not arbitrary but restricted to take on a particular form, we will define and specify it accordingly); whether or not job-dependent release times are given; and finally, specifications regarding job duration times, *e.g.*, all jobs possess *unit duration times* ($\tau_j = 1$), $\tau_j \in \{1, 2\}$, etc. As before, an unreferenced attribute is intended to signify that the relevant condition is not imposed. For example, no reference to duration times implies that such values are arbitrary; no specification for *prec* is taken to mean that jobs are not constrained by any precedence structure.

In addition, we will generally assume that a job (resp. operation, task, etc.) starts as soon as possible subject only to natural delay caused by precedence or by processing capacity. With regard to the latter, we will assume a capacity that limits processing to at most one job at a time. Then as soon as a processor becomes available and if a job's technological predecessors (if any) have been completed, the job starts; no superfluous idle time on a processor is created. Schedules formed in this way are often referred to as *semi-active.* Important to note (and easy to see) is that when operating in this context, the distinction between "sequences" and "schedules" is eliminated.

1.2.3 Optimality Criteria

The third and final field is used for specifying measures of performance and, as a consequence, is generally self-evident. We need only be clear on notation. The *completion time* of a job is denoted by c_j and so if the measure of performance is to minimize a schedule's total completion time the entry in the third field would be Σc_j. On the other hand if the aim is to minimize the completion time of all jobs, we are really seeking to minimize the maximum completion time (sometimes called *schedule length* or *makespan*). This is denoted as $C_{\max}$.

The *lateness* of a job is given by L_j and is measured as $c_j - d_j$. When only nonnegative lateness is important, our measurement pertains to job *tardiness* denoted by T_j and is given as $\max(0, L_j)$. Total lateness or tardiness as well as maximum values for each are defined in the obvious manner and are denoted as $\Sigma L_j, \Sigma T_j, L_{\max}$, and $T_{\max}$ respectively. Other useful measures of schedule performance exist as well and will be defined when the need arises.

Relative to schedule performance generally (and where most relevant, *i.e.*, **Chapters 3, 4,** and **5**), our interest will remain confined to *regular measures.* Formally, these are measures expressible as nondecreasing functions f of job completion times such that

$$f(c_1, c_2, \ldots, c_n) < f(c_1', c_2', \ldots, c_n') \Rightarrow c_j < c_j'$$

for at least one j. The aim, naturally enough, is to produce schedules that exhibit minimum or near-minimum values of a given regular measure. Happily, the class of regular measures of schedule performance is quite rich including many that square with legitimate aims often arising in practical contexts. It is easy to see that the measures suggested above are all regular.

To illustrate overall, the problem denoted by $P3|prec|C_{\max}$ is a three-machine, parallel-machine problem with all machines identical and where the objective is to produce a minimum length schedule subject to precedence constraints on the order of job processing. This is exactly the bicycle assembly problem described earlier. On the other hand, a specification of the form $1||T_{\max}$ refers to a single-processor problem without precedence restrictions where the maximum job tardiness is to be minimized. The classic, minimum makespan, two-machine flowshop problem would appear as $F2||C_{\max}$.

1.3 Outline of the Book

In **Chapter 2** we provide background material the intent of which is to facilitate accessibility regarding the scheduling results that follow. The assumption underlying this aim is that a so-called "average user" of the book will find the coverage helpful (to be sure, some users may find it necessary while others, of course, can skip it altogether). While the heaviest treatment in the chapter pertains to the subject of *computational complexity*, we also provide enough coverage of standard notions in *graph theory* in order to more easily present and deal with existing results. We conclude the chapter with a modest discussion of *branch-and-bound.*

Our coverage of scheduling results commences in **Chapter 3** and proceeds in rather standard fashion beginning with single-machine problems. In **Chapter 4**, we consider parallel-processor models while in **Chapter 5**, flow shop, job shop, and more general models are examined. Scheduling problems that in some sense tend to resist our convenient classification format are addressed in **Chapter 6**, which includes models of the classroom scheduling problem described earlier, staffing or workforce scheduling problems, and last, timetabling problems. **Chapter 7** takes up issues in the area of *project scheduling.* Topics included are ones related to issues in activity network construction, basic scheduling calculations, and problems of time-cost optimization. We conclude with **Chapter 8** dealing with traversals. Considered are Eulerian and Hamiltonian traversals and their practical manifestations in the form of the Chinese postman and traveling salesman problems.

CHAPTER 2

MATHEMATICAL PRELIMINARIES

In this chapter, we present material that was described previously as background for the purpose of making more accessible the results in the remainder of the text. The (intended) implication is that this background will also provide enough support so that even fairly sophisticated literature on topics in scheduling can be negotiated. Of course, readers already possessing such a background need not dwell on the subject matter in this chapter, but can use it only (if at all) as a reference. On the other hand, those without a suitable background as well as those who are only marginally comfortable with the material presented here are advised to become so acquainted in order that their progression through the rest of the book be facilitated.

The first and by far the lengthiest section in this chapter pertains to the subject of computational complexity. Key in this introductory coverage is the concept of problem reduction and the notions underlying the formation of fundamental equivalence classes for problems. Following this we include a section on graphs (both directed and undirected). We do so to provide enough support so that particular results covered later can be appreciated at more than a superficial level. Last, we provide a very brief, elementary treatment of partial enumeration or branch-and-bound. We include this coverage since such approaches are often the principal general-purpose recourse for treating scheduling problems.

2.1 Computational Complexity

Let us begin with an example. Suppose we consider a set of n jobs that are to be processed on a single machine. In addition, processing is not constrained in any meaningful way other than that at most one job can be worked on at a time by the machine and each job, once started, is processed without interruption. Each job possesses a fixed, nonnegative, and integer-valued processing time as well as a due-date. The objective is a simple one: create a sequence of the jobs that minimizes the total (over all jobs) simple difference between job completion times and their respective due-dates.

Now, this particular problem, recognized in our adopted notation as $1||\Sigma L_j$, is a very easy one to solve; all that is required is a permutation of the jobs arranged in nondecreasing processing time order. That is, such an ordering is always correct regardless of the job due-dates; the proof (given in Chapter 3) of this fact is easy. In addition, the effort (shortly, we will be more precise in describing how this is measured) required for implementing this strategy is minimal, involving only a sort on job processing times.

On the other hand, suppose we consider a simple modification to the stated problem. Rather than minimizing the linear sum of differences of job completion times from due-dates, let us now seek an ordering that minimizes total job tardiness. Recall that a job incurs tardiness in the length of time beyond which the due-date is exceeded by the job's completion time. Alternately, a job is not tardy if it is completed on or prior to its due-date. In our shorthand, this is the problem $1||\Sigma T_j$.

Now, it is not a challenge to find examples that demonstrate that the correct ordering strategy employed for the problem $1||\Sigma L_j$ does not extend to the total tardiness version. With only a little more effort, it also becomes apparent that other ordering strategies (ordering on due-dates, etc.) do not seem to work either. Moreover, there is recent evidence that suggests that it is highly unlikely that the problem $1||\Sigma T_j$ will *ever* be resolved by the sort of approach used in the former case. Here then is a pair of problems, differing only in their objectives (and seemingly, only slightly so) that apparently possess very different solvability properties.

But is it evident that these two particular problems should respond differently to attempts at their solution? Indeed, on what basis does one make such a bold statement regarding the bleak prospects for an "efficient" solution in the second case? And even if the ultimate outcome is doomed insofar as an effective resolution is concerned, this does not cause the second problem to evaporate or to become less important. How then should a practitioner of scheduling proceed? Negotiating answers to these sorts of questions constitutes much of the activity in the field of study known generally as *complexity theory*.

Loosely, modern complexity theory deals with the search for fundamental distinctions in the tractability of problems. While it is true that these sorts of issues have been the source of concern for mathematicians and theoretical computer scientists for many years, their study in the realm of combinatorial optimization in general, and scheduling theory in particular, can be traced to approximately 1970. Following, we provide an overview of some of the basic notions of complexity, reminding the reader again that our coverage is introductory at best. Greater detail can be found in the seminal work of Garey and Johnson (1979). Also suitable, but not at the level of Garey and Johnson, is Chapter 2 in Parker and Rardin (1988).

2.1.1 Problems

In the illustration used earlier, two *problems*, $1||\Sigma L_j$ and $1||\Sigma T_j$, were described. For either of these problems, a particular choice of n, of duration times, and/or due-dates would define an *instance* of the problem. That is, a problem is a (usually infinite) collection of instances of like mathematical form but that differ in size and in the values of parameters in the problem form.

Often, our tendency is to relax this distinction between problems and instances, especially when the context is clear. It is also the case that we often employ a fairly liberal interpretation of the notion of a problem. When one talks of an "integer programming problem" there is often no compulsion to require specificity other than perhaps a qualification as to whether it is a "binary problem" or a "knapsack problem," etc. On the other hand, in the area of scheduling, this relaxation can be easily abused. Surely, one could not speak of a "scheduling problem" and expect not to be pressed further for details, *e.g.*, distinguishing between the "problems" in our prior illustration is important because what we know about each is apparently quite different.

In fact, it would be difficult to think of another field where the need for these sorts of distinctions has been more profound than in scheduling. Delineations within structures can be so subtle that they are often neglected and yet they have to be made because they contribute directly (although we may not fully understand why) to the problem's status in terms of solvability. Often, we even have to be precise regarding parametric information; a problem might be very easy when job durations are assumed to have only value 1 but very hard when durations can be 1 *or* 2. A problem may be very easy if processing involves two machines but very hard when three machines are required.

2.1.2 Computational Orders

As stated, a specific problem includes instances of different *size*, and however measured, we could probably agree that the magnitude of *computational effort* required to treat instances of a given problem would generally increase with size. Here it will suit our purposes to take computational effort to be the number of elementary steps required, in the *worst case*, by an algorithm to actually solve (*i.e.*, stop with a correct outcome) a specific problem instance. The relationship then between effort and size is captured by an algorithm's *time-complexity function.* For example, if size is denoted by n then an algorithm might possess a time-complexity function that is linear in n. On the other hand, another algorithm might require effort that grows faster (*i.e.*, is a slower algorithm), as, say, n^3. Still another might require effort growing exponentially as $n!$. Of course, the actual forms of these functions might involve constants (*e.g.*, $2.5n^2$) or may contain lower order terms (*e.g.*, $n^4 + n$). Our convention, however, is to generally neglect these, preserving only the dominant components. Our vehicle for this is to refer to broad computational orders in the context of so-called "big oh" notation. Hence, our latter functions would be order n^2 and order n^4, written $O(n^2)$ and $O(n^4)$ respectively.

An interesting activity, especially for novices in the subject of complexity, is to examine the behavior of various time-complexity functions. Accordingly, let us consider a selection of such functions, taken from Garey and Johnson (1979) and shown below:

Time Complexity Function	Size, n 10	20	30	40	50	60
n	.00001sec	.00002sec	.00003sec	.00004sec	.00005sec	.00006sec
n^2	.0001sec	.0004sec	.0009sec	.0016sec	.0025sec	.0036sec
n^5	0.1sec	3.2sec	24.3sec	1.7min	5.2min	13min
2^n	.001sec	1.0sec	17.9min	12.7days	35.7years	366cent.

Thus, an $O(n)$ algorithm operating on an instance of size $n = 10$ requires effort, upon conversion to time units, of 0.00001 seconds. If instance size is increased sixfold ($n = 60$), this algorithm requires six times the previous effort, or 0.00006 seconds. Alternately, a slower algorithm, say the $O(n^5)$, one would solve the size 10 instance in 0.1 seconds while requiring approximately 13 minutes for the one of size 60, *i.e.*, $(60)^5$ computations $\times 0.00006$ seconds per 60 computations. While this effort is certainly less appealing than that of the linear-time alternative it is not nearly so severe as that required by the user of the $O(2^n)$ procedure! Indeed, instructors who assign homework requiring the application of the latter procedure, and especially on instances of moderate size, should be prepared for some late submissions.

But do the entries in this table really mean anything? After all, the time-complexity functions are constructed from a worst-case perspective so won't typical instances require substantially less effort? Perhaps. The problem is that we can't predict such a thing and the worst-case perspective, even if pessimistic in its outcome, does eliminate uncertainty in its intent. Even so, substantial reductions in the times shown in the table might not be very meaningful in any event, *e.g.*, $O(2^{n/2})$ rather than $O(2^n)$ does not constitute a debatable difference for reasonable values of n.

Then, what about improvements in technology? Does a faster (serial) computer make a difference? The computations in the next table (also from Garey and Johnson, 1979) give an indication of the answer. Shown are the respective sizes of the largest instances of a problem that could be treated in a fixed period of time when increasing computer power 100- or even 1000fold. As we see, for exponential time-complexity functions, faster machines seem to have little effect.

Complexity Function	Size on Present Computer	×100	×1000
n	A	100A	1000A
n^2	B	10B	31.6B
n^5	C	2.5C	3.98C
2^n	D	D+6.64	D+9.97

2.1.3 Problem Size and Encoding

To this point, we have been intentionally vague in our interpretation of instance "size." But, a quick examination of the literature will reveal that this relaxation carries over even at more sophisticated levels where n is taken to represent, for example, the number of variables in a problem, the number of vertices in a graph, etc. So, if this sort of looseness is permissible, it must be the case that such transgressions in preciseness leave unchanged the ultimate lines of delineation between tractability classes. It turns out that this is indeed the case.

Certainly, when an instance is specified, care must be exercised and rules have to be followed. For example, when creating a specification for a graph we might adopt a convention where vertex labels are first written down as, say, $v[1], v[2], \ldots, v[n]$ followed by the edges given by the corresponding vertex pairs, *e.g.*, $\{v[1], v[3]\}, \{v[3], v[7]\}$, etc. We also would require delimiters to denote conditions such as when the vertex listing stops and the edge listing begins, the end of the instance specification, and so forth. In short, we *encode* our instance using symbols drawn from some machine alphabet. This in turn leads to the more formal notion of size as the *length* of this instance encoding.

But certainly the same instance could be encoded in alternative ways, yielding different lengths accordingly. For example, a graph could be encoded by its adjacency matrix: n rows, each with n elements either 0 or 1 depending on whether or not a pair of vertices are adjacent. The rows would have to be distinguished in some fashion, perhaps using the symbol /. Clearly, the same graph when encoded by each of these two schemes results in encodings of different lengths. It should also be clear, however, that each scheme is expressible as a polynomial function of the other and as a polynomial function of the number of vertices in the original graph. And this is the point; if we possess an algorithm for a problem with time-complexity function expressible as, say, some polynomial in the number of vertices in a graph, then no matter what our encoding (so long as our choices are polynomially related in the sense above) the general time-complexity status (polynomial, exponential, etc.) of our algorithm remains unaffected.

What about numbers in the encoding (vertex labels, edge weights, etc.)? Can they be expressed in any fashion? Not quite. Indeed, almost any fixed-based number system (binary, octal, etc.) will work so long as the length of the resultant encoding grows logarithmically with the magnitude of the numbers. For example, a binary encoding requires $log_2(k+1)$ (binary) digits to specify any positive integer k in the encoding. On the other hand, if our base is 1 (*i.e.*, a number k is encoded by k 1s), the corresponding growth of a number's encoding length is linear in the number's magnitude. This could lead to the unsettling outcome where an algorithm for a problem experienced polynomial (including linear) growth under this base-1 encoding but was exponential under, say, a binary encoding. This base-1 encoding is often referred to as *unary encoding* and plays prominently in complexity theory if for no other reason than that it creates some apparently anomalous behavior. Algorithms that are polynomially bounded with respect to a unary encoding are called *pseudopolynomial algorithms.*

2.1.4 Problem Reductions

One of the most important notions in complexity theory, certainly from the perspective of its application, is the concept of problem reduction. We say that a problem P *reduces* to another problem P' if any algorithm that solves P' also solves (possibly by a suitable conversion) problem P. In particular, we are interested in polynomial reductions. That is, we say that P *polynomially reduces* to P' if a polynomial time algorithm for P' implies the existence of a polynomial time algorithm for problem P. There is a less than subtle observation here in that if this sort of reducibility property exists between a pair of problems P and P', we can safely interpret problem P to be a "special case" of problem P'. Hence, if problem P is, in some sense, difficult then P' must be as well.

The trick of course is in creating the stated reduction. Sometimes this is easy (even trivial) and sometimes it is not. In many cases, including some prominent ones in scheduling, we know of no such reductions even though we suspect they must exist. In any event, our strategy is this: create from an arbitrary instance of P, a specific instance of P' such that any suitable algorithm for the P' instance would always produce an outcome that exhibited a correct solution for P *vis-a-vis* the solution on the contrived instance. If this contrived instance can be constructed and solved in polynomial time, we would have a meaningful result. The best way to show this is with an illustration.

Example 2.1 Polynomial Reduction Consider an n-job, one-machine scheduling problem like that described at the outset of Section 2.1 but where now we assume the existence of a set of penalties, one for each job. If a job is tardy, its penalty is incurred (independent of the length of tardiness), whereas if the job is not tardy, there is no penalty assessment. The aim is to find an ordering of the jobs that produces the smallest total tardiness penalty. For ease, let us refer to this probem as P_S.

Now, consider the well-known KNAPSACK problem stated below, in equality form, and that we denote by P_K:

Given positive integers $a_1, a_2, \ldots, a_t$ and b, does there exist a subset $S \subseteq \{1, 2, \ldots, t\}$ such that $\sum_{i \in S} a_i = b$?

We will show that *any* instance of P_K can be transformed into a valid instance of P_S that when solved produces an outcome (interpretable as a solution to P_S) yielding a solution to P_K directly.

To proceed, we need a precise description of a way to convert any set of input data for the P_K instance to one for P_S. First, we set $A = \sum a_i$. Next, let the processing times and penalty values for a t-job scheduling instance be given by $\tau_i = p_i = a_i$ respectively for $1 \leq i \leq t$, and finally assume that the due-date for all t jobs is b. Clearly, the size of the instance constructed for P_S is polynomial in the length of the P_K instance.

Now, suppose there exists an appropriate subset of as which correctly solves P_K. Then, it is easy to see that a schedule can be formed from this by simply packing the respective jobs (the ones indexed as the hypothesized as in the knapsack instance and given by S) in the interval $[0, b]$. This leaves $A - b$ units of processing time which contribute to tardiness penalty and this penalty value is exactly $A - b$. In fact, it is also easy to see that this value is a lower bound on the tardiness penalty for any schedule. Thus, the schedule just formed is optimal. The construction is illustrated in Figure 2.1.

Conversely, suppose we solve the particular scheduling instance and find the optimal penalty value to be, accordingly, $A - b$. Now, for this to be the

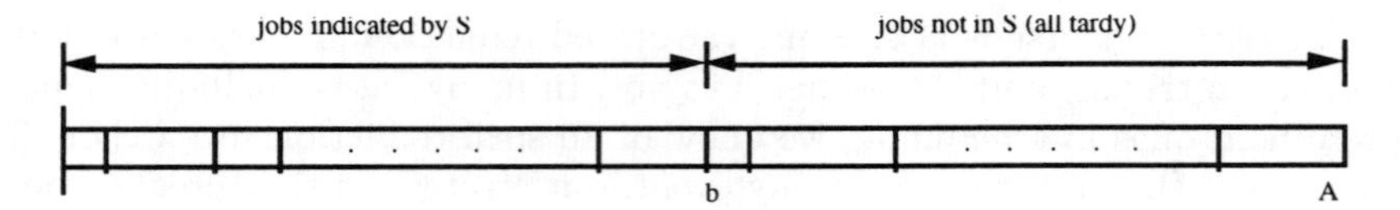

Figure 2.1. *Schedule construction from knapsack solution*

case, some job starts its processing exactly at time b, which means that some job completes at exactly b, or equivalently that some set of jobs are processed completely in the interval $[0, b]$. Otherwise, our penalty would be strictly greater than $A - b$. So, if we denote the set of indices for these jobs by S, it is apparent that we have produced a subset of as that constitutes a solution on the original instance of P_K.

Thus, the knapsack problem possesses a solution if and only if the solution to our specially contrived scheduling instance has penalty value exactly $A - b$. This is the desired result; any algorithm for solving P_S can be used to solve P_K. That is, we simply form, from an instance of the latter, one of the former which is then solved by the algorithm. The solution value of the optimal schedule is compared to $A - b$ and the outcome regarding the knapsack instance is given as a by-product accordingly. □

Polynomial reducibility of a problem P to another one, P', is denoted as $P \propto P'$. In the illustration above $P_K \propto P_S$. Clearly, the relation $\propto$ is *reflexive* in that a problem polynomially reduces to itself. It is also *transitive* such that $P \propto P'$ and $P' \propto P''$ imply $P \propto P''$. Whenever we can demonstrate *symmetry* (*i.e.*, $P \propto P'$ implies $P' \propto P$) $\propto$ is an equivalence relation that will ultimately form a nice tool for grouping problems in terms of tractability.

2.1.5 Fundamental Problem Classes

In contrast to a problem stated in an optimization context (*e.g.*, find a schedule having minimum tardiness penalty, etc.), let us now concentrate on *decision problems* or more formally, *language recognition problems*. These are problems where a solution corresponds to a correct *yes* or *no* response. For example, in the above reduction illustration, where a schedule with minimum total tardiness penalty was sought, the analogous decision problem would be: Given a set of t jobs with the stated duration times, due-dates, and tardiness penalties, does there exist an admissible schedule having total tardiness penalty no greater than $A - b$?

In general, we convert optimization problems to their decision analogs by posing the question of whether or not there *exists* a feasible solution to a given problem having objective function value equal or superior to a specified *threshold* where the latter is taken as part of the instance.

The Class $\mathcal{P}$. In terms of tractability, the status of decision problems exhibits substantial variance. At one extreme (which is not of interest to us here) there are decision problems that are *undecidable.* That is, they are known to be unsolvable-by any algorithm. At the other end of the tractability spectrum are problems that are *well-solved.* These are problems for which an algorithm is known to exist and that will always stop on the correct outcome, doing so by consuming, in the worst case, effort bounded by a polynomial function of instance size. In some quarters these algorithms have been referred to as *good* algorithms (*cf.* Edmonds, 1965a).

The set of all problems that admit polynomial time algorithms belong to the *Class* $\mathcal{P}$. Well-known members of this class include the decision problem analogs of linear programming, matching, shortest path, and various scheduling problems. The first single-machine illustration used in this section $(1||\Sigma L_j)$ is in Class $\mathcal{P}$. On the other hand, the total tardiness penalty version just discussed is not.

In the world of complexity theory, membership in $\mathcal{P}$ is quite prestigious. Indeed, it has become almost automatic to interpret a problem's inclusion in $\mathcal{P}$ as evidence that large instances can be routinely solved. On the other hand, one shoud be reminded that while an $O(n^{10})$ algorithm would qualify as "good," its defense as a practical algorithm or its role in rendering a problem "well-solved," is tenuous at best. Still, even cases akin to the one just raised at least serve to *fix* a problem's status. This in turn can heighten the intensity of effort on the problem's behalf, possibly giving rise ultimately to a substantially improved (*i.e.*, lower order polynomial) algorithm which, in fact, is of real practical value. As testimony in this regard, we need look no further than the well-documented experiences in linear programming (*cf.* Khachian, 1979; Karmarkar, 1984) and matroid parity (*cf.* Lovász, 1981; Gabow and Stallmann, 1986; Orlin and Vande Vate, 1990).

The Complement of Problems in $\mathcal{P}$. Remaining with the total tardiness penalty problem, suppose we leave the input exactly as before but now ask: Does there exist no admissible schedule having tardiness penalty less than or equal to $A - b$? Observe that instances for which the response was previously *yes* are now answered with *no* and vice versa.

As it turns out, this change from an existence problem to one of nonexistence is inconsequential for problems in $\mathcal{P}$. That this is so is fairly obvious following from our definition of $\mathcal{P}$. That is, the certificate of membership in Class $\mathcal{P}$ is an algorithm that always stops in polynomial time with the correct answer. The same algorithm will respond to the complementary problem if we simply read *yes* when the procedure responds *no*, and conversely. As we shall see, however, this rather simple justification for dealing with Class $\mathcal{P}$ problem complements, indeed their inclusion in $\mathcal{P}$ itself, does not easily extend.

The Class $\mathcal{NP}$. To be sure, very stern requirements have to be met for membership in Class $\mathcal{P}$. To be included, an algorithm that actually solves a decision problem in polynomially bounded time is needed. Suppose, however, that we relax our standards and require only that a correct solution (one for which we would respond *yes*) be verifiable in polynomially bounded time. For the tardiness problem used in the reduction illustration, a suitable schedule could certainly be checked for feasibility in polynomial time as could the value of its tardiness penalty. This rather simple notion, of easy to verify but not necessarily easy to solve, is at the heart of the *Class* $\mathcal{NP}$.

$\mathcal{NP}$ stands for *nondeterministic polynomial* where the terms also invite an alternative (though rather loose) interpretation of the class. Crudely, problems in $\mathcal{NP}$ can be thought of as those that could be solved in polynomial time if there is a suitable random proof generator (or guessing device) that can, with positive probability, "guess" a polynomial-length proof of *yes* when the correct answer is *yes*. Most conventional decision problems are in $\mathcal{NP}$.

Clearly, Class $\mathcal{NP}$ contains $\mathcal{P}$ and the interesting question is whether or not this containment is proper. Presently, a great number of problems (decision problems) are known to be in $\mathcal{NP}$ but not known to be in $\mathcal{P}$ and so the issue is whether or not this is a permanent condition. We return to this relationship between these very important classes later.

Complements of Problems in $\mathcal{NP}$. The class of decision problems for which a *yes* response corresponds to a *no* relative to a member of $\mathcal{NP}$ is referred to as $\mathcal{C}o\mathcal{NP}$. We just saw that polynomial-time algorithms for decision problems that are in $\mathcal{P}$ are easily adaptable to suitable algorithms for solving the corresponding complementary *no* versions. Thus, letting those complements be given by $\mathcal{C}o\mathcal{P}$, it follows that $\mathcal{P} = \mathcal{C}o\mathcal{P}$. The situation with $\mathcal{NP}$ complements is different, however.

By its very definition, membership in $\mathcal{NP}$ is predicated upon polynomial verifiability of a correct *yes* outcome. But what if the correct outcome is *no*? That is, rather than verification of the *existence* of a solution, we might be interested in verifying *nonexistence*. In this case, it is possible that we would have to enumerate every potential solution, concluding *no* only if we are certain that none could lead to *yes*. But this enumeration is inherently exponential in nature and so it may not be possible to even verify correct *no* answers in a polynomially bounded effort. This means that there are likely to be many members of $\mathcal{C}o\mathcal{NP}$ that do not belong to $\mathcal{NP}$. Clearly, class $\mathcal{P}$ lies at the intersection of $\mathcal{NP}$ and $\mathcal{C}o\mathcal{NP}$.

$\mathcal{NP}$-Hardness and $\mathcal{NP}$-Completeness. Certainly, an ambitious aim would be to seek membership in $\mathcal{P}$ for those problems from $\mathcal{NP}$ not known to be so contained. In reality, this would not be a well-advised undertaking however. First, there are many such problems, and secondly, most of them have been heavily studied for many years-by very talented researchers. At the very least, this provides circumstantial evidence that our efforts might be tedious in the best case and doomed in the worst.

To see why this is so, let us return to the notion of polynomial reduction. Problems to which all members of $\mathcal{NP}$ polynomially reduce are referred to as *$\mathcal{NP}$-Hard.* Thus, a suitable algorithm for an $\mathcal{NP}$-Hard problem would yield one for every member of $\mathcal{NP}$ as well. In a real sense, an $\mathcal{NP}$-Hard problem is at least as hard as any member of $\mathcal{NP}$. But are there any $\mathcal{NP}$-Hard problems? Actually, there are many. Indeed, most of the interesting problems that one might encounter in combinatorial optimization, scheduling in particular, are $\mathcal{NP}$-Hard.

The seminal result in this area is due to Stephen Cook (1971), who demonstrated that every problem in $\mathcal{NP}$ reduces (polynomially) to a particular problem in logic known as the *Satisfiability Problem* which we denote as SAT. An instance of the satisfiability problem (loosely stated) consists of a Boolean expression formed as a finite conjunction (read "and") of disjunctions (read "or"), each containing some configuration of Boolean variables x and/or their complements, $\bar{x}$ (read "not x"). These variables take on values of *true* or *false* (if a variable is *true*, its complement is always *false*) and the requirement is to decide if there exists an assignment of such values to all the variables that renders the entire expression *true* or *satisfiable.* The rules are easy; if at least one true element exists in a disjunctive grouping or *clause*, then we count the clause true and if all the clauses are simultaneously true, we count the entire expression satisfiable. Formed from Boolean variables $\{x_1, x_2, x_3, x_4\}$, the expression below is satisfiable by setting $x_1 = x_2 = x_3 =$ true and $x_4 =$ false.

$$(x_1 \vee x_2 \vee x_4) \wedge (\bar{x}_1 \vee x_2 \vee \bar{x}_3) \wedge (\bar{x}_2 \vee \bar{x}_4)$$

Observe that in this notation, $\wedge$ and $\vee$ stand for "and" and "or" respectively.

Hence, satisfiability was the first $\mathcal{NP}$-Hard problem. But observe that a given assignment of true-false values to the variables/complements of a satisfiable expression is easily (*i.e.*, polynomially) checked in order to see if the expression is satisfied by that specific assignment accordingly. That is, the satisfiability problem is also in $\mathcal{NP}$.

Problems that are in Class $\mathcal{NP}$-Hard and are also in $\mathcal{NP}$ are called *$\mathcal{NP}$-Complete.* So, if P and P' are any two $\mathcal{NP}$-Complete problems, we have that $P \propto P'$ and $P' \propto P$. That is, the relation $\propto$ is symmetric. But we already know that it is reflexive and transitive as well and we see that $\mathcal{NP}$-Complete problems form an equivalence class, in turn providing a powerful tool for categorizing problem tractability. If any member of the Class $\mathcal{NP}$-Complete admits a polynomial-time algorithm then so must all members of the class. Similarly, if a single $\mathcal{NP}$-Complete problem is ever proved to be permanently intractable, then this must be the status of all other $\mathcal{NP}$-Complete problems as well. In either case, we would have resolved the issue raised earlier regarding the relationship between $\mathcal{P}$ and $\mathcal{NP}$.

It is easy to see how one should go about proving that a problem P is $\mathcal{NP}$-Complete (it may be quite difficult to actually find the proof-if it is there at all). Required is a given $\mathcal{NP}$-Complete problem, say P^*, and a reduction, $P^* \propto P$, along with a valid argument that P is in $\mathcal{NP}$. Once accomplished, the new $\mathcal{NP}$-Complete problem, P, can then be used to establish the $\mathcal{NP}$-Completeness of other problems. Whereas in the beginning there was only the satisfiability problem, there are now hundreds of $\mathcal{NP}$-Complete problems, including many in scheduling. The knapsack problem used previously is $\mathcal{NP}$-Complete and it is not hard to see that the decision version of the tardiness penalty problem employed in Example 2.1 is in $\mathcal{NP}$. This, coupled with the valid reduction shown by the example, establishes the latter problem's $\mathcal{NP}$-Complete status as well.

Generally, the class $\mathcal{NP}$-Complete contains no optimization problems. Every member is a decision problem. However, optimization problems possess $\mathcal{NP}$-Complete decision problem analogs and are thus $\mathcal{NP}$-Hard. For example, a polynomial algorithm for finding a schedule for the aforementioned problem having minimum total tardiness penalty certainly provides a polynomial resolution for the corresponding decision version relative to a given threshold value. For this reason, it is often the case that problems are left in their optimization context and are accordingly described as being $\mathcal{NP}$-Hard.

2.1.6 What to Do with $\mathcal{NP}$-Hard ($\mathcal{NP}$-Complete) Problems

It is important to know if a problem is $\mathcal{NP}$-Hard/$\mathcal{NP}$-Complete. At least its status is clarified in such cases and if our supervisor assigned us the task of producing a polynomial-time algorithm for it anyway, we can offer pretty incriminating evidence that such an undertaking might be monumental if not altogether hopeless. The supervisor might fire us, but as suggested in Garey and Johnson (1979), the particular problem's certificate of apparent intractability renders a search for our replacement fairly moot; chances are that nobody else will find such an algorithm.

But a problem does not go away simply because it is known to be $\mathcal{NP}$-Hard. In fact, there are some options that our supervisor can consider. First of all, it must be understood that a proof of $\mathcal{NP}$-Hardness is a worst-case argument, *i.e.*, there are instances that can force our best algorithms to perform very poorly. However, this does not necessarily mean that such algorithms perform poorly all the time. It may well be that instances encountered in a particular setting are, in some loose sense, manageable. For example, the knapsack (optimization) problem is a well-known $\mathcal{NP}$-Hard problem and yet algorithms have existed for some time that can solve instances involving thousands of variables. Indeed, the knapsack problem has often been referred to as one of the easiest "hard" problems where the latter adjective follows from its $\mathcal{NP}$-Hard status and the former arises from the ease of its solution in "real-world" contexts.

It may also be the case that a generally $\mathcal{NP}$-Hard problem exhibits interesting special cases that are polynomially solvable. This phenomenon occurs often in scheduling where, for example, certain problems possess polynomial time algorithms in the cases where instances have no more than two processors but are $\mathcal{NP}$-Hard when three or more processors are specified. Certainly, if our scheduling environment was confined to the former setting, it would not be particularly bothersome to know that the general problem was $\mathcal{NP}$-Hard.

Finally, it may be that we simply have to be less ambitious and settle for solutions to a problem that are not guaranteed to be optimal. That is, we may have to give up something in solution quality to gain something in computational efficiency. This point of view is adopted often in scheduling and leads to the development of approximation or heuristic approaches. Of course, our interest is in polynomial approximations, since in giving up optimality, the least we should get in return is polynomiality.

Fortunately, there are some very clever and effective approximation strategies for otherwise hard scheduling problems. Sometimes, however, we do not have a very strong sense of how effective (in terms of solution quality) these approaches are and in some cases, performances can be substantially worse than anticipated including performance that is arbitrarily poor. As we shall see later in this book, heuristic approaches can even lead to anomalous or counterintuitive behavior (recall the bicycle assembly illustration in Chapter 1).

In order to assess the quality of a heuristic algorithm's performance, a common tactic is to examine the strategy's *performance ratio*; for a given instance, this is the ratio of the value of a heuristically obtained solution to the optimum value. Often, the literature reports empirical statements regarding this ratio for a particular heuristic, *e.g.*, grand mean of a large number of such ratios generated from a sample set of instances. Sometimes, however, more stringent standards than this are enforced, two of which predominate. Most often, the notion of a *performance guarantee* is employed. This is a value that establishes an upper bound or worst-case limit on the performance ratio for any instance of a problem to which the heuristic might be applied. Often these ratios do not square closely with the empirically based statistics. That is, it is not uncommon for heuristics to exhibit ratios of the latter variety that are very close to 1 (let us assume our problem seeks minimum value solutions) but that have realizable (instances exist that achieve the ratio) performance guarantees of say $3/2, 2$, or even ∞. The other option is to produce a formal *expected performance guarantee*. By definition, we would expect these values to reflect a less pessimistic view than the worst-case result and, in that regard, provide insight more representative of a heuristic's so-called typical performance. In fact, this is often the case. Unfortunately, the determination of these expected-case results can be a complicated undertaking and as a consequence, fewer results of this variety are known than in the worst-case context.

2.2 Graph Theory

Our intention in this section is to provide enough coverage of some basic results in graph theory so that their application in scheduling is not only facilitated but also appreciated. Readers desiring a serious treatment of this much studied topic for its own sake are advised to consult any of a number of excellent publications. Particularly appropriate in this regard is Bondy and Murty (1976).

2.2.1 Fundamental Definitions

A graph G is a pair (V, E) where V is a set of *vertices* and E a set of two-element (not necessarily distinct) subsets of V called *edges*. Letting $|V| = n$, we sometimes say that the graph is of *order* n. In a graph $G = (V, E)$, we say that a pair of distinct vertices are *adjacent* if they define an edge, and that an edge is *incident* to its defining vertices. The *degree* of a vertex is the number of edges incident to it.

An edge defined by nondistinct vertices is called a *loop*, and *multiple edges* result when more than one copy of the same edge is allowed; the resulting structure is called a *multigraph*. On the other hand, a graph without loops or multiple edges is said to be *simple*. A simple graph is given on the right in Figure 2.2 while on the left, a multigraph is shown.

When the defining vertices for edges are ordered, a graph is said to be *directed* and is often referred to as a *digraph* denoted as $G = (V, A)$. Here, A is the set of directed edges or *arcs*. The notion of vertex degree for digraphs

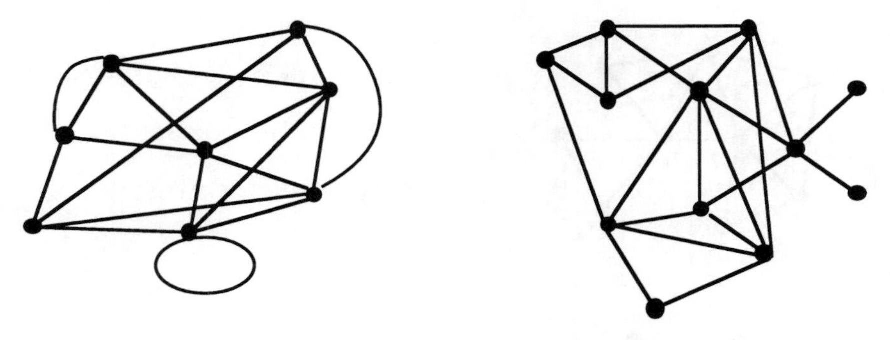

Figure 2.2. *Graphs (right) and multigraphs (left)*

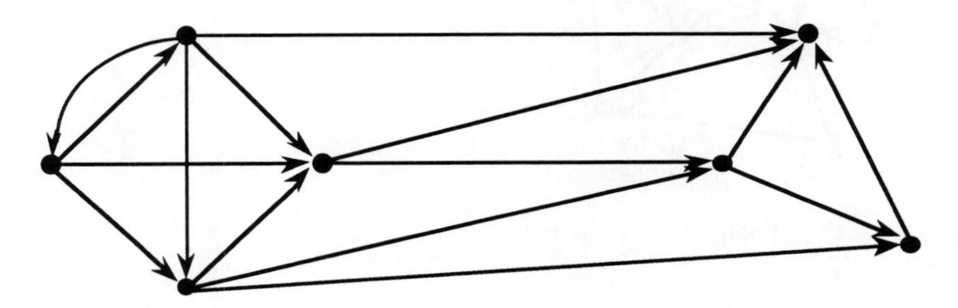

Figure 2.3. *A sample digraph*

is relativized in an obvious way where the number of arcs directed out of a vertex is the *outdegree* and the number of arcs directed into the vertex is the *indegree*. A digraph is shown in Figure 2.3.

A *walk* in an undirected graph is an alternating sequence of vertices and incident edges that begins at some vertex s and ends at another, t. When all vertices are distinct, the walk is a *path*. When only the first and last vertices are the same, the walk is a *cycle*. In digraphs, a path must conform to direction while the corresponding notion ignoring direction gives rise to a *chain*. Similarly, a cycle need not conform to direction but when direction is consistent, a *circuit* results. Any graph without circuits (cycles in the undirected case) is termed *acyclic*. Illustrations follow in Figure 2.4. Observe that we have denoted edges (arcs) individually rather than by vertex pairs. Often, paths, cycles, circuits, etc. are denoted by vertex or edge (arc) sequences when the context is clear.

A graph is *connected* if there exists a path between every pair of vertices in the graph. A digraph is connected if its underlying undirected structure is connected (equivalently, if every pair of vertices is joined by a chain). The

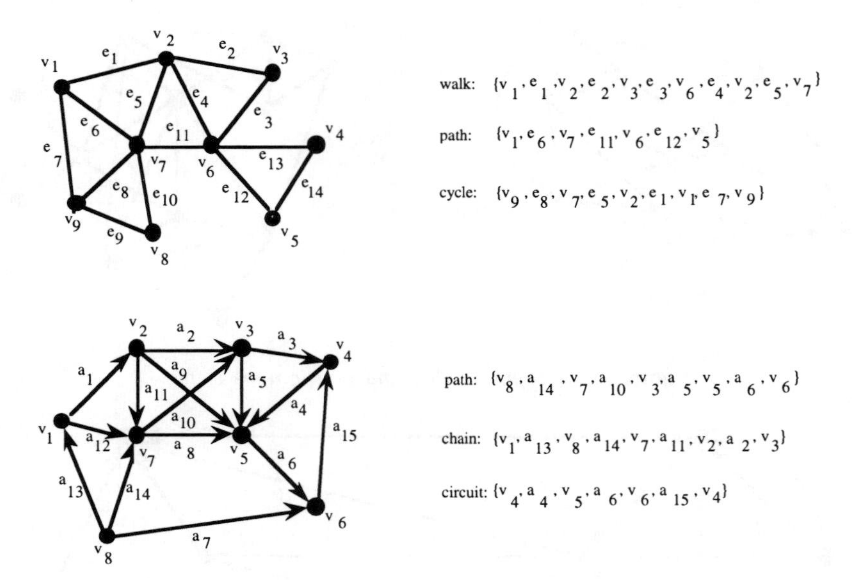

Figure 2.4. *Walks, paths, chains, cycles, and circuits*

vertex connectivity of a graph is the minimum number of vertices whose removal leaves a disconnected or *trivial* graph (a graph consisting of a single vertex and no edges) while *edge connectivity* is defined in the same fashion relative to edge removal. These two parameters are related in a formal way in that a graph's vertex connectivity is always bounded from above by its edge connectivity, which in turn is bounded from above by the graph's minimum vertex degree. Moreover, it is easy to show that these relationships can be strict.

Given two graphs $G = (V, E)$ and $G' = (V', E')$ we call G' a subgraph of G if $V' \subseteq V, E' \subseteq E$, and V' contains the set of vertices met by members of E'. G' is a *spanning subgraph* of G if $V = V'$. For some nonempty subset $H \subseteq V$, the subgraph of $G = (V, E)$ *induced* by H is the subgraph having vertex set H and all edges of E that are incident only to vertices in H. An *edge-induced* subgraph is defined in a similar manner relative to an edge subset of G. The graphs in Figure 2.5 illustrate these types.

If the vertex set of a graph can be partitioned into two subsets V_1 and V_2 such that every edge in the graph is incident to a vertex in one subset and a vertex in the other, the graph is said to be *bipartite*. Equivalently, a graph is bipartite if and only if it is free of odd cycles (cycles consisting of an odd number of vertices).

A connected graph having the same degree r at every vertex is called *regular* in degree r or simply *r-regular*, and a simple graph where every pair of vertices is adjacent is a *complete* graph. Similarly, a *complete bipartite* graph possesses edges connecting every vertex in V_1 with every vertex in V_2. We denote the complete graph on n vertices as K_n and by $K_{m,n}$ the complete bipartite graph having bipartition components of order m and n respectively. The first graph in Figure 2.6 is 3-regular, the second is K_6, and the third is a complete bipartite graph $K_{4,5}$.

Given a pair of graphs G and G', we say that G and G' are *isomorphic* if there exists a one-to-one mapping between their vertex sets, preserving adjacency. Graphs (a) and (b) in Figure 2.7 are isomorphic while neither is isomorphic to the graph in (c).

G and G' are *homeomorphic* if both can be obtained from the same graph by a sequence of edge subdivisions (*i.e.* insertion of vertices on edges). Equivalently, a pair of graphs are homeomorphic if they are isomorphic to within vertices of degree 2. All graphs in Figure 2.7 are homeomorphic, formed by edge subdivisions of K_4.

2.2.2 Trees, Forests, Arborescences, and Branchings

A *forest* is a graph without cycles. If the forest is connected, it is called a *tree*. A tree (resp. forest) is said to be *spanning* if it includes every vertex of a graph G. A *branching* is a digraph with no cycles and no vertex with indegree in excess of 1. If the branching is connected, it is an *arborescence*. As we will observe later in the text, these sorts of special structures play important roles in scheduling, particularly in the complexity hierarchy where various, otherwise hard, problems are well solved when instances are restricted to such structures.

In the graphs of Figure 2.8, the marked edges (resp. arcs) form a spanning tree (arborescence). Removing any corresponding edge (arc) of the tree (arborescence) leaves a forest (branching) that is not a tree (arborescence). Adding any edge (arc) creates a cycle and thus precludes the resulting subgraph from being a forest (branching).

2.2.3 Covering, Matching, and Independence

A subset of vertices in a graph is *independent* or *stable* if no two vertices in the subset are adjacent. The corresponding notion of edge independence is a *matching*. That is, a matching is a subset of edges no two of which are incident to the same vertex. If the matching spans, it is said to be *perfect* (perfect matchings are sometimes called *1-factors*). Interestingly, the problem of finding maximum cardinality matchings in arbitrary graphs is

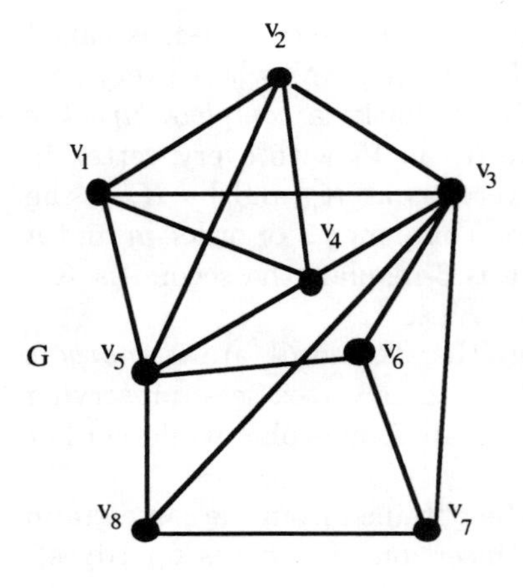

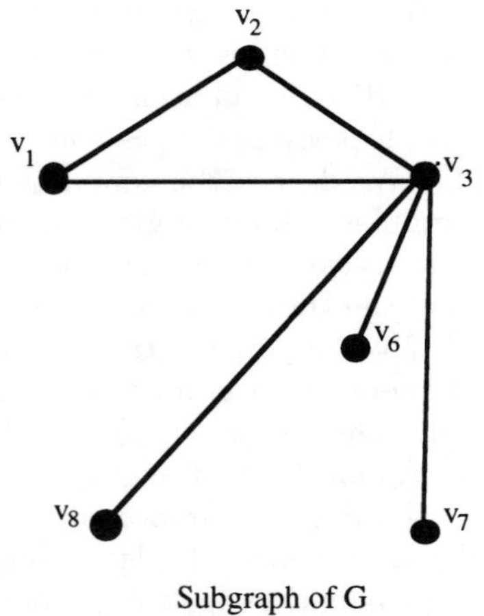

Subgraph of G
(nonspanning)

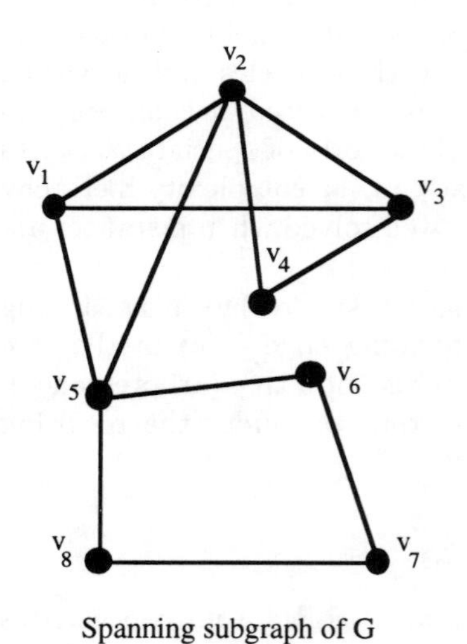

Spanning subgraph of G

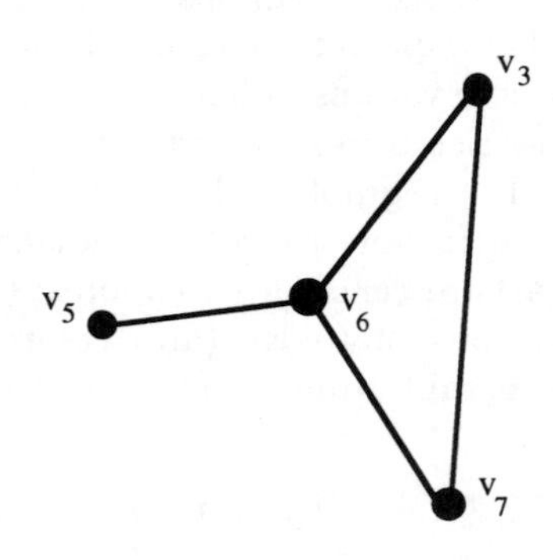

Subgraph induced by
$H = \{v_3, v_5, v_6, v_7\}$

Figure 2.5. *Subgraphs*

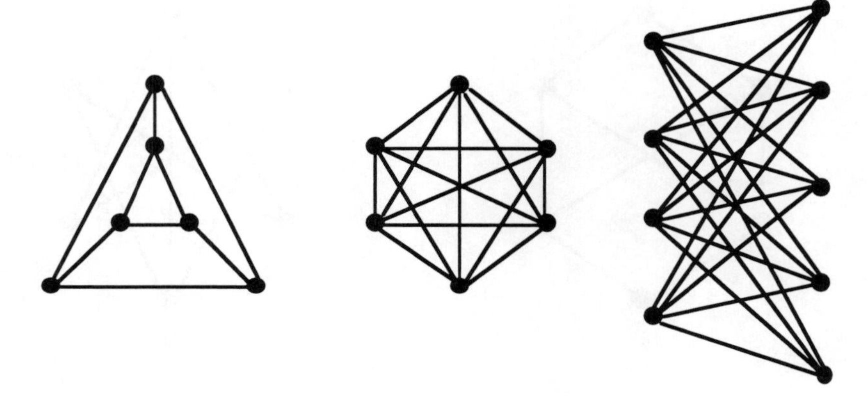

Figure 2.6. *Regular and complete graphs*

well solved following an ingenious and well known strategy by Edmonds (1965a). On the other hand, the problem of seeking a maximum cardinality independent set is $\mathcal{NP}$-Hard in general but polynomially solvable on bipartite graphs as well as some other special structures (*e.g.*, see Borie *et al.*, 1992).

Given a graph $G = (V, E)$, any subset of V having the property that every edge in E is incident to at least one vertex in V is called a *cover* in G. Finding minimum cardinality vertex covers is a hard problem on arbitrary graphs but, as with independent sets, is solvable on bipartite graphs. On the other hand, the analogous problem of finding least cardinality *edge covers* (subset of edges with the property that every vertex is incident to at least one edge in the set) is well solved on any graph. Note that any perfect matching constitutes a minimum edge cover but not the converse.

An important (and useful) relationship between vertex covers and independent sets holds that the respective minimum and maximum cardinalities of each always sum to the order of the relevant graph. In general, if V' is a vertex cover, then $H = V \backslash V'$ is an independent set.

A *clique* is a subset of vertices that induces a complete graph. If $G = (V, E)$ is simple, then clearly any independent set in G is a clique in the edge complement of G, *i.e.*, the graph having the same vertex set as G but only edges connecting vertices that are not adjacent in G. It follows from the relationship between minimum covers and maximum independent sets that specification of maximum cliques would also yield the former two subsets indirectly. That is, if the status of any one of these three problems is resolved then the status of the other two is resolved accordingly.

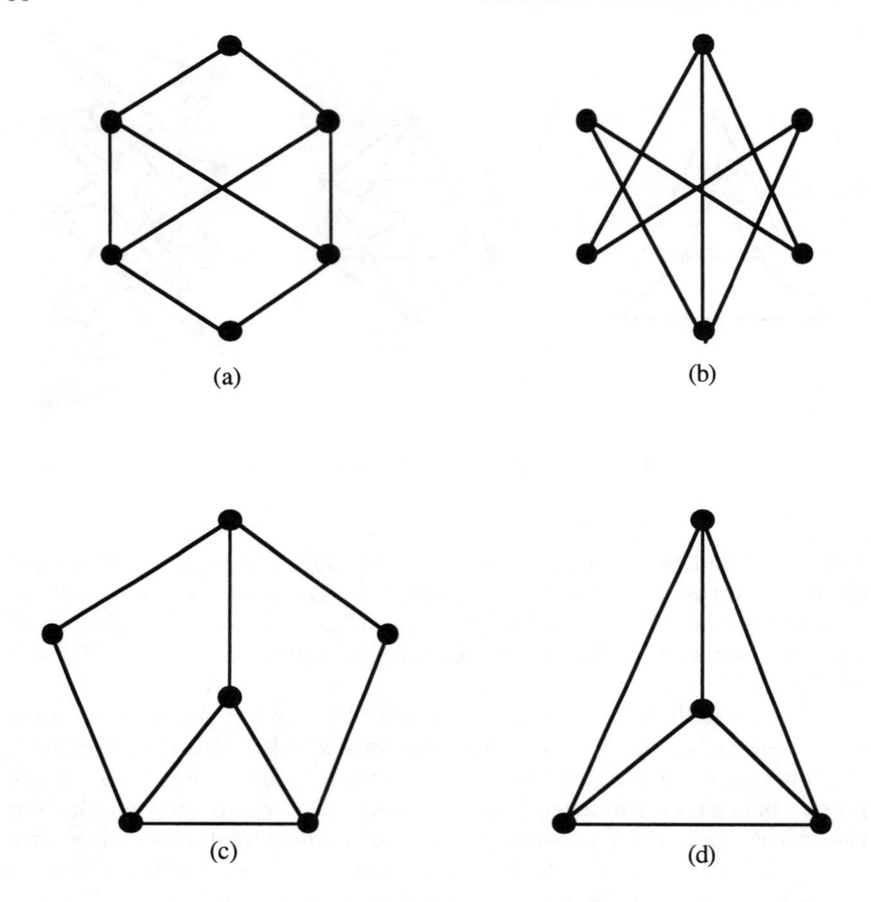

Figure 2.7. *Isomorphism and homeomorphism*

In Figure 2.9, we illustrate covers, matchings, independent sets, and cliques.

2.2.4 Vertex Colorings, Edge Colorings, and Perfect Graphs

A *proper vertex coloring* of a graph G is an assignment of colors to the vertices of G in such a way that no two adjacent vertices have the same color. If k colors are used, we say the assignment constitutes a *k-vertex coloring*. The *chromatic number* of a graph is the smallest k for which it can be k-colored. Determining if a graph is k-colorable is hard and remains so even for k fixed at 3.

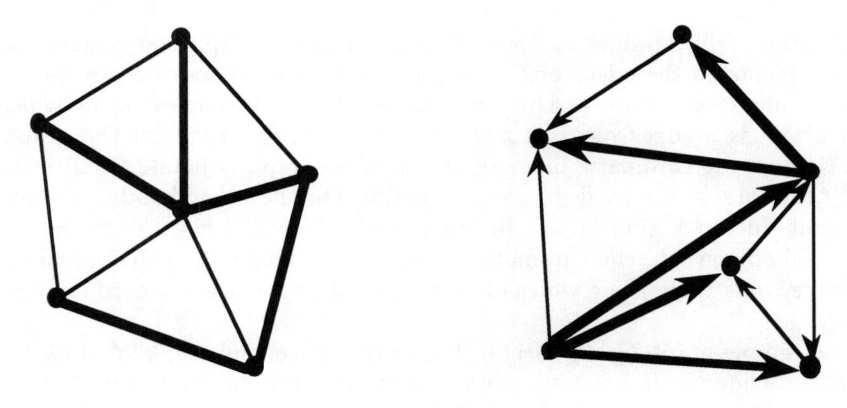

Figure 2.8. *Trees and arborescences*

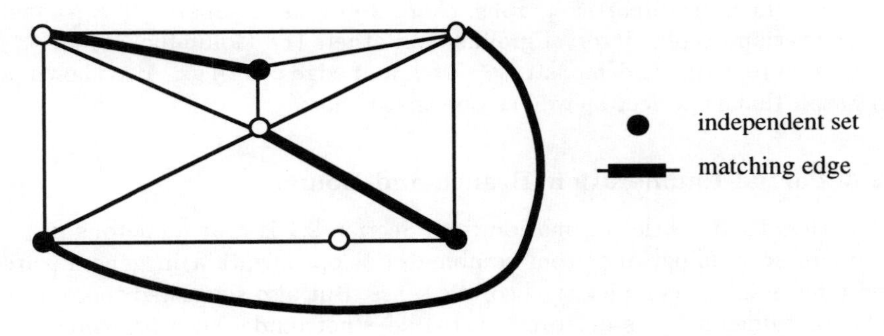

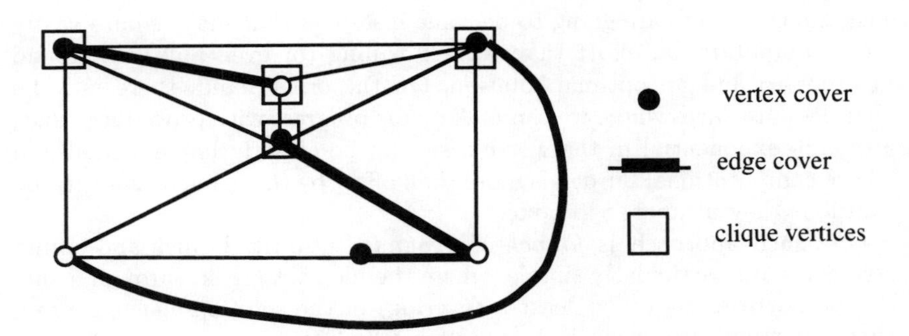

Figure 2.9. *Covers, independent sets, matchings, and cliques*

Alternately, a proper *k-edge coloring* of a loopless graph is an assignment of k colors to the edges of the graph such that no pairs of edges having a common vertex are assigned the same color. The smallest k for which a graph is k-edge colorable is called the *chromatic index* of the graph. Trivially, the chromatic index of any loopless graph is bound from below by the largest vertex degree in the graph. On the other hand, a famous result due to Vizing (1964) (independently, Gupta, 1966), holds that in any simple graph, the chromatic index is never larger than the maximum degree plus 1. Deciding which of these two values is correct is hard (Holyer, 1981).

A simple graph G is *perfect* if, for every induced subgraph H of G, the largest clique in H has cardinality equal to the chromatic number of H. At this writing, the problem of recognizing perfect graphs remains a celebrated open problem. However, various classes of graphs are known to be perfect among which are bipartite graphs, chordal graphs, comparability graphs, permutation graphs, interval graphs, and others (*cf.* Golumbic, 1980).

In Figure 2.10, we demonstrate vertex and edge colorings. Also shown is a graph that is perfect as well as one that is not.

2.3 Partial Enumeration/Branch-and-Bound

The (less than subtle) intimation from Section 2.1 is that numerous problems in scheduling, and from combinatorial optimization in general, are very hard–in the complexity-theoretic sense. But also suggested there, this sort of "difficult by association" status does not render such problems unworthy of some form of analysis; to suggest otherwise, especially to the practitioner, is silly and, some would claim, even borders on zealotry.

Still, knowing that a problem is $\mathcal{NP}$-Hard/$\mathcal{NP}$-Complete is to know that any approach to the problem has to be tempered with a dose of reality. It suggests that there are going to be some instances that may require vastly more computational effort than we are willing (or even able) to expend in order to find an optimal solution. On the other hand, there may be countless instances where we can make some progress with procedures that, although exponential in the worst case, can be cleverly implemented and whose computational burden is more than offset by the value of the optimal solutions they ultimately produce.

One such approach is *branch-and-bound.* In spirit, branch-and-bound strategies are particularly simple, where the idea is to seek, through a successive partitioning (or at least a covering) of the solution space, a proof that a particular solution is optimal (this "solution" might be a dummy, signaling an empty solution space). From this perspective, it is easy to

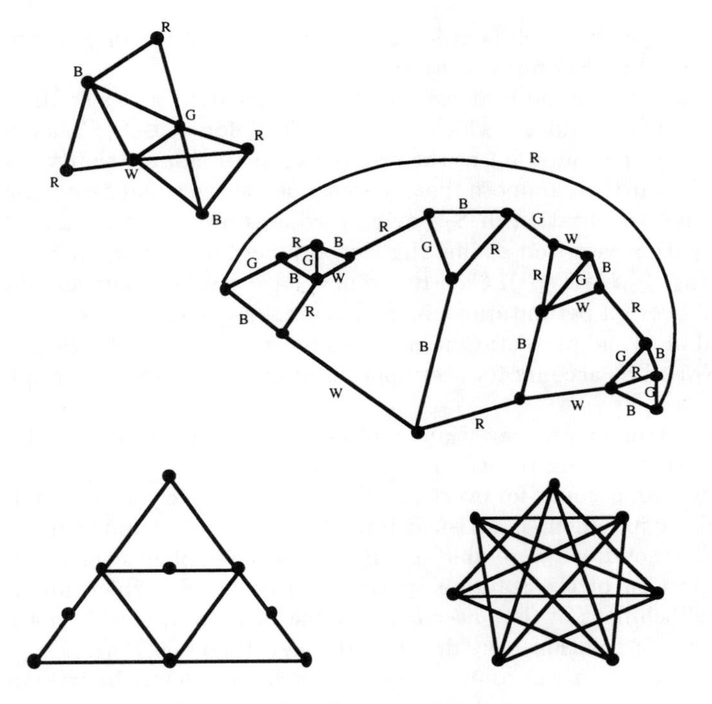

Figure 2.10. *Colorings in graphs*

see why branch-and-bound is often described as a "divide-and-conquer" strategy.

Of course, it's not difficult to argue why such an approach, which amounts to a search over the solution space, needs to be implemented in an intelligent manner. Indeed, one clear (but certainly unintelligent) partitioning strategy would be to enumerate every solution. Selecting the best from among these provides an air-tight proof of optimality. But we already know that *total* enumeration is not viable. Lest the uninitiated reader forget, recall that the solution space for an instance of $1||\Sigma T_i$ consists of $n!$ permutations and as we would deduce from the first table in Section 2.1.2, the task of examining each of these would be most time-consuming—even for uninteresting values of n.

On the other hand, suppose we were to carve up the solution space for the stated tardiness problem (with respect to any instance) in the following way. Denoting the entire space of permutations by S, let S_1 contain those permutations in S that have in common the attribute that job 1 is ordered first. Similarly, let S_2 be the set where job 1 is not first. Trivially, $S_1 \cup S_2 = S$

and $S_1 \cap S_2 = \phi$ (a job is ordered in a specific position or it is not); that is, the pair $\{S_1, S_2\}$ forms a *partition* of S.

Now, let us assume that we know that a solution π^* exists in S_1 with tardiness $T(\pi^*)$ and for which $T(\pi^*) \leq T(\pi_i)$ for $\pi_i \in S_1$. That is, π^* is an optimum permutation in the set S_1 (we shall assume that we actually know π^*). Further, suppose that by some means, we could assert that there existed no permutation in S_2 having tardiness any less than $T(\pi^*)$. More formally, this assertion claims that for any permutation π_i in S_2, we have that $T(\pi_i) \geq \beta \geq T(\pi^*)$. Clearly, we now have a valid argument that π^* is optimal over all permutations in S. This is self-evident; π^* is known to be optimal in S_1, no permutation in S_2 can have a value smaller than this and since S_1 and S_2 account for every permutation in S, we have our optimality proof.

Two outcomes are essential to observe in the illustration. First, it is obvious that claims regarding the solution space S are not meaningful unless we can account for every member in S, either explicitly or implicitly. But this requirement is satisfied immediately by the condition that $S_1 \cup S_2 = S$. Second, we were able to eliminate from explicit consideration the entire portion of the solution space captured by S_2, which follows from the establishment of the *lower bound* value β. Indeed, if π^* is not optimal then either a solution has "dropped through the cracks" (*i.e.*, S_1 and S_2 do not cover S) or the computation producing the lower bound value β is not valid.

But suppose that $T(\pi^*)$ is not bounded from above by β as indicated. Then π^* may or may not be optimal in S. In fact, the best solution in S_2 might actually be inferior to π^*, but without further exploration of the set S_2 we can make no such claim. In this case, we would simply reconstitute the search, this time over the subspace S_2. As before, we could create two new sets by "branching" to form a partition of the set of solutions in S_2. Suppose we call these S_{21} and S_{22}, where in the former we would collect all those permutations that expressly exclude job 1 from the first position, but fix it in the second. The set S_{22} accounts for all the other solutions (in S_2); permutations prohibiting job 1 in *either* the first or the second position. Lower bound values would be computed relative to optima in each of these new subsets and the overall state of our search would be re-evaluated. Our options in this regard would include: a conclusion that π^* was indeed optimal after all, the discovery of a better solution, say π' (which now becomes the standard against which to test further), or lingering inconclusiveness which would require further searching within some unaccounted for region of the entire space. Clearly, this process has to stop sooner or later (possibly, much later) since the solution space is finite.

The Fundamental Branch-and-Bound Approach. We will not give a detailed or "algorithmic" specification of the general strategy of branch-and-bound, but rather, we direct the reader to a host of suitable treatises in that regard such as Parker and Rardin (1988), Nemhauser and Woolsey (1988), and Papadimitriou and Steiglitz (1982). Following, we will present a description of the general notion which should suffice in order to support material presented later.

At any state of our search, we may consider the entire solution space, S, to be accounted for in terms of a family of subsets $S_0, S_1, S_2, \ldots, S_q$ such that $S = S_0 \cup S_1 \cup S_2 \cup \ldots \cup S_q$. Here, S_0 consists of a subset of solutions that have already been considered in the search. Each subset S_k characterizes a *candidate problem* derived from the original problem instance by some well-defined restriction of the solution space. The strategy is to select some candidate problem from the list, examine it, and decide whether or not any solution to that candidate need be considered further in identifying an optimal solution over S. If we conclude that we do not, we can consider the restricted problem defined over S_k to be *fathomed*. In this case, the problem is removed from the list of active candidates and we set $S_0 \leftarrow S_0 \cup S_k$. On the other hand, this evaluation may fail in that we are not able to establish or prove that the respective candidate can be fathomed. In this case, the set S_k is replaced (in the candidate list) by one or more new candidate problems each of which is derived from the one defined over S_k by appending additional restrictions.

Clearly, the evaluation alluded to in the fathoming test is crucial in the overall process; here is where we seek and ultimately obtain a proof of optimality. One possibility, of course, is to test for simple feasibility. Certainly, if it can be demonstrated that no feasible solution exists at all over some S_k, then there is obviously none that is optimal (from S_k) in S accordingly. Another tactic is to evaluate solutions over some S_k relative to any known admissible solutions from S. Some of the latter may be known beforehand while others may be generated during the overall search. Regardless, suppose we have in hand a best solution discovered so far in the search. This is called an *incumbent solution.* If its value is ν^*, then any candidate can be fathomed if it can be established that no solution accordingly can have value less than ν^*. This is accomplished by showing that some easily (*i.e.*, polynomial) computed lower bound (throughout, we have assumed problems of the minimization form) on the value of an optimal solution over the respective S_k is at least as great as ν^*. Of course, any feasible solution provides an initial incumbent value or alternately, we could start the process with a dummy value of $+\infty$.

Naturally, a specific implementation of the branch-and-bound strategy requires substantial attention to details. For a given problem, there will be alternative ways to perform branching *vis-a-vis* the creation of the subsets $S_1, S_2, \ldots, S_q$ and numerous ways to form lower bounds and accomplish the fathoming process. Note further that different choices from among these alternatives can lead to substantial differences in the computational behavior of the resulting branch-and-bound application.

Example 2.2 We shall demonstrate an application of branch-and-bound by considering again the problem $1||\Sigma T_i$ and specifically, an instance having four jobs with $\tau = (8, 6, 10, 7)$ and $d =$ (14,9,16,16). Now, while the general problem is $\mathcal{NP}$-Hard (otherwise, its solution by branch-and-bound would require some tall explaining), the special case where all jobs have the same due-date is easily solved by arranging the jobs in SPT (shortest processing time) order, *i.e.*, in nondecreasing order of processing times. The proof of this result is left as an exercise in Chapter 3. In any event, suppose we modify the data of the original instance so that all jobs are assumed to have due-dates equal to the maximum of the original (actual) due-dates. Then it should be obvious that the tardiness of the SPT ordered sequence provides a lower bound on the tardiness value of any permutation (and hence, on an optimal one) subject to the true due-dates. As we shall see, this trivial observation provides us with an easy source of lower bounds that can be used throughout the entire search.

As for branching, the following strategy is also easy to see. Employing somewhat of a reversal of the notion suggested earlier, let us create candidates by fixing sequential positions from last to first. That is, a given candidate problem will be denoted by a partial solution characterized by a fixed subsequence of jobs in positions $n, n-1, n-2, \ldots, 2$. We form candidates in this fashion since it provides an explicit measure of completion times, and hence tardiness value contributed by the fixed subsequence, leaving an estimate of additional contribution by the remaining "unfixed" jobs easily measurable by utilizing the common due-date result just described.

The entire process can be represented as a tree where nodes in the tree correspond to candidate problems and where branches are formed by the nature of the restrictions that are imposed in order to create the candidates. So then, abiding by the notational convention adopted in the earlier discussion, the full space S of solutions for this sample instance consists of 24 permutations. Now, we partition S into four subsets, each representing a possibility for the final job in a permutation. No matter which job is last, the completion time of any one is a constant, fixed at the sum of the four processing times since it is obvious we will never insert idle time in an optimal solution. Let these candidate problems be defined on subspaces S_1, S_2, S_3, and S_4 respectively where the subscript and the job in the final position are in correspondence.

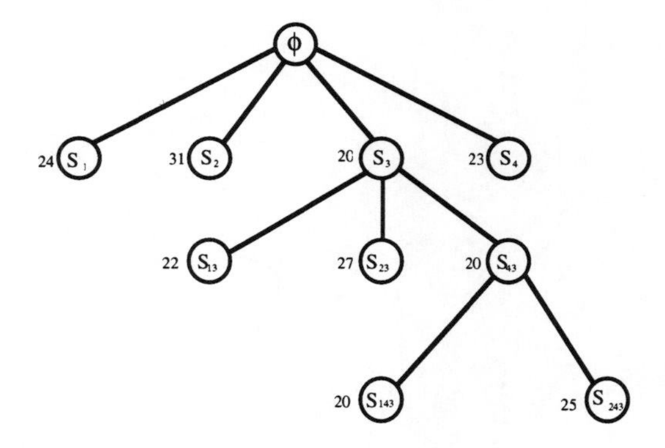

Figure 2.11. *Branch-and-bound tree for Example 2.2*

Consider S_1. Here, job 1 completes at time 31 (in all permutations in S_1) and is tardy by 17 units. Of the remaining jobs (2, 3, and 4), no matter how they are ultimately ordered, we will sustain total tardiness of no less than 7 units. This follows from the SPT order where the three jobs are assumed to have a common due-date of 16 (the maximum of d_2, d_3, and d_4). Overall, the value of 24 results as a valid bound on the optimal solution value relative to any permutation in S_1. Repeating this simple notion for each of the other three sets produces lower bound values of 31, 20, and 23 respectively. If we adopt a so-called *least-bound branching* strategy, we would select the candidate problem corresponding to S_3 and proceed with the enumeration.

Branching from S_3 will produce three replacement candidates. Defined by restrictions on the third sequential position, let us denote (for ease) these problems by sets S_{13}, S_{23}, and S_{43}. Again, the lower bounds are calculated (for S_{43}, we have $T_3 = 15, T_4 = 5$, and the SPT ordered value relative to jobs 1 and 2 with common due-date of 14 is 0 for the total of 20) and upon comparison we select the candidate associated with S_{43} for continued search. A binary branching is performed creating sets S_{143} and S_{243} with bounds of 20 and 25 respectively.

At this point we are done, having produced an optimal ordering of $\{2, 1, 4, 3\}$ with tardiness vaue of 20. The proof of optimality follows since every other branch of the tree is fathomed, *i.e.*, recall that by definition, the lower bounds are nondecreasing in value. The entire computation is summarized by the tree in Figure 2.11. Observe that the root of the tree is labeled with ϕ to signify the full space of permutations, *i.e.*, no restrictions are applied. □

CHAPTER 3

SINGLE-PROCESSOR PROBLEMS

We begin our coverage of scheduling problems by examining results that pertain to scheduling models involving only one processor. In this context, claims that there are more than a few interesting results are sometimes met with a fair amount of skepticism, especially by the uninitiated. But in fact, even more intimate observers of our subject are frequently surprised, as well, at the depth (and elegance) of what is known in the single-processor case. Our aim in this chapter is to describe some of these results and in so doing, provide exposure to and an appreciation of this depth.

3.1 Independent Jobs

We take up first the class of single-processor models in which jobs are assumed to be free of precedence restrictions. When jobs are unconstrained in this sense, they are often said to be *independent.*

3.1.1 Completion Time Models

Suppose we have a set of jobs where these may possibly be "weighted," in some sense, by parameters w_i reflecting, presumably, some notion of preference or priority among the jobs. Then a classic problem results if our objective is to arrange the jobs in such a way as to form a sequence having minimum total weighted completion time. Assuming the weights to be arbitrary integer values, we have:

Lemma 3.1 *Let S be a sequence that has a pair of adjacent jobs i and j with i prior to j and where $\tau_i/w_i > \tau_j/w_j$. Then the sequence $\bar{S}$ created from S by interchanging jobs i and j produces a weighted completion time value strictly less than that of S.*

Proof. Let S be given by $\{\theta_1, i, j, \theta_2\}$ where θ_1 and θ_2 are (possibly empty) sets of jobs. Let us denote the weighted completion times of S and $\bar{S}$ as $C_w(S)$ and $C_w(\bar{S})$ respectively. Now, if the completion time of the last job in θ_1 is $t \geq 0$, then we have $c_i(S) = t + \tau_i$ and $c_j(S) = t + \tau_i + \tau_j$. For $\bar{S}$, these values are $c_j(\bar{S}) = t + \tau_j$ and $c_i(\bar{S}) = t + \tau_i + \tau_j = c_j(S)$. Then,

$$C_w(\bar{S}) - C_w(S) = w_i c_i(\bar{S}) + w_j c_j(\bar{S}) - w_i c_i(S) - w_j c_j(S)$$

and upon expanding and eliminating like terms, we obtain

$$C_w(\bar{S}) - C_w(S) = w_i \tau_j - w_j \tau_i < 0$$

and $\bar{S}$ is strictly better than S. □

With this lemma, we are led to the following fundamental result:

Theorem 3.2 *The single-processor problem* $1||\Sigma w_i c_i$ *is solved by a sequence* S^* *with jobs arranged in nondecreasing* τ/w*-order.*

Proof. Any sequence either conforms to the stated ordering or does not. But by the earlier lemma, any one that does not can be strictly improved by a switch of an existing pair of adjacent jobs (not conforming to τ/w-order). Thus, no sequence that is in violation of the stated order can be optimal and conversely, any that does so conform must be. □

Note that the sense in which the converse direction in the theorem holds is easy which follows since multiple orderings created by ties in τ/w ratios have the same value. We leave this detail to the reader.

Example 3.1 Let job duration times and the respective weights be given by τ and w as follows: $\tau = (5, 4, 7, 10, 12, 9, 6, 3, 9, 4)$, $w = (4, 1, 8, 2, 3, 2, 3, 1, 5, 4)$. Applying the ordering suggested by the theorem yields an optimal sequence $S^* = \{3, 10, 1, 9, 7, 8, 2, 5, 6, 4\}$. Existing ties (*e.g.*, $\tau_2/w_2 = \tau_5/w_5$) indicate alternative optima. □

If all jobs possess the same weight (equivalently, $w_i = 1$ for all i), the next result follows trivially:

Corollary 3.3 *The problem* $1||\Sigma c_i$ *is solved by a sequence* S^* *with jobs arranged in nondecreasing order of duration times.* □

For ease we often refer to these orderings as WSPT (weighted shortest processing time) and SPT (shortest processing time) respectively.

Example 3.2 Operating on the same instance as above with all weights the same yields the SPT sequence $S^* = \{8, 2, 10, 1, 7, 3, 6, 9, 4, 5\}$. □

Clearly, minimizing total completion time (resp. weighted completion time) is equivalent to minimizing mean job completion time (resp. mean weighted completion time). But while trivial to observe, when viewed in this latter context, it is perhaps easier to see that optimal permutations accordingly (*i.e.*, SPT, WSPT) directly affect measures related to practical concerns such as in-process inventory.

Now, suppose that jobs are not required to be simultaneously available for processing. That is, a job may possess some nonzero release time denoted by r_i which, in turn, acts as a lower bound on the respective job's start time. Unfortunately, it is now not so clear how jobs are to be arranged so as to minimize total completion time. In fact, we have just turned an easy problem into a very difficult one (the reader is prompted to recall the earlier discussion pertaining to the fine distinction that can often exist between so-called easy scheduling problems and apparently intractable ones). The following result provides evidence. First, we state the well-known number-theoretic problem, 3-PARTITION:

> Given positive integers n, B, and a set of integers $A = \{a_1, a_2, \ldots, a_{3n}\}$ with $\sum_{i=1}^{3n} a_i = nB$ and $B/4 < a_i < B/2$ for all i, does there exist a partition of A as $\{A_1, A_2, \ldots, A_n\}$ such that $|A_j| = 3$ for $1 \leq j \leq n$ and where $\sum_{a \in A_j} a = B$?

Following a fairly complicated transformation, we have:

Theorem 3.4 *The (recognition version) single-processor problem $1|r_i \geq 0|\Sigma c_i$ is $\mathcal{NP}$-Complete by a reduction from 3-PARTITION.* □

We provide no details of the proof of this result, directing interested readers instead to Lenstra *et al.* (1977). On the other hand, it is important to observe that the apparent difficulty of problem $1|r_i \geq 0|\Sigma c_i$ transcends even a pseudopolynomial resolution. This follows since 3-PARTITON is an especially hard problem, exhibiting $\mathcal{NP}$-Completeness in the *strong sense*, *i.e.*, it remains hard even under unary encoding.

3.1.2 Lateness Models

Recall that job lateness L_i is simply the difference between a job's completion time and its due-date. Since due-dates are fixed parameters, the next result follows directly from previous ones.

Theorem 3.5 *The single-processor problems $1||\Sigma w_i L_i$ and $1||\Sigma L_i$ are solved by sequences formed by WSPT and SPT orderings respectively.*

Proof. The proofs are exactly those of the previous theorem and corollary which follows since $L_i = c_i - d_i$ with d_i a constant. □

Also easy is the problem of minimizing maximum lateness. The argument is again one based on pairwise switching. We have:

Theorem 3.6 *The single-processor problem $1||L_{\max}$ is solved by a sequence S^* with jobs arranged in nondecreasing order of due-dates (EDD or earliest due-date).*

Proof. Any sequence in violation of the stated order can be altered to create a sequence that does so conform and that is no worse. This is sufficient to establish the result. □

Observe that the theorem above does not claim that any non-EDD sequence could not be optimal. Unlike the total completion time results, the pairwise switch may not strictly improve–we only claim that it will not be worse.

Example 3.3 Suppose due-dates are supplied for the jobs in the earlier example. If these are given by $d = (20, 45, 15, 11, 34, 25, 40, 62, 62, 50)$ then EDD ordering would produce $S^* = \{4, 3, 1, 6, 5, 7, 2, 10, 9, 8\}$ with $L_{\max}(S^*) = L_5 = L_7 = 9$. □

3.1.3 Tardiness Models

If a job completes at a time strictly greater than its due-date, we say that the job is *tardy*. Letting this be denoted by T_i, we have $T_i \triangleq \max(0, c_i - d_i)$. Accordingly, some results (but not many) carry over as before. The proof of the next theorem is the same as that in the $L_{\max}$ case:

Theorem 3.7 *The single-processor problem* $1||T_{\max}$ *is solved by EDD-order.* □

Unfortunately, our problems now begin to get somewhat more difficult. To see how this might be so, consider the total tardiness problem $1||\Sigma T_i$. Clearly, SPT ordering will not work as the following instance establishes: $\tau_1 = t, \tau_2 = kt$ with $k, t > 1, d_1 = (k+1)t$, and $d_2 = (k+1)t - 1$. Here, SPT produces the sequence $\{1, 2\}$ with strictly positive total tardiness ($\{2, 1\}$ is optimal). Similarly, EDD is also not a correct strategy. For this, consider an instance with $n = t+1 \geq 3$ and assume $\tau_1 = t$ with all other jobs having unit duration times. Let $d_1 = t$ and $d_i = t + \epsilon$ for $2 \leq i \leq t + 1$ where $\epsilon > 0$. Then EDD produces the family of $t!$ sequences which have job 1 first followed by the other jobs in any order. But the total tardiness in all cases is $t(t + 1)/2$. On the other hand, the (non-EDD) sequences, which place job 1 last, have total tardiness value t.

In fact, the prospects of finding any fast algorithm that solves problem $1||\Sigma T_i$ appear to be gloomy. Indeed, the following result resolves what stood for some time as one of the more interesting open problems in single-processor scheduling. We sketch the proof of the following.

Theorem 3.8 *The (recognition version) single-processor problem* $1||\Sigma T_i$ *is $\mathcal{NP}$-Complete.*

Proof. (Sketch) A not uncommon trick in complexity proofs is to establish the $\mathcal{NP}$-Completeness of an intermediate problem which, in turn, is used as the basis for the intended reduction. This is precisely the case here. Let us consider the following, EVEN-ODD PARTITION problem which is known to be hard:

Given a set of integers $B = \{b_1, b_2, \ldots, b_{2t}\}$ where $b_i \epsilon \mathbb{Z}^+$ and such that $b_i > b_{i+1}$ for each $1 \leq i < 2t$, is there a bipartition of B into subsets B_1

and B_2 such that $\sum_{b_i \epsilon B_1} b_i = \sum_{b_i \epsilon B_2} b_i$ and where for each $1 \leq i \leq t$, B_1 (and thus B_2) contains exactly one of the pair $\{b_{2i-1}, b_{2i}\}$?

Now, we can pose a modified version of this problem where the elements to be partitioned are restricted. Then by a reduction from the unrestricted version above, it can be shown that the so-called RESTRICTED EVEN-ODD PARTITION problem, which we make formal below, is also $\mathcal{NP}$-Complete:

Given a set of integers $A = \{a_1, a_2, \ldots, a_{2t}\}$ where $a_i \epsilon \mathbb{Z}^+$ and such that $a_i > a_{i+1}$ for each $1 \leq i < 2t$, $a_{2j} > a_{2j+1} + \delta$ for each $1 \leq j < t$, and $a_i > t(4t+1)\delta + 5t(a_1 - a_{2t})$ for each $1 \leq i \leq 2t$, where $\delta = 1/2 \sum_{1 \leq i \leq t}(a_{2i-1} - a_{2i})$, is there a bipartition of A as $\{A_1, A_2\}$ such that $\sum_{a_i \epsilon A_1} a_i = \sum_{a_i \epsilon A_2} a_i$ and where for each $1 \leq i \leq t$, A_1 (and thus A_2) contains exactly one of the pair $\{a_{2i-1}, a_{2i}\}$?

The proof now proceeds by establishing a reduction from this restricted partition version. Following, we give the explicit mapping. Let $A = \{a_1, a_2, \ldots, a_{2t}\}$ constitute the instance of the relevant partition problem and set $\bar{A} = 1/2 \sum_{1 \leq i \leq 2t} a_i$. Observe that

$$\bar{A} = \sum_{1 \leq i \leq t} a_{2i-1} - \delta = \sum_{1 \leq i \leq t} a_{2i} + \delta$$

and also, that we may safely assume that $a_{2i-1} \leq a_{2i} + \delta$ for $1 \leq i \leq t$. The latter follows since otherwise there is no solution for A. Now, let us form a tardiness instance having $n = 3t + 1$ jobs which we shall denote as $\{v_1, v_2, \ldots, v_{2t}\} \cup \{w_1, w_2, \ldots, w_{t+1}\}$. These "$v$- and w-jobs" have duration times given as

$$\tau_{v_i} = a_i, \quad 1 \leq i \leq 2t$$

$$\tau_{w_i} = b = (4t+1)\delta, \quad 1 \leq i \leq t+1$$

$$d_{v_i} = \begin{cases} (j-1)b + \delta + (a_2 + a_4 + \ldots + a_{2j}) & \text{if } i = 2j-1, \\ d_{v_{2j-1}} + 2(t-j+1)(a_{2j-1} - a_{2j}) & \text{if } i = 2j. \end{cases}$$

$$d_{w_i} = \begin{cases} ib + (a_2 + a_4 + \ldots + a_{2i}) & \text{if } 1 \leq i \leq t, \\ d_{w_t} + \delta + b & \text{if } i = t+1. \end{cases}$$

and, finally, we set a threshold on tardiness as

$$k = \bar{A} + t\left[(t+1)b + 2\bar{A}\right] - t(t-1)b/2 - t\delta - \sum_{1 \leq i \leq t} (t-i+1)(a_{2i-1} + a_{2i}).$$

Then, we can show that the given instance of RESTRICTED EVEN-ODD PARTITION possesses a solution exactly when the constructed instance of $1||\Sigma T_i$ has an admissible n-job sequence with total tardiness no greater than k. This establishes the result of the theorem. □

The sense in which we have characterized the proof of Theorem 3.8 as "sketched" follows from the degree of technical detail that accompanies the full argument presented in Du and Leung (1990). In particular, we have left out of this description some technical lemmas, where key among these is a result establishing that it is sufficient to confine any search for an optimal tardiness solution to a reduced, but highly structured set of so-called *canonical schedules.* Then, concentrating on the latter, it can be shown that the threshold value given by k above is actually a lower bound on the total tardiness of any canonical schedule, and thus on an optimal one. Moreover, and particularly important, this bound is shown to be tight exactly when an appropriate partition of the "v-jobs" exists where the latter is related to a correct partition of the elements in A.

Prior to the result of Theorem 3.8, the strongest complexity result available was an $\mathcal{NP}$-Completeness proof for the weighted version. This is now an easy corollary that follows by setting weights at 1 to obtain the simple tardiness problem of the previous theorem.

Corollary 3.9 *The (recognition version) single-processor problem* $1||\Sigma w_i T_i$ *is* $\mathcal{NP}$*-Complete.* □

Note, however, that under a particularly limiting restriction on duration times, problem $1||\Sigma w_i T_i$ is polynomially solvable. Specifically, if duration values are identical, say all unit times, then the problem is solvable as a bipartite matching problem—for any set of weights. The construction is easy: define a complete bipartite graph $G = (X, Y, E)$ where the bipartition $\{X, Y\}$ is such that X corresponds to jobs and Y to sequential positions $1, 2, \ldots, n$. Edges (i, j) are defined so as to correspond to the assignment of job i to order position j. These edges are weighted as $c_{ij} = w_i \max(0, j - d_i)$. That is, if job i is placed in position j, it completes at time j (or kj if duration times are different than 1) and incurs tardiness $w_i T_i$. A minimum weight, perfect matching on G solves the problem.

Example 3.4 Suppose five unit execution time jobs have the due-dates and weights given by:

$$d = (4, 3, 2, 3, 4),$$
$$w = (3, 5, 6, 4, 7).$$

The corresponding bipartite graph is shown in Figure 3.1 and an optimum matching is denoted in bold. Note that edges with zero weight have been left out for ease of depiction. The corresponding optimum sequence is $\{3, 2, 4, 5, 1\}$ with weighted tardiness value 3. □

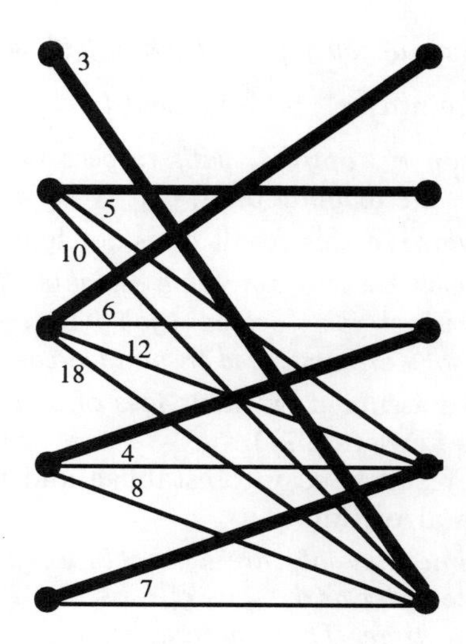

Figure 3.1. *Bipartite graph of Example 3.4*

A_{LT} : Lawler's Approach to the Tardiness Problem

Prominent in the context of procedures for certain tardiness/lateness problems are pseudo-polynomial algorithms and their extensions to *fully polynomial approximation schemes.* Following, we describe the development of one of these approaches, namely the treatment of the total tardiness case by an algorithm due to Lawler. We then show explicitly its conversion to the stated approximation scheme.

In fact, this particular pseudo-polynomial algorithm of Lawler is applicable to the weighted tardiness case but only under an assumption that job weights are *agreeable*, *i.e.*, $\tau_i < \tau_j \Rightarrow w_i \geq w_j$. Of course, that such a limitation exists is anticipated by the known, strong sense $\mathcal{NP}$-Completeness status of the problem with arbitrary weights (see Exercise 3-34). On the other hand, when weights are identical, Lawler's result is general since the "agreeability" condition is trivially satisfied by any instance. Our description below assumes that weights are arbitrary (but agreeable) while the development of the approximation scheme assumes unit weights at the outset.

To generate a sense for the procedure, three results are useful. For the first two, we shall simply state the respective outcomes without proof. Following Lawler (1977) we have:

Theorem 3.10 *Let jobs possess arbitrary weights w_j and denote by π any ordering that is optimal with respect to due-dates $d_1, d_2, \ldots, d_n$. Let c_j de-*

note the completion time of job j in π. Pick any d'_j satisfying

$$\min(d_j, c_j) \leq d'_j \leq \max(d_j, c_j).$$

Then any ordering, say π', optimal with respect to due-dates d'_j is also optimal with respect to the original ones, d_j. □

Observe that the converse of this result does not hold.

Theorem 3.11 *Let jobs possess agreeable weights. Then there exists an optimal ordering π in which job i precedes job j if $d_i \leq d_j$ and $\tau_i < \tau_j$ and in which all nontardy jobs are arranged in nondecreasing due-date order.* □

Note also that we have assumed, without loss of generality, that duration times are distinct (see Lawler, 1977).

Now, the following result is easy to establish and plays prominently in the development of an algorithm.

Theorem 3.12 *Assume that jobs are agreeably weighted and numbered in nondecreasing due-date order as $d_1 \leq d_2 \leq \cdots \leq d_n$. Further, suppose job k has the largest duration time. Then there exists an integer $\delta, 0 \leq \delta \leq n-k$ such that there is an optimal ordering π in which job k is preceded by jobs $j \leq k+\delta$ and followed by jobs indexed as $i > k+\delta$.*

Proof. Consider, among optimal orderings with respect to due-dates $d_1, d_2, \ldots, d_n$, one where job k completes as late as possible. Let this completion time be c'_k. Now, let π be an optimal ordering under due-dates $d_1, d_2, \ldots, d_{k-1}, d'_k, d_{k+1}, \ldots, d_n$ where $d'_k = \max(c'_k, d_k)$. Suppose further that π satisfies the conditions of Theorem 3.11 relative to the stated due-dates and let c_k be the completion time of k in π. From Theorem 3.10, π is also optimal under the original due-dates. But by hypothesis, $c_k \leq d'_k$. Also, k is not preceded in π by any job j for which $d_j > d'_k$ for if so, job j is nontardy, denying the conditions of Theorem 3.11. Job k must be preceded by all jobs with due-dates no greater than d'_k. So, if we pick δ as the largest integer for which $d_{k+\delta} \leq d'_k$ we have the stated result. □

The strategy should now begin to reveal itself. That is, Theorem 3.12 suggests (relative to the stated hypotheses regarding job weights and due-date numbering) that we can produce an optimal sequence for a set of jobs if we know that particular subsequences are optimal. More precisely, if k is a job having maximum duration time, then the theorem claims that there is some $\delta, 0 \leq \delta \leq n-k$ and for which there exists an optimal ordering of jobs in the form shown in Figure 3.2 with job k positioned as indicated. Here, job k completes at time $t + \sum_{j \leq k+\delta} \tau_j$. Then the overall sequence is optimal only if the jobs processesd before and after job k are optimally ordered (relative to starting times of t and $t + \sum_{j \leq k+\delta} \tau_i$ respectively). Recursively then, the

Figure 3.2. *Sense of ordering from Theorem 3.12*

optimal solution for some job set S starting at time t can be determined from the optimal solutions to subproblems defined by job subsets $S^* \subset S$ with start times $t^* \geq t$. That is, we can solve the problem by dynamic programming.

To set up the necessary recursion, denote by $S_{ij}(k)$ the set of jobs contained in the closed interval $[i, j]$ and which have duration strictly less than τ_k. Then if we let $T(S_{ij}(k), t)$ denote the weighted tardiness of an optimal permutation of the jobs in $S_{ij}(k)$ with starting time t, we have

$$T(S_{ij}(k), t) = \min_{\delta}\{T(S_{i,k+\delta}(k'), t) + w_{k'}T_{k'} + T(S_{k'+\delta+1,j}(k'),\ c_{k'}(\delta))\}$$

which follows from the result of Theorem 3.12 and by direct appeal to the principle of optimality. Note that $T_{k'}$ is the tardiness of job k' given by $\max(0, c_{k'}(\delta) - d_{k'})$ where as before, $c_{k'}(\delta) = t + \Sigma\tau_j$ summing over jobs in $S_{i,k+\delta}(k')$. In addition, k' is the job having the largest duration over jobs in the set $S_{i,j}(k)$. We can initialize the recursion by

$$\begin{aligned} T(\phi, t) &= 0 \\ T(\{j\}, t) &= w_j \max(0, t + \tau_j - d_j) \end{aligned}$$

Clearly, the order of computation under this recursive strategy requires no more than $O(n^3)$ subsets of the stated form $S_{i,j}(k)$ and there need be no more than $\sum \tau_j \leq n\tau_{\max}$ values of t to examine. Each evaluation of the functional equation requires a search over at most n alternatives with $O(n)$ effort yielding an overall complexity time bound of $O(n^5\tau_{\max})$.

Now, to produce the aforementioned approximation scheme we can employ a clever scaling trick described in Lawler (1982). Given the original instance defined by processing times and due-dates τ_j and d_j respectively, let us determine a set of "reduced" integer duration times by $q_j = \lfloor \frac{\tau_i}{K} \rfloor$, where K is an appropriately defined scaling factor to be described shortly. Let us also replace the old due-dates by scaled, real-valued ones given by $d'_j = \tau_j/K$. Then, operating on the instance defined by these new processing times and due-dates, the notion is to simply apply the recursive procedure described above with the ordering produced acting as our approximation of the optimum solution on the original instance. The next example illustrates the strategy. Before demonstrating the approach, however, let us define a suitable scaling factor and establish bounds for the approximation's performance as well as on its computational order.

Let S' be the ordering produced when operating on the data given by q_j and d'_j and let its tardiness value be ν'. Now, let the value of S' relative to the original due-dates d_j and with processing times $p_j = Kq_j \leq \tau_j$ be ν_p. Clearly, $\nu' \leq \nu_p$. But also, ν_p is bounded from above by the optimal tardiness value, T^*. Now, recall the definition of q_j. Clearly, $Kq_j \leq \tau_j < K(q_j + 1)$ and denoting the value of S' with the original data as ν_A, we have the inequality

$$\nu_A < \nu_p + K[n(n+1)/2]$$

and thus

$$\nu_A - T^* < K[n(n+1)/2].$$

Let us now choose the value of K. Accordingly, we may set

$$K = [2\varepsilon/n(n+1)]T_{\max}$$

which yields

$$\nu_A - T^* < \varepsilon T_{\max} \leq \varepsilon T^*$$

and we have the desired performance bound. For the required complexity function, we observe that in the application of the pseudo-polynomial procedure, our search over values t can be confined to at most $nT_{\max}$ choices rather than $\sum \tau_j$ (*cf.* Lawler, 1982). Hence, the time bound specified earlier can now be replaced by $O(n^5 T_{\max}/K)$. Upon substituting the given value for K, we obtain a bound of $O(n^7/\varepsilon)$ and the claim of a fully polynomial approximation scheme is substantiated, *i.e.*, the bound is polynomial in n and ε^{-1}.

We now consider a modification to the total (weighted) tardiness version where rather than a penalty measurement that grows linearly as tardiness, we assume that a single penalty charge is assessed for *any* nonzero amount of tardiness. Formally, let us associate a binary variable u_i with each job where u_i is 1 if job i is tardy and 0 otherwise. Then problem $1||\Sigma w_i u_i$ seeks to minimize the total weighted number of tardy jobs. The problem is hard.

Theorem 3.13 *The (recognition version) single-processor problem* $1||\Sigma w_i u_i$ *is* $\mathcal{NP}$*-Complete.*

Proof. The reduction is from the knapsack problem P_K, the proof of which was described in the previous chapter. □

But suppose we restrict the problem of Theorem 3.13 so that jobs have unit duration times. Certainly such instances are amenable to the bipartite matching tactic employed in the $\Sigma w_i T_i$ case but there is, however, a better way. Let us proceed as follows: simply arrange jobs in nonincreasing w-order and from this list, select jobs in order, rejecting a choice only if it or a previously chosen job must be tardy. We can demonstrate with the following illustration.

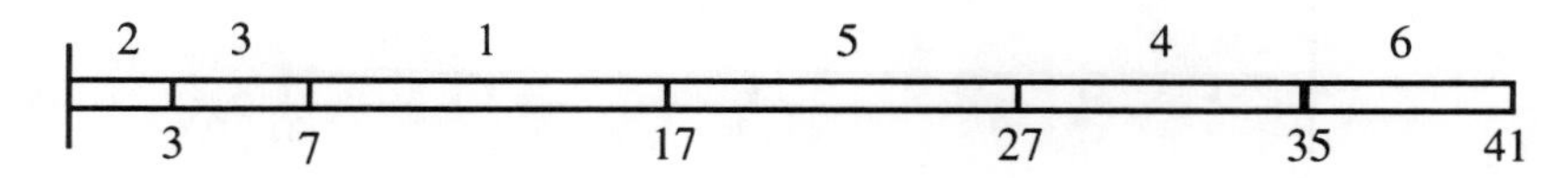

Figure 3.3. *EDD ordering in Example 3.6*

Example 3.5 Consider four jobs with $d = (2, 1, 2, 3)$ and $w = (9, 8, 7, 6)$. We would select job 1 first, job 2 next, reject job 3, and finally pick job 4 to be processed third. Only job 3 (processed last) is tardy with weight 7. □

Interestingly, the algorithm described above should catch our fancy by its inherently greedy nature. Indeed, the stated problem can be described in the context of a *transversal matroid* where independence testing is an easy consequence of the matching structure of the underlying bipartite graph described in the $1|\tau_i = 1|\Sigma w_i T_i$ case (*cf.* Lawler, 1976b).

Returning to the general $\Sigma w_i u_i$ problem, the following result of Moore (1968) is also well known for the subcase of identical weights (but with arbitrary duration times).

A_M : Moore's Algorithm for $1||\Sigma u_i$

Step 0: Form the EDD sequence S. If the number of tardy jobs is 0 or 1 stop.

Step 1: Locate the first tardy job in S, say x. Find any job among those in S up to and including x that has maximum duration. Letting this job be y, then set $S \leftarrow S \backslash y$ and repeat this step until all jobs in S are nontardy. The final sequence is given by $\{S, \bar{S}\}$ where $\bar{S}$ is the complement of S. □

Example 3.6 Consider six jobs having duration times and due-dates given by $\tau = (10, 3, 4, 8, 10, 6)$ and $d = (15, 6, 9, 23, 20, 30)$. Now, the EDD sequence results as $\{2, 3, 1, 5, 4, 6\}$ and the timing diagram is given as shown in Figure 3.3. Here the first job tardy is job 1. But job 1 is also the one (compared to those earlier in the sequence) having the largest duration time as required in the algorithm. We thus remove job 1 from its position and consider the updated schedule in Figure 3.4. Now, job 4 is the first job that is tardy but job y per the statement of the algorithm corresponds to job 5. Hence, we now have $S = \{2, 3, 4, 6\}$ with all jobs shown nontardy and the optimal sequence for the full problem is $\{2, 3, 4, 6, 1, 5\}$ ($\{2, 3, 4, 6, 5, 1\}$ works as well). □

The correctness of Moore's algorithm is established by the next theorem:

Theorem 3.14 *Algorithm A_M will, in polynomial time, produce a correct sequence for the single-processor problem $1||\Sigma u_i$.*

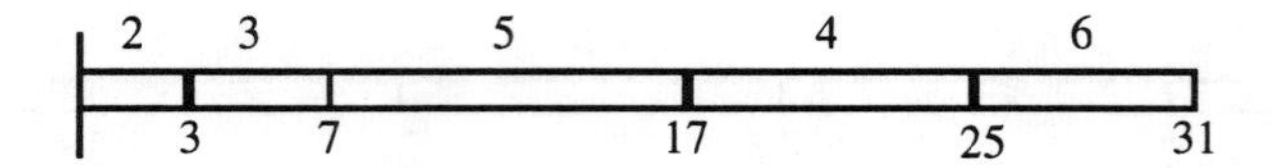

Figure 3.4. *Outcome of iteration 2 in Example 3.6*

Proof. Before proving the theorem, let us state a lemma due to Jackson (1955) that plays a key role. We have:

Lemma 3.15 *Given a set of jobs with duration times τ_j and due-dates d_j, there exists an ordering of the jobs in which none are tardy if and only if this outcome is possible under EDD ordering.* □

That is, if the EDD ordering produces some nonzero number of tardy jobs, then no other ordering (non-EDD) could produce zero. Trivially, if EDD yields zero or one tardy jobs the sequence is optimal. In the following we shall therefore assume that at stopping, at least two tardy jobs result.

Now, to establish the correctness of Moore's algorithm, we shall follow Sturm (1970) and show that when a correct application of the algorithm stops it does so with a sequence yielding θ tardy jobs and that there is no other ordering that exhibits fewer.

At some point, the algorithm (after starting with an EDD ordering) has produced a subset of jobs $\bar{S}$ that are unavoidably tardy and that can, moreover, be processed in any order at the end of the final sequence given by jobs in S. Let us assume these jobs, y, which are "removed" in Step 1, are denoted and indexed as $\beta_1, \beta_2, \ldots, \beta_h$ where β_i signifies the ith job selected as y in the process. Also, for notational ease, we will assume that these β-jobs are not actually removed (as per the algorithmic statement yielding the construction of $\bar{S}$) but remain in sequence with processing simply skipping over the stated jobs.

Similarly, the first tardy job encountered in successive sets S of the algorithm and given by x in Step 1 will be denoted sequentially as $\alpha_1, \alpha_2, \ldots, \alpha_k$. In the illustration earlier, $\alpha_1 = 1$ and $\alpha_2 = 4$ while $\beta_1 = 1$ and $\beta_2 = 5$.

Relativizing, the algorithm locates (if possible) a job in the current S say α_ℓ such that

$$d_{\alpha_\ell} < \sum_{j=1}^{\alpha_\ell} \tau_j - \sum_{j=1}^{\ell-1} \tau_{\beta_j}$$

and

$$d_i \geq \sum_{j=1}^{i} \tau_j - \sum_{\beta_j \leq i} \tau_{\beta_j} \text{ for } 1 \leq i \leq \alpha_\ell - 1.$$

That is, α_ℓ is the first job completing beyond its due-date (recall also that x is chosen prior to y). Now, the next β-job is selected as β_ℓ such that

$$\tau_{\beta_\ell} = \max\{\tau_j | 1 \leq j \leq \alpha_\ell \text{ and } j \neq \beta_i, 1 \leq i \leq \ell - 1\}$$

and if this is done, we have

$$\begin{aligned} d_{\alpha_\ell} &\geq d_{\alpha_{\ell-1}} \geq \sum_{j=1}^{\alpha_{\ell-1}} \tau_j - \sum_{j=1}^{\ell-1} \tau_{\beta_j} \\ &= \sum_{j=1}^{\alpha_\ell} \tau_j - \sum_{j=1}^{\ell-1} \tau_{\beta_j} - \tau_{\alpha_\ell} \\ &\geq \sum_{j=1}^{\alpha_\ell} \tau_j - \sum_{j=1}^{\ell-1} \tau_{\beta_j} - \tau_{\beta_\ell} \\ &\geq \sum_{j=1}^{\alpha_\ell} \tau_j - \sum_{j=1}^{\ell} \tau_{\beta_j}. \end{aligned}$$

Thus, job α_ℓ will no longer be tardy (unless $\alpha_\ell = \beta_\ell$, *i.e.*, $x = y$) after skipping (removing) job β_ℓ. If no such job α_ℓ is found, the algorithm stops and the claim is that the resultant order of processing is optimal.

Let us suppose otherwise. That is, let us assume the solution produced by the algorithm has $\theta > 1$ tardy jobs given by $\beta_1, \beta_2, \ldots, \beta_\theta$ but that there exists another ordering having fewer, say $\theta - 1$. Let this alternative have its tardy jobs denoted by $\epsilon_1, \epsilon_2, \ldots, \epsilon_{\theta-1}$. Then by hypothesis, for all i

$$d_i \geq \sum_{j=1}^{i} \tau_j - \sum_{\epsilon_j \leq i} \tau_{\epsilon_j}$$

where we assume without loss of generality that ϵ_i and β_j are distinct for all i and j.

Now Moore's algorithm found a first job α_1 that was tardy, so clearly $\epsilon_j \leq \alpha_1$ for some ϵ_j. But there is also a job ϕ_1 for which

$$\tau_{\phi_1} = \max_{\epsilon_j \leq \alpha_1} \{\tau_{\epsilon_j}\}$$

and by the definition of β_1, we know

$$\tau_{\phi_1} \leq \tau_{\beta_1}.$$

So, assume there are at least $\ell < \theta$ jobs ϵ_i for which $\epsilon_i \leq \alpha_\ell$ and that from among these, we have selected $\phi_1, \phi_2, \ldots, \phi_\ell$ where $\tau_{\phi_i} \leq \tau_{\beta_\ell}, 1 \leq i \leq \ell$.

On the other hand, the algorithm produced a job $\alpha_{\ell+1}$ with

$$d_{\alpha_{\ell+1}} < \sum_{j=1}^{\alpha_{\ell+1}} \tau_j - \sum_{j=1}^{\ell} \tau_{\beta_j} \leq \sum_{j=1}^{\alpha_{\ell+1}} \tau_j - \sum_{j=1}^{\ell} \tau_{\phi_j}.$$

But from the relationship for d_i given earlier, we have that

$$d_{\alpha_{\ell+1}} \geq \sum_{j=1}^{\alpha_{\ell+1}} \tau_j - \sum_{\epsilon_j \leq \alpha_{\ell+1}} \tau_{\epsilon_j},$$

implying that there are at least $\ell+1$ jobs ϵ_i and for which $\epsilon_i \leq \alpha_{\ell+1}$. Hence, we can select a job $\phi_{\ell+1}$ for which

$$\tau_{\phi_{\ell+1}} = \max\{\tau_{\epsilon_i} | \epsilon_i \leq \alpha_{\ell+1} \text{ and } \epsilon_i \neq \phi_1, \phi_2, \ldots, \phi_\ell\}.$$

With the algorithm, we located job $\beta_{\ell+1}$ with

$$\tau_{\beta_{\ell+1}} \geq \max \left\{ \begin{array}{r} \tau_{\epsilon_i} | \epsilon_i \leq \alpha_{\ell+1} \text{ and } \epsilon_i \neq \beta_1, \beta_2, \ldots, \beta_\ell; \\ \epsilon_i \neq \phi_1, \phi_2, \ldots, \phi_\ell \end{array} \right\} = \tau_{\phi_{\ell+1}}$$

But this means there are at least $\ell+1$ ϵ-jobs having $\epsilon_i \leq \alpha_{\ell+1}$ and $\ell+1$ of these given as $\phi_1, \phi_2, \ldots, \phi_{\ell+1}$ for which $\tau_{\phi_i} \leq \tau_{\beta_i}$ for $1 \leq i \leq \ell+1$. Proceeding inductively, the algorithm finds a job α_θ and so there must be θ ϵ-jobs having $\epsilon_i \leq \alpha_\theta$ which leads to the desired denial that $\theta - 1$ jobs are tardy in the hypothesized alternative. This completes the proof. □

We conclude this entire section with an interesting result that will be very useful later. It pertains to minimization problems of the form $1||f_{\max}$ where $f_{\max} \triangleq \max_i(f_i(c_i))$ and $f_i(c_i)$ is any continuous, nondecreasing function of job completion time. Consider the following algorithm due to Lawler (1973):

A_{LFm}: Lawler's Algorithm for $1||f_{\max}$

Step 0: Set $S = \{1, 2, \ldots, n\}$ and let $k = n$.

Step 1: Set $\lambda \leftarrow \sum_{j \in S} \tau_j$ and determine x such that $f_x(\lambda) \leq f_i(\lambda)$ for all $i \in S$.

Step 2: Place job x in the kth order position, set $S \leftarrow S \backslash x$, $k \leftarrow k-1$, and return to step 1 until $|S| = 0$. □

Example 3.7 Consider the following instance on four jobs:

$\tau = (2, 4, 6, 4)$

$f(\cdot) : (c_1^2, 4c_2, .5c_3^3, 25c_4^{3/2})$

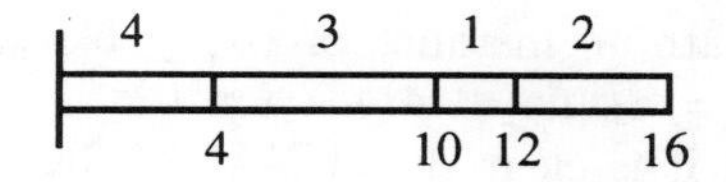

Figure 3.5. *Final schedule for Example 3.7*

Then, $S = \{1, 2, 3, 4\}$ and $\lambda = 16$ with $f_2(16)$ minimum. Thus, job 2 is scheduled last. Next, $S = \{1, 3, 4\}$ with $\lambda = 12$ and $f_1(12)$ is minimum so job 1 is processed third in the sequence. Continuing, we see that job 3 is chosen next and finally, job 4 is selected (it goes first in the final schedule). The schedule that results is shown in Figure 3.5 and has $f_{\max} = f_3 = 500$. □

Following, we establish the validity of algorithm $\boldsymbol{A_{LF_m}}$.

Theorem 3.16 *Algorithm* $\boldsymbol{A_{LF_m}}$ *correctly solves the single-processor problem* $1||f_{\max}$.

Proof. Let S be the sequence produced by the algorithm and denote its value by $\nu(S) = \max_j(f_j(c_j(S))) = f_x(c_x(S))$. Now, assume there exists another sequence $\bar{S}$ such that $\nu(\bar{S}) < \nu(S)$. Then job x must appear earlier in $\bar{S}$ than in S. Also, there must exist a job y such that $c_y(\bar{S}) \geq c_x(S)$. But then

$$f_y(c_y(\bar{S})) \geq f_y(c_x(S)) \geq f_x(c_x(S)),$$

where the first part follows by the nondecreasing nature of f and the second by a correct application of the algorithm. Thus,

$$\nu(\bar{S}) \geq f_y(c_y(\bar{S})) \geq f_x(c_x(S)) = \nu(S),$$

which is a contradiction. □

Observe that the earlier results regarding EDD ordering for the $1||L_{\max}$ and $T_{\max}$ cases now follow as simple corollaries of Theorem 3.16. That is, the last job sequenced has a largest due-date, the next to last, a next largest, and so forth. Clearly, lateness and tardiness functions are nondecreasing in job completion time and so the application of the algorithm is valid.

If we introduce nonzero release times then just as before (in the Σc_i case), we again see a marked difference in complexity status. The following result provides testimony in the $L_{\max}$ case (cases ΣT_i and $\Sigma w_i u_i$ are already hard with $r_i = 0$).

Theorem 3.17 *The (recognition version) problem* $1|r_i \geq 0|L_{\max}$ *is* $\mathcal{NP}$*-Complete.*

Proof. The reduction is easy, following a transformation from KNAPSACK (recall the stated form given in Example 2.1). As shown in Lenstra *et al.*,

(1977), we simply create an instance of $1|r_i \geq 0|L_{\max}$ on $n = t+1$ jobs for which $r_i = 0$, $\tau_i = a_i$, $d_i = A+1$ for $1 \leq i \leq t$; $r_n = b$, $\tau_n = 1$, and $d_n = b+1$. Then it is clear that KNAPSACK solves exactly when a schedule exists having $L_{\max} = 0$. □

On the other hand, if we impose the restriction of unit duration times, we preserve the efficient solvability status observed previously. That is, problems $1|r_i \geq 0, \tau_i = 1|\Sigma w_i T_i$ and $1|r_i \geq 0, \tau_i = 1|\Sigma w_i u_i$ can be solved in polynomial time by approaches akin to those described earlier. The reader is asked to verify these outcomes as an exercise.

3.2 Dependent Jobs

We now consider an environment where instances are subject to a nonempty, irreflexive partial order $\prec$ defined on the job set. Specified as *prec* and as suggested earlier, these partial orders correspond to operational precedence constraints, and are represented as acyclic digraphs. Moreover, it may be (and often is) the case that various, otherwise hard problems behave very differently when these digraphs exhibit particular structures. In the next section, we examine some of these.

3.2.1 Special Precedence Structures

Precedence Chains

Particularly simple are so-called *precedence chains.* Here, subsets of jobs are constrained by orderings, which when depicted graphically are given by digraphs where each vertex has indegree and outdegree of at most 1. An illustration is provided in Figure 3.6.

Forests

An obvious generalization of the precedence chain can be created by relaxing the bound on the indegree and/or outdegree requirement of the digraph. If outdegree is required to be at most 1 (indegree not restricted) an *in-forest* results and if indegree is at most 1 (outdegree not restricted), we have an *out-forest.* If these digraphs are connected, the structure is an *in-tree* or an *out-tree* respectively (also called *arborescences*). And if the digraph is the disjoint union of an in-forest and an out-forest, we refer to the precedence structure as an *opposing forest.* These digraphs are illustrated in Figure 3.7.

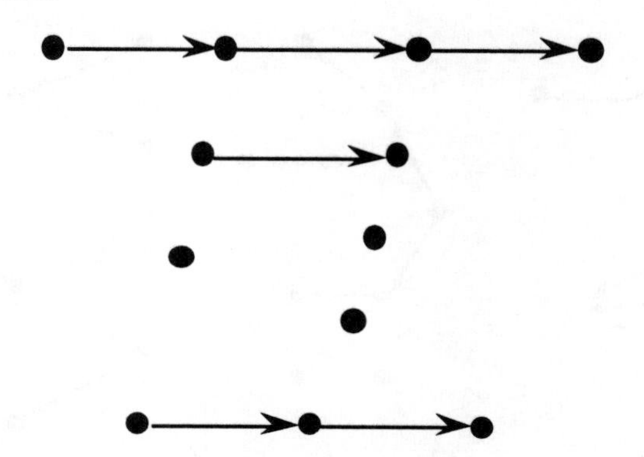

Figure 3.6. *Precedence chains*

Series-Parallel Digraphs

More general yet, and certainly more interesting, is the class of digraphs known (somewhat loosely) as *series-parallel* graphs. These graphs are recursively defined, a fact that in itself provides a glimpse into how problems might be solved on such structures. There are two forms often described: *vertex series-parallel* and *edge series-parallel.* We begin with the first of these.

Vertex series-parallel digraphs. Following Lawler (1978), let us define first a class of digraphs referred to as *vertex transitive series-parallel.* We have:

(i) A digraph $G = (V, A)$ consisting of a single vertex and $A = \phi$ is vertex transitive series-parallel.

(ii) For the vertex transitive series-parallel graphs $G_1 = (V_1, A_1)$ and $G_2 = (V_2, A_2)$ where $V_1 \cap V_2 = \phi$, $G_3 = (V_3, A_3)$ is vertex transitive series parallel if

$$\begin{aligned} V_3 &= V_1 \cup V_2 \\ A_3 &= A_1 \cup A_2 \cup V_1 \times V_2. \end{aligned}$$

Here, G_3 is formed by *series-composition.*

(iii) For the vertex transitive series-parallel graphs $G_1 = (V_1, A_1)$ and $G_2 = (V_2, A_2)$ where $V_1 \cap V_2 = \phi$, $G_3 = (V_3, A_3)$ is vertex transitive series-parallel if

$$\begin{aligned} V_3 &= V_1 \cup V_2 \\ A_3 &= A_1 \cup A_2. \end{aligned}$$

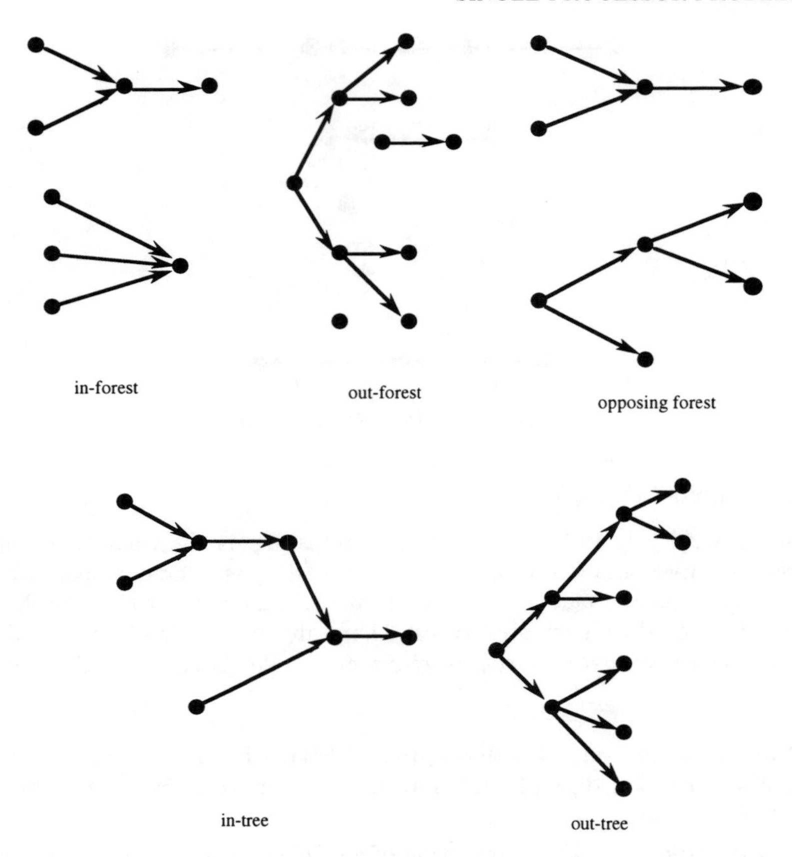

Figure 3.7. *In/out forests/trees*

Here, G_3 is formed by *parallel-composition.*

All graphs formed by a finite application of (i)–(iii) are vertex transitive series-parallel.

Example 3.8 The graph G in Figure 3.8 is shown to be constructable by the operations (i)–(iii) just given, and hence is vertex transitive series-parallel. Observe that s and p signify series and parallel composition respectively. □

Now, let $G = (V, A)$ be a digraph. Then $G' = (V, A')$ is the *transitive closure* of G such that there is an arc (i, j) in A' if and only if there is a directed path from i to j in G. A digraph then is said to be *vertex series-parallel* if and only if its transitive closure is vertex transitive series-parallel.

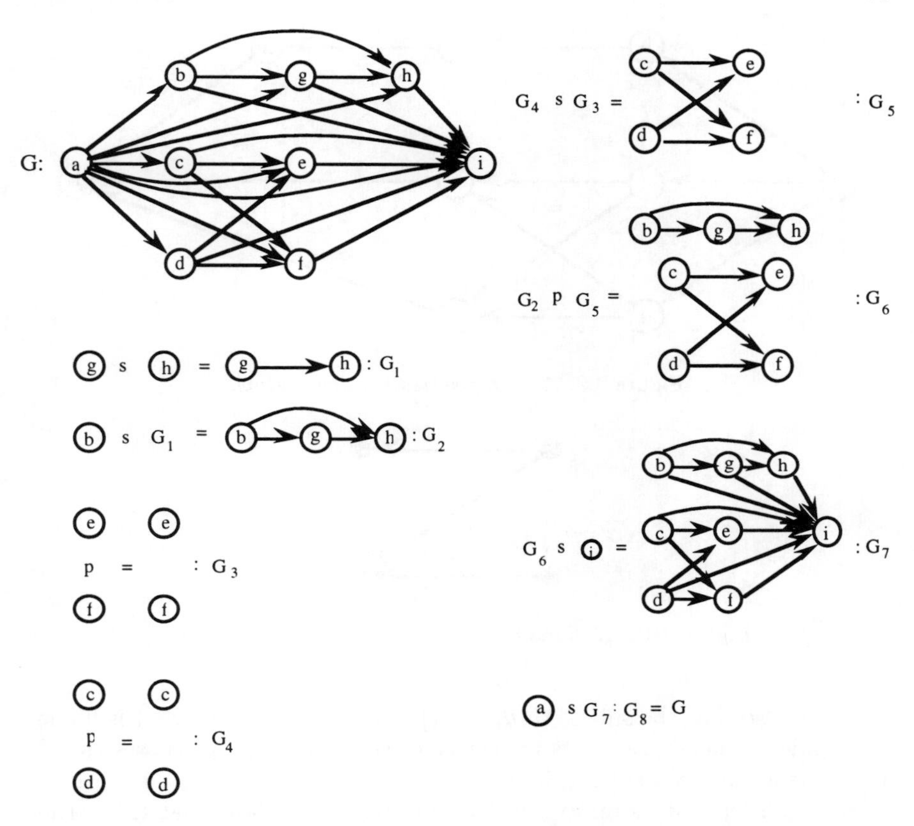

Figure 3.8. *Construction of a vertex transitive series-parallel graph*

Example 3.9 The digraph in Figure 3.9 is vertex series-parallel; its transitive closure was formed in the previous example. □

On the other hand, the digraph in Figure 3.10 is not vertex series-parallel; the structure shown is its own transitive closure and it is easy to see that it cannot be constructed using the series or parallel composition rules.

In fact, the digraph just exhibited is the smallest digraph that is not vertex series-parallel. However, it plays a much more prominent role than this. Referring to the given structure as a "Z-graph" we have the following forbidden subgraph characterization which we state without proof.

Theorem 3.18 *A digraph is vertex series-parallel if and only if its transitive closure possesses no Z-graph as an induced subgraph.* □

The digraph in Figure 3.11 is not vertex series-parallel since the subset of vertices $\{a, b, d, e\}$ clearly induces a subgraph of the forbidden form.

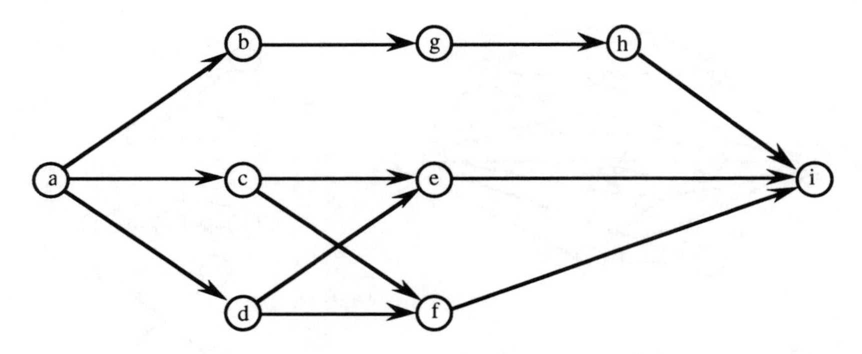

Figure 3.9. *A vertex series-parallel digraph*

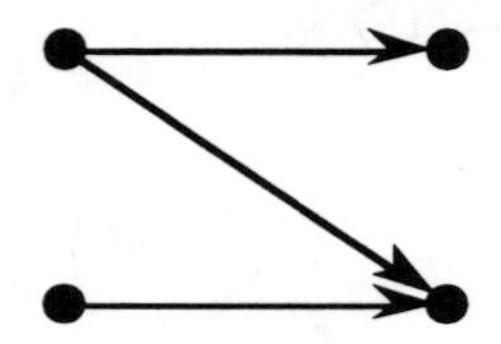

Figure 3.10. *A digraph that is not vertex series-parallel*

Observe also that the subset $\{b, d, f, g, i\}$ induces a subgraph that is homeomorphic to the Z-graph; its transitive closure, however, possesses the Z-graph as an induced subgraph.

We shall leave it as an exercise for the reader to show that the earlier structures (chains, in- and out-trees) are vertex series-parallel (in the case of chains and forests, we need only apply the definitions to the connected components or alternately create a connected structure by adding a dummy terminal vertex). Rather, we now remind ourselves of the original point: vertex series-parallel digraphs apparently yield something positive insofar as treating an otherwise hard problem is concerned. But, it should presently not be difficult to imagine how this could be, for if two graphs G_1 and G_2 are joined by a binary series or parallel operation then we may be able to determine an optimal solution for our problem on the composed graph, say G_3, assuming we have optimal solutions on G_1 and G_2. Recursively, if G_3 is the original digraph, we would have solved the full problem accordingly.

Still, even if this recursive strategy is valid for some problem restricted to the class of vertex series-parallel digraphs, it is clearly of little use if we are not able to *recognize* membership in the class, *i.e.*, we need to be able to decide (quickly) if a digraph is vertex series-parallel or not. Fortunately, however, this can be done.

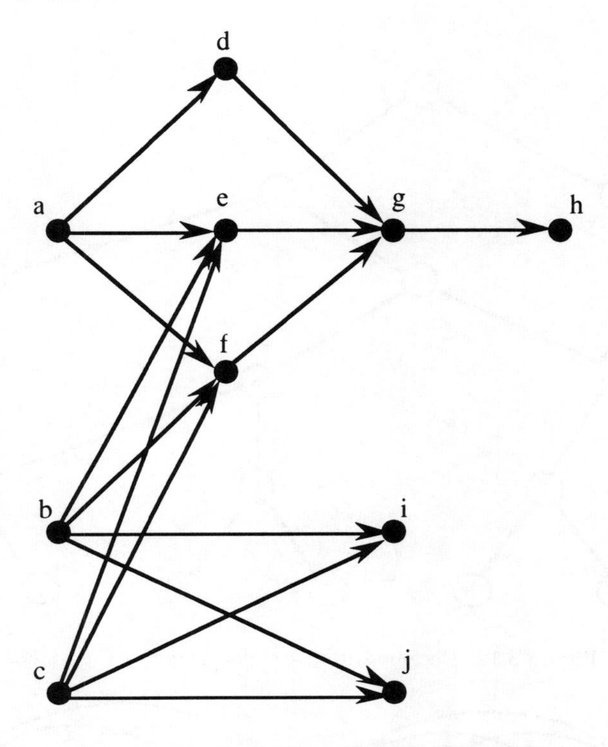

Figure 3.11. *Forbidden subgraph recognition*

A *decomposition tree* of a vertex series-parallel digraph is a rooted tree where the leaves are jobs and each internal node in the tree corresponds to a series or parallel (but not both) operation. Identified with each node is a graph formed by the respective composition rule (the leaves are graphs consisting of a single vertex) with the instance graph identified with the root. The decomposition tree for the graph given earlier is shown in Figure 3.12.

The important result here is that the decomposition tree of a vertex series-parallel digraph can be exhibited quickly (*i.e.*, polynomially) and as a consequence can be considered as part of the instance (Valdes, Tarjan, and Lawler, 1982). In fact, the recognition of vertex series-parallel graphs follows by either exhibiting a decomposition tree or by discovery (during the process) of a forbidden subgraph which produces the correct conclusion that the graph is not vertex series-parallel. Note also that the decomposition tree of a vertex series-parallel digraph is not unique. This is illustrated in Figure 3.13.

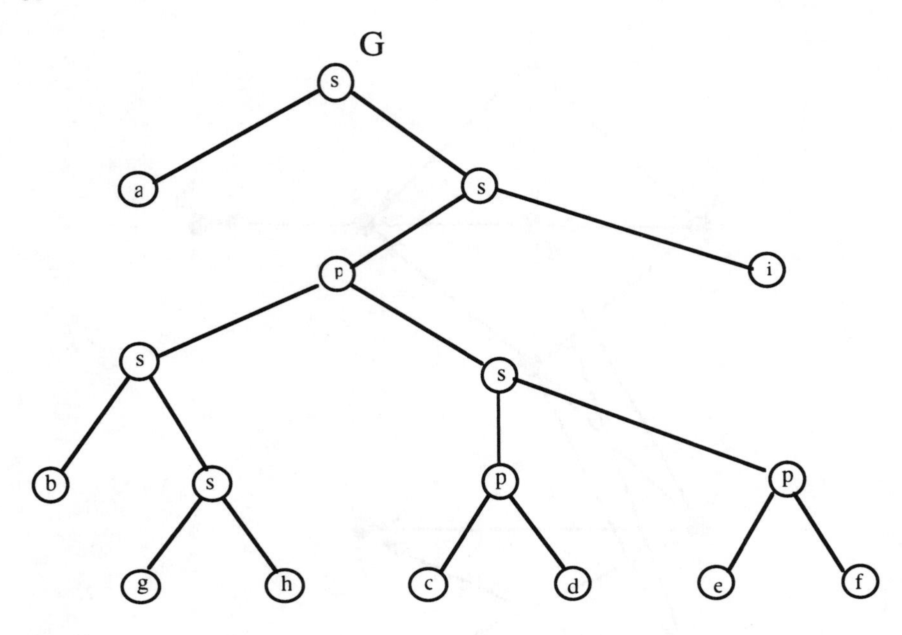

Figure 3.12. *Decomposition tree for graph of Figure 3.9*

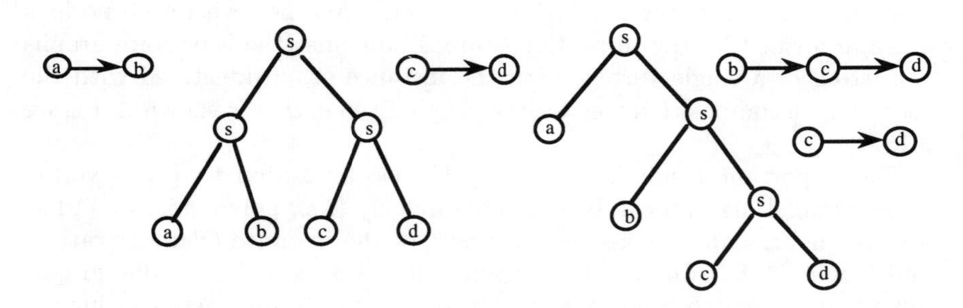

Figure 3.13. *Two decomposition trees of the same graph*

Edge series-parallel graphs. Probably more familiar, in that appeal to traditional series and parallel compositions is immediate, is the class of so-called *edge series-parallel digraphs.* Here, the primitive components are directed edges (arcs) rather than vertices, with each vertex of the arc referred to as a *terminal*; t_1 is the label for the vertex of outdegree 1 while t_2 is the terminal label for the other (indegree = 1) vertex. If e and f are two arcs with terminals labeled in this way, then the arcs are *series* composed by matching t_2 of e with t_1 of f producing a new digraph having terminals t_1 and t_2, where the first is preserved from e while the second is preserved from f. The common vertex (formerly t_2 from e and t_1 from f) relinquishes status as a terminal after e and f have been joined by the series operation. The *parallel* composition matches like terminals from e and f, both of which are preserved as terminals in the resulting composed graph. Similarly, subsequent graphs are composed in a series or parallel fashion in this way, yielding after each composition a new graph with two terminals. In this regard, graphs formed in this fashion are often called *2-terminal*, edge series-parallel digraphs (since the context is clear, we will hereafter drop the reference to "2-terminal"). A decomposition tree for an edge series-parallel structure is shown in Figure 3.14 where terminals are denoted by darkened vertices.

It should be clear that these series and parallel constructions do not form circuits and since arcs (the primitive graphs) are acyclic, any and all digraphs formed in the stated fashion are also acyclic. In addition, edge series-parallel digraphs are particularly easy to recognize. Given an acyclic digraph, all that is required is that we successively perform series and/or parallel *reductions* [*i.e.*, series reduction: replace a 2-arc path of the form (i, j, k) by a single arc (i, k); parallel reduction: replace two arcs both existing as (i, j) by a single arc, (i, j)] until we can reduce no further. At stopping, if we have a single arc, we know the original graph was edge series-parallel and if we have something different than an arc, it is not. That is, we have the following characterization:

Theorem 3.19 *A digraph G is edge series-parallel if and only if it can be reduced to single arc by series and parallel reduction.* □

The reader may want to test the result of the theorem by applying it to the previous digraph.

There is also a forbidden subgraph result for edge series-parallel digraphs.

Theorem 3.20 *A digraph is (2-terminal) edge series-parallel if and only if it possesses no subgraph homeomorphic to the 2-terminal digraph in Figure 3.15.* □

The darkened vertices in the digraph of the theorem are terminals.

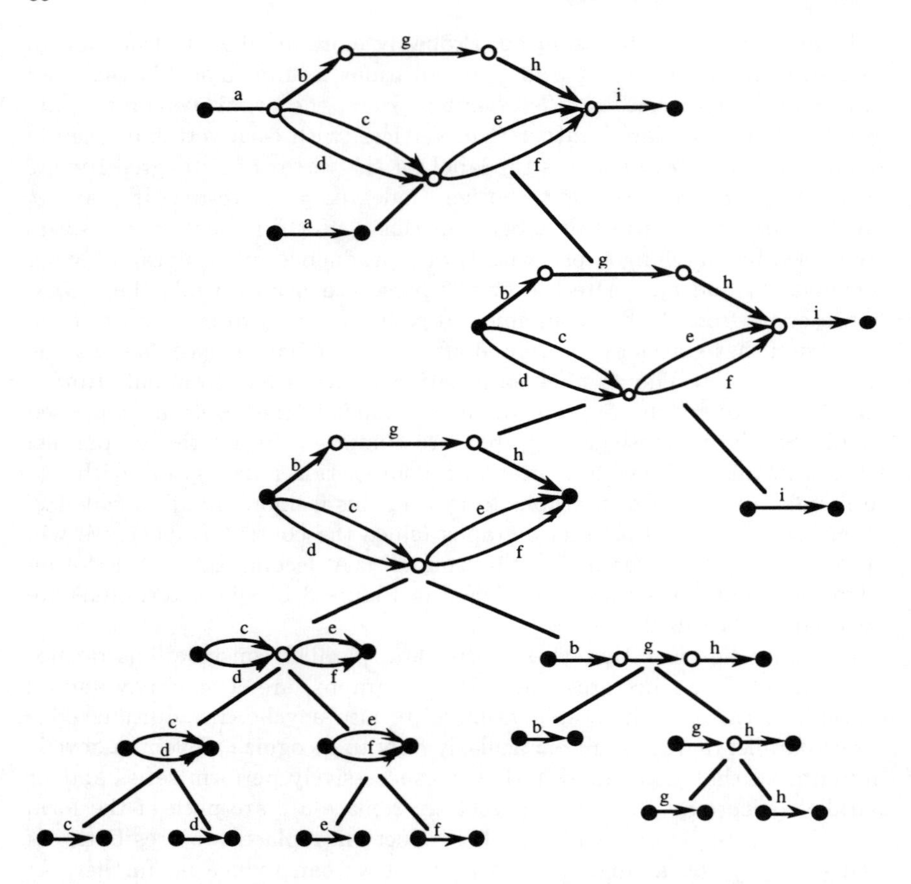

Figure 3.14. *Edge series-parallel decomposition*

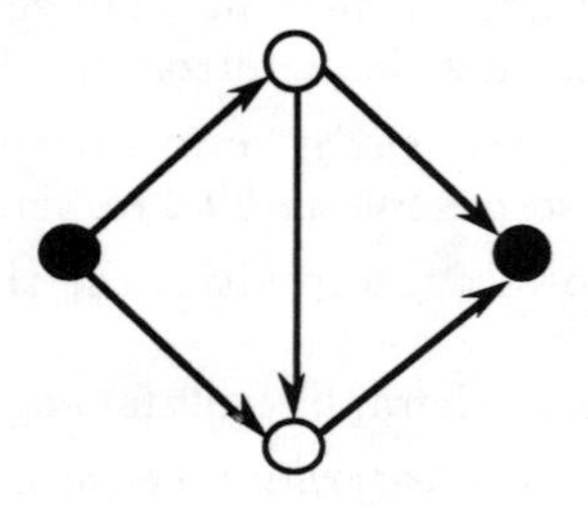

Figure 3.15. *Forbidden subgraph of 2-terminal, edge series-parallel digraphs*

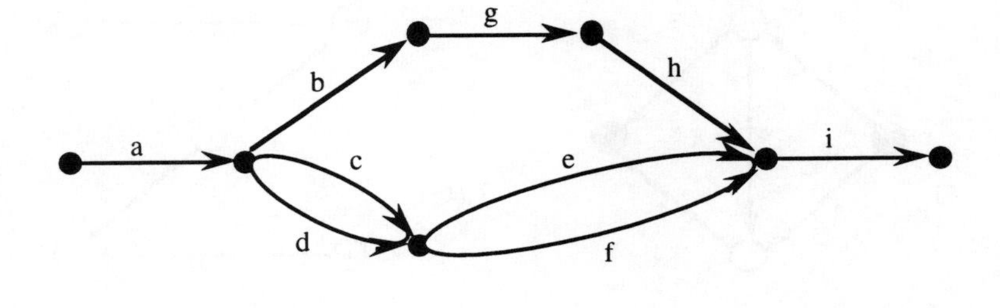

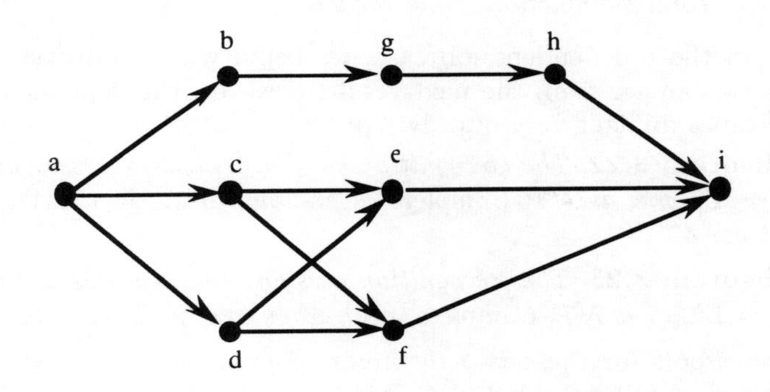

Figure 3.16. *Line graph construction*

But are vertex and edge series-parallel digraphs related? To see that they are, let us consider the notion of a *line graph.* The line digraph of a directed acyclic graph G is the digraph $L(G)$ with a vertex $f(e)$ for each arc e of G and an arc $(f(e_1), f(e_2))$ for each pair of arcs e_1, e_2 in G of the form $e_1 = (u,v), e_2 = (v,w)$. A graph and its line graph are shown in Figure 3.17. The observant reader will be quick to note that the graphs in Figure 3.16 are precisely the edge and vertex series-parallel digraphs illustrated previously. In fact, we are thus led to the following result.

Theorem 3.21 *A digraph G is edge series-parallel if and only if its line graph, $G' = L(G)$, is vertex series-parallel.* □

As one might expect, the forbidden subgraphs in each case are related in an obvious way which we demonstrate in Figure 3.17. Observe that $L(G)$ in the figure is a homeomorph of the forbidden Z-graph; its closure contains the Z-graph as an induced subgraph.

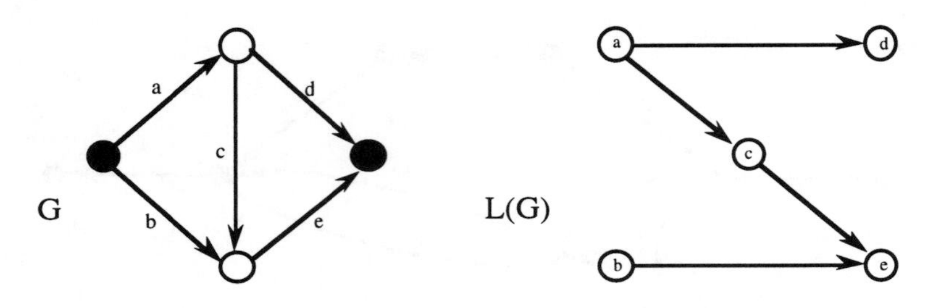

Figure 3.17. *Relationship between forbidden graphs*

3.2.2 Total Completion Time Models

As in the independent job case, we begin with completion time models. As we can see from the next result, however, the dependent job problem becomes difficult very quickly. We have:

Theorem 3.22 *The (recognition version) single-processor problem* $1|$ *prec,* $\tau_i = 1|\Sigma w_i c_i$ *is* $\mathcal{NP}$*-Complete, even when* $w_i \in \{\delta, \delta+1, \delta+2\}$ *for any integer* δ. □

Theorem 3.23 *The (recognition version) single-processor problem* $1|$ *prec,* $w_i = 1|\Sigma_i c_i$ *is* $\mathcal{NP}$*-Complete, even when* $\tau_i \in \{1,2,3\}$. □

The proofs for these two theorems, both of which follow by a reduction from the LINEAR ARRANGEMENT problem, are given in Lawler (1978).

On the other hand, when the precedence structure is vertex or edge series-parallel, problem $1|prec|\Sigma_i w_i c_i$ is solvable in polynomial time. Following, we present an algorithm due to Lawler (1978) (we shall assume the vertex series-parallel case in the presentation).

Conceptually, the procedure is simple. Starting with single jobs identified by leaves of the instance graph's decomposition tree, we progress up the limbs of the tree, merging optimal job subsequences as prescribed by the respective series or parallel composition operations identified with internal (non-leaf) nodes of the decomposition tree. The sequence obtained at the tree's root solves the problem. Of course, a key algorithmic issue to be negotiated in this scheme is precisely how the aforementioned merging proceeds *vis-a-vis* the series and parallel operations.

Parallel composition: Of the two rules, this is the easiest. Let M_1 and M_2 be sets of jobs each with elements arranged in nondecreasing τ/w-order. Then the *parallel composition* of M_1 and M_2 is simply the union $M = M_1 \cup M_2$ with elements in M also arranged in nondecreasing τ/w-order. Clearly, the sequence of jobs given by M is feasible and nondecreasing τ/w-order remains optimal (for M) given that it is so for M_1 and M_2. Note, however, that for this ordering to be consistent, we will adopt a generalization of the notion of a job to include ones defined as *composite*. These are artificial jobs given by subsets of two or more real jobs combined and treated as a singleton for purposes of the ratio-based ordering (observe that real elements comprising a composite job may or may not be sequenced in nondecreasing τ/w-order).

Series composition: On the surface, it would seem that the merger of sets M_1 and M_2 by *series composition* would be uninteresting, requiring a simple concatenation of the two subsequences. It is the case, however, that simple concatenation *per se* fails to preserve the nondecreasing τ/w-ordering. For example, suppose M_1 and M_2 each consist of single jobs, say i and j respectively. Then the graph resulting from the series composition of these two sets (jobs) is the arc directed from vertex i to vertex j, in turn reflecting the precedence relationship between job i and job j. Now if $\tau_i/w_i \leq \tau_j/w_j$ this same ordering is preserved trivially in $M = M_1 \cup M_2$. On the other hand, with $\tau_i/w_i > \tau_j/w_j$, it is not and the reversal of i and j is, by definition, inadmissible. Herein is the role of the composite job described above. We may simply replace real jobs i and j by a new one which is treated subsequently as a single job and which has duration time and weight equal to the corresponding sums of these parameters for the original jobs. Moreover, this strategy can be repeated where new composite jobs are ultimately created by the contraction of previously formed ones. We will not pursue the notion further, but there are some technical details involved in justifying this composite job formation strategy and the interested reader is directed accordingly to Lawler (1978) as well as Sidney (1975). Following, we will simply state the computational process for the series composition. Note that as a convenience, we will take the liberty to refer to ratios τ_i/w_i where the subscript can refer to real jobs as well as composite ones. In addition, to get around references to empty sets, we will assume that M_1 and M_2 possess dummy elements with ratio value $-\infty$ and $+\infty$ respectively. The algorithm for series composition is:

Step 1: Locate a maximum ratio (τ/w) element in M_1 and one of minimum value in M_2. Calling these i and j respectively, then if $\tau_i/w_i \leq \tau_j/w_j$, set $M = M_1 \cup M_2$ and stop. If $\tau_i/w_i \geq \tau_j/w_j$, remove i from M_1, remove j from M_2, and form the composite job $k = [i, j]$.

Step 2: (2.1.) Locate a maximum ratio element in M_1, say i. If $\tau_i/w_i \leq \tau_k/w_k$, go to Step 2.3.

(2.2) Remove i from M_1 and form composite job $k \leftarrow [i, k]$. Return to Step 2.1.

(2.3.) Find a minimum ratio element j in M_2. If $\tau_k/w_k \leq \tau_i/w_i$, set $M = M_1 \cup M_2 \cup \{k\}$ and stop.

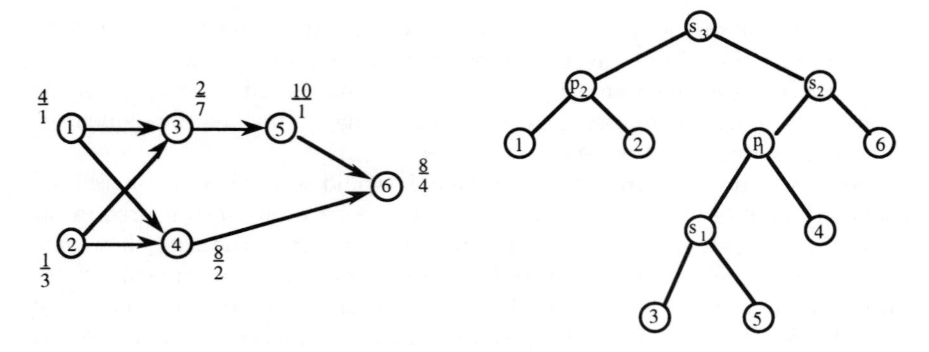

Figure 3.18. *Vertex series-parallel graph and its decomposition tree*

(2.4.) Remove j from M_2 and create composite job $k \leftarrow [k, j]$. Return to Step 2.1. □

The reader should have little difficulty in observing how the iterative composite job creation strategy stops with a set M having elements that when arranged in nondecreasing τ/w-order preserve feasibility. Below, we demonstrate the entire strategy with an illustration.

Example 3.10 Consider the (vertex) series-parallel graph in Figure 3.18 where the ratio next to each vertex signifies duration time and weight as τ_i/w_i. The decomposition tree for the graph is also shown to the right. Note that we have indexed the series (s_i) and parallel (p_i) operations so that the ordering of their application can be recorded. Beginning with s_1 and operating on $M_1 = \{3\}$ and $M_2 = \{5\}$ (for ease, we will not exhibit the dummy elements $\pm\infty$) we have $M = \{3, 5\}$ since $\tau_3/w_3 < \tau_5/w_5$. Similarly, letting $M_1 = \{3, 5\}$ and $M_2 = \{4\}$, then under p_2 we obtain $M = M_1 \cup M_2$ and by ordering on τ/w ratios, we produce $M = \{3, 4, 5\}$. The computation proceeds by next composing under s_2 the outcome of which is given along with the remainder of the solution, in the list below.

$$\begin{array}{llll} s_2: & M_1 = \{3,4,5\} & M_2 = \{6\} & M = \{3,[4,5,6]\} \\ p_2: & M_1 = \{1\} & M_2 = \{2\} & M = \{2,1\} \\ s_3: & M_1 = \{2,1\} & M_2 = \{3,[4,5,6]\} & M = \{2,[1,3],[4,5,6]\} \end{array}$$

The reader may want to verify the formation of the two composite jobs created under s_2 and s_3.

The final ordering is shown by the timing diagram in Figure 3.19. The weighted completion time value is 244. □

As described in Lawler, a careful implementation of the above strategy can be accomplished so that an n-job instance requires $O(n \log n)$ effort.

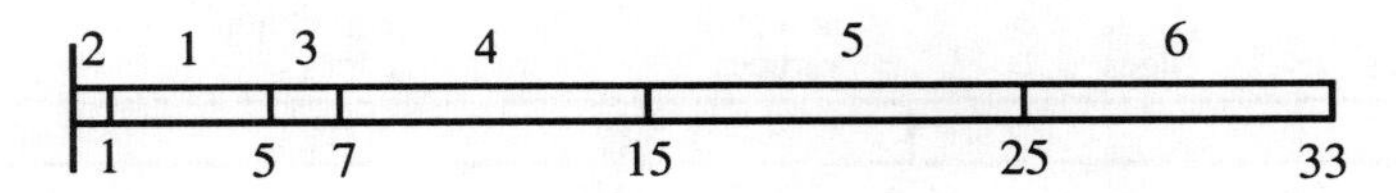

Figure 3.19. *Final sequence*

This time bound assumes that the decomposition tree is part of the instance.

3.2.3 Due-Date Problems

Just as we observed in the unconstrained case, most due-date problems remain hard when the dependencies are introduced. As testimony, we have:

Theorem 3.24 *The (recognition version) single-processor problem* $1|$ *prec,* $\tau_i = 1|\Sigma T_i$ *is* $\mathcal{NP}$*-Complete.*

Proof. Our proof follows Rinnooy Kan (1976) using a reduction from k-CLIQUE. Accordingly, consider a finite graph $G = (V, E)$ and an integer k that comprises an instance of the clique problem and let $n_0 = |V|, e_0 = |E|, n_1 = n_0 - k, e_1 = e_0 - f$ where $f = k(k-1)/2$. Note that f is the edge cardinality of a complete graph on k vertices. Now, let us define an instance of the stated scheduling problem in the following way:

$n = n_0 + n_0 e_0,$

$d_i = n_0 + f n_0, 1 \le i \le n_0,$

$d_{(j+1)n_0} = k + f n_0, 1 \le j \le e_0,$

$d_{jn_0+t} = n_0 + n_0 e_0, 1 \le j \le e_0, 1 \le t \le n_0 - 1,$

$i \prec jn_0 + 1$ if vertex i is incident to edge j, $1 \le i \le n_0, 1 \le j \le e_0,$

$jn_0 + t \prec jn_0 + t + 1, 1 \le j \le e_0, 1 \le t \le n_0 - 1,$

$y = \frac{1}{2} n_0 e_1 (e_1 + 1) + e_1 n_1.$

From the construction shown, for every edge in E, a corresponding precedence chain is created. There are n_0 jobs in each chain, the last of which has due-date $k + fn_0$. Also, from the precedence relationships created by the vertex-edge incidence, whenever a chain of these "edge-jobs" is scheduled, it must be preceded by the two vertices to which the particular edge e is incident in G. Clearly, at most f edge-chains can be scheduled within the interval $[0, k + fn_0]$. But this means that at least $e_1 = e_0 - f$ chains must be scheduled "tardy" where the only tardy job in any such chain is the last one. We then have for these tardy jobs, indexed by i:

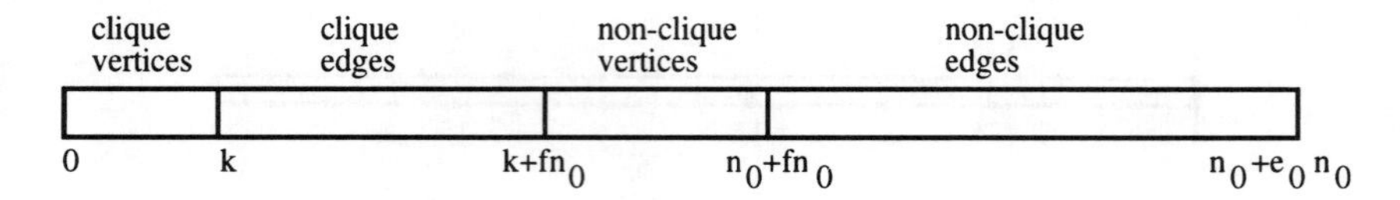

Figure 3.20. *A suitable schedule from a clique of size k*

$$\begin{aligned}\sum_i T_i &\geq \sum_{j=1}^{e_1}[(fn_0 + n_0 + jn_0) - (k + fn_0)]\\ &= e_1 fn_0 + (2n_0 + 3n_0 + \cdots + (e_1+1)n_0) - e_1(k + fn_0)\\ &= \frac{1}{2}n_0 e_1(e_1+1) - n_0 + (e_1+1)n_0 - e_1 k\\ &= \frac{1}{2}n_0 e_1(e_1+1) + n_0 e_1 - e_1 k\\ &= \frac{1}{2}n_0 e_1(e_1+1) + e_1 n_1 = y,\end{aligned}$$

and y is indeed a lower bound on minimum total tardiness.

Now, suppose G possesses a k-clique. Then it is easy to see that a suitable schedule can be formed and that has total tardiness y. We demonstrate this assertion in Figure 3.20. Here, the first two segments on the timing diagram pertain to jobs defined by the (existing) clique vertices and edges respectively. The third segment schedules jobs defined by vertices not in the given k-clique, and the final segment, those defined by nonclique edges.

Conversely, assume that no k-clique in G exists. Then the maximum number of edges in any k-vertex induced subgraph of G is $f-1$. It is easy to see that for any schedule, we must have no fewer than e_1+1 jobs tardy and so $\Sigma_i T_i \geq y + 1 > y$. This completes the proof. □

On the other hand, some results do carry over from the unconstrained case. By using a generalization of EDD we can solve the problems $1|$ *prec* $|T_{\max}(L_{\max})$.

Theorem 3.25 *The* $1|$ *prec* $|T_{\max}(L_{\max})$ *problems are solved by arranging jobs in nondecreasing* $\bar{d}$*-order where* $\bar{d}_j \triangleq min(d_j, min\{d_k|k$ *is a successor of* j *in* $\prec\})$. □

We leave the proof of Theorem 3.25 as an exercise (see Exercise 3-29). Still, it is helpful to observe that the same "last to first" sequence construction scheme employed previously for problem $1||f_{\max}$ also extends in a natural way to the present case. That is, from among those jobs that can finish last (*i.e.*, jobs without successors or, in general, ones without un-

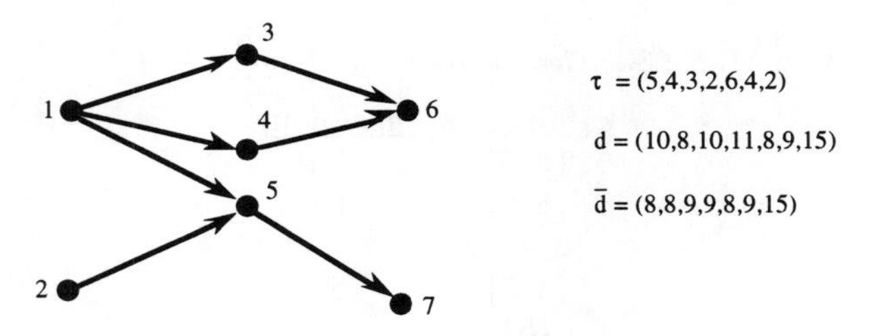

Figure 3.21. *Graph of Example 3.11*

scheduled successors) we need simply select one that exhibits a minimum difference $c_j - d_j$ where c_j is the sum of job duration times. This job is placed last in the sequence; its duration is subtracted from c_j, and then from among jobs able to complete at this new time, we repeat the process, selecting a job for the second to last sequential position. Continuing in this way, we ultimately form a complete sequence that solves the problem. It is also easy to see that the sequence produced will satisfy the conditions of Theorem 3.25. We illustrate both approaches by the following.

Example 3.11 Consider the precedence structure given by the graph in Figure 3.21. Duration times, original due-dates, and the modified ones are indicated accordingly. Now, ordering on $\bar{d}_j$ values produces a sequence {2,1,5,3,4,6,7}. Other sequences, both feasible as well as infeasible, exist and satisfy the stated ordering. Important, but easy to guarantee, is that the latter be avoided.

On the other hand, if we employ the strategy of filling sequential positions from last to first, our initial set of candidates (ones capable of being processed last) would be jobs 6 and 7. Evaluating $c_6 - d_6$ and $c_7 - d_7$ requires a comparison of the respective due-dates upon which our selection places job 7 last. This leaves jobs 5 and 6 as the new candidates for processing and when applying the same test, we see that $d_6 > d_5$ and so job 6 is processed in the 6th position. Next, jobs 3, 4, and 5 become candidates and we select job 4 which is ordered 5th. The procedure continues in this fashion and is summarized in Table 3.1. The (optimal) value of $T_{\max}$ (and $L_{\max}$) is seen to be $T_6 = 15$. □

3.2.4 Sequence-Dependent Setup Problems

We conclude this chapter by briefly describing a class of single-processor problems that take into account the (very real) possibility that there may be some unavoidable and nonnegligible amount of time required between the completion and start times of pairs of adjacent jobs in a sequence. Moreover,

Table 3.1. *Computation for Example 3.11*

candidates	*decision*
$\{6,7\}$	7
$\{5,6\}$	6
$\{3,4,5\}$	4
$\{3,5\}$	3
$\{5\}$	5
$\{1,2\}$	1
$\{2\}$	2

Figure 3.22. *Timing diagram relative to an instance of* $1|seq.dep.|C_{\max}$

it is natural to adopt an assumption that these so-called *changeover* or *setup* times can vary as a function of job pairs; that is, they are *sequence-dependent.*

Now, suppose our interest is in minimizing makespan. In our adopted notation, the problem is $1|seq.dep.|C_{\max}$ and for a given job ordering, a schedule would appear as shown in Figure 3.22. Observe that the shaded portions on the timing diagram indicate the changeover portions of total time. It should be evident that since the actual job processing times are constant, the objective of minimizing makespan is equivalent to minimizing total changeover time.

Two things should be apparent. First, the problem just described is only a (thinly) disguised version of a particularly well-known combinatorial problem, and secondly, its difficulty, therefore, is immediate. Consider a graph $G = (V, A)$ and let us take the vertices in V to correspond to jobs in the given instance of $1|seq.dep.|C_{\max}$. Let members $(i, j) \in A$ signify the processing order requiring job i immediately prior to job j and let arc (i, j) be weighted by the respective changeover value. Then clearly, a least total changeover job ordering corresponds to a least-weight Hamiltonian path in G. We can even "attach" an artificial vertex if we want to model the notion of startup prior to the first job in a sequence and a final shutdown after the last job is processed. In this case, our abstraction would then require a least-weight Hamiltonian cycle/circuit in the resultant graph. Regardless, in either context, it should be evident that we are solving a traveling salesman problem (TSP). The bottling illustration employed in Chapter 1 demonstrated the idea.

It follows then that what is worth knowing about TSPs is equivalently so for $1|seq.dep.|C_{\max}$ and to be certain, a great deal is known about the former. To repeat the claim made in the first chapter, a more studied problem in combinatorial optimization, if not operations research in general, would be hard to find. Accordingly, there is a voluminous literature dealing with the traveling salesman problem, including even a book of the same name: *The Traveling Salesman Problem: A Guided Tour of Combinatorial Optimization*, by Gene Lawler, Jan Karel Lenstra, Alexander Rinnooy Kan, and Dennis Shmoys, eds. (Lawler *et al.*, 1985).

As a consequence, we devote no coverage, at this point, to the problem $1|seq.dep.|C_{\max}$ *per se*. On the other hand, the interested reader should have no difficulty locating material; indeed, a good starting point for any literature search would be the lengthy list of references in the book just named. On the other hand, in Chapter 8 of this book, we take up the notion of traversals in graphs with the aim of appealing to slightly broader interests in the area of routing. There, we deal with a few issues pertaining to Hamiltonian cycles/circuits as well as the conventional TSP.

3.3 Exercises

3-1 Find an optimal sequence for the instance of $1||\Sigma w_i c_i$ below. Repeat the exercise for the measure Σc_i.

$$\tau = (7,4,6,3,2,8,7,5,3,4,1)$$
$$w = (2,8,1,4,7,9,2,1,8,3,6)$$

3-2 Suppose we wanted to *maximize* total weighted completion time. Modify Theorem 3.2 accordingly and apply the result to the instance in Exercise 3-1.

3-3 Give a fast algorithm for finding a sequence that minimizes the weighted start time of the last job in a single–processor, unconstrained problem.

3-4 Repeat Exercise 3-3 by finding a fast procedure for minimizing the weighted start time of the last k jobs where k is fixed.

3-5 Find a sequence that solves the following instance of $1|\tau_i = 1|\Sigma w_i T_i$.

$$w = (7,4,6,3,2,1,8,5,4,9)$$
$$d = (4,3,7,8,5,7,8,4,9,6)$$

3-6 Consider the following PARTITION SCHEDULING problem:

Given a set of n jobs on a single processor, a partition of these as $\{E, T\}$, and an integer $k \geq 0$, does there exist a sequence of jobs where those in E are nontardy, those in T are tardy, and total tardiness does not exceed k?

Either give a fast algorithm for this problem or prove it to be $\mathcal{NP}$-Complete.

3-7 Give an instance of $1||T_{\max}(L_{\max})$ that shows that EDD ordering is not necessary.

3-8 Prove each of the following propositions regarding $1||\Sigma T_i$ (*cf.* Baker, 1974):

- If an EDD ordering results in at most one tardy job, then this sequence minimizes ΣT_i.
- If every job has the same due-date, then ΣT_i is minimized by SPT ordering.
- If every ordering produces n tardy jobs, then ΣT_i is minimized by SPT ordering (n is the total number of jobs in the instance).
- If SPT ordering produces no nontardy jobs, then the given sequence minimizes ΣT_i.

3-9 Solve the following instance of $1||\Sigma T_i$ by the dynamic programming approach of Lawler.

$$\tau = (5, 6, 3, 7, 2, 4)$$
$$d = (12, 10, 14, 9, 15, 12)$$

3-10 Give an instance of $1||\Sigma w_i T_i$ that establishes that the "agreeability" condition cannot be relaxed.

3-11 Resolve the instance of Exercise 3-9 by a partial enumeration approach. Be clear on a subproblem creation strategy and a procedure for determining bound values.

3-12 Solve the following instance of $1|\tau_i = 1|\Sigma w_i u_i$ by applying the greedy algorithm illustrated in Example 3.5.

$$w = (7, 6, 4, 5, 6, 3, 8)$$
$$d = (4, 2, 3, 5, 6, 5, 3)$$

3-13 Consider the following strategy for problem $1|\tau_i = 1|\Sigma w_i u_i$:
Construct a bipartite graph G having bipartition $\{X, Y\}$ with vertices in X corresponding to sequential positions $1, 2, \ldots, n$ and those in Y corresponding to jobs labeled in nonincreasing w-order. For every $j \in Y$, create an edge (i, j) if job j is nontardy when sequenced in position i. Now, taking vertices in Y in order, construct a maximum cardinality matching in G as follows: for given $j \in Y$, place in the current matching, edge (i, j) where i is maximum over unmatched vertices incident to j. We demonstrate this idea on the following instance:

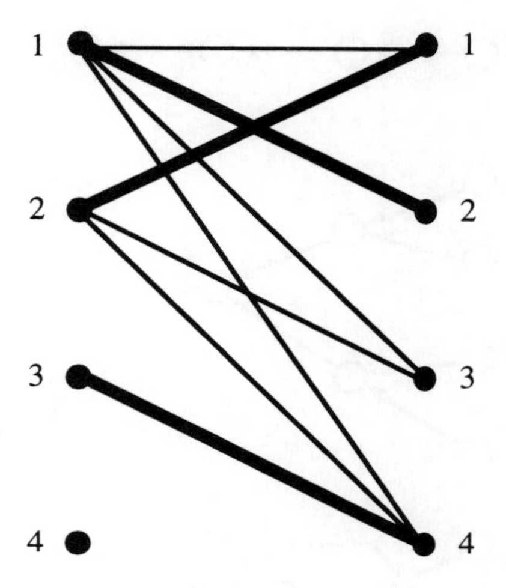

Figure 3.23. *Exercise 3-13*

$w = (6, 5, 3, 2)$
$d = (2, 1, 2, 3)$

The corresponding graph is shown in Figure 3.23 and the edges in the matching are indicated in bold. The sequence produced is $\{2, 1, 4, 3\}$. First, apply this procedure to the instance of Exercise 3-12. Then, give a proof of its correctness *or* exhibit a counterexample.

3-14 Give an instance of $1||\Sigma w_i u_i$ for which $\boldsymbol{A_M}$ fails.

3-15 Apply $\boldsymbol{A_M}$ to the following instance of $1||\Sigma u_i$.

$\tau = (3, 5, 2, 4, 2, 6)$
$d = (10, 12, 14, 10, 15, 9)$

3-16 Apply Lawler's procedure, $\boldsymbol{A_{LF_m}}$, to the following instance of $1||f_{\max}$.

$\tau = (5, 3, 4, 6, 2, 4)$
$f = (c_1^2, 5c_2, c_3^{3/2}, 10c_4^{1/2}, \frac{c_5^3}{100}, c^6)$

3-17 Give an instance of $1||f_{\max}$ that establishes that the montonicity condition on $f_j(c_j)$ is meaningful.

3-18 Prove that EDD-ordering solves $1||T_{\max}(L_{\max})$ by employing the result of Theorem 3.16.

3-19 Show that each digraph in Figure 3.24 is vertex series-parallel by producing the decomposition tree for each.

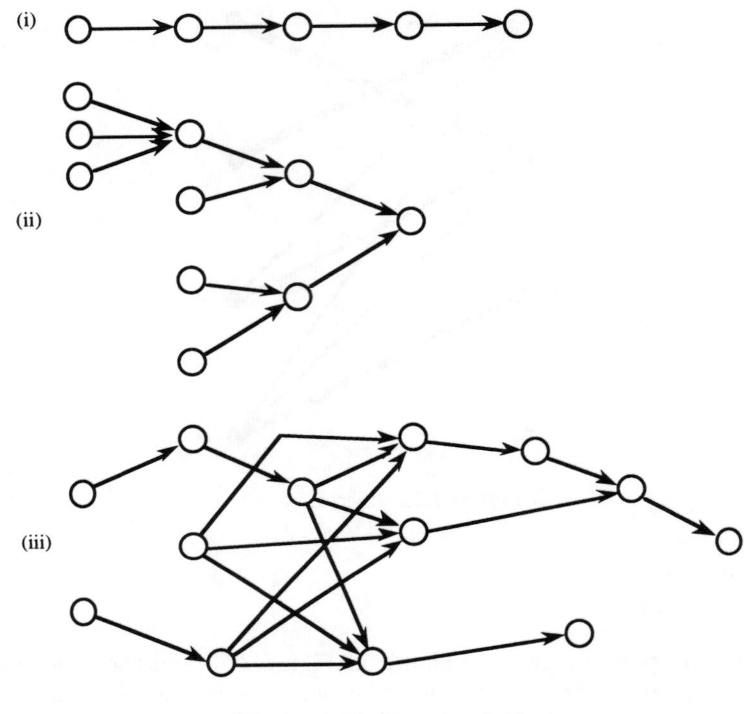

Figure 3.24. *Exercise 3-19*

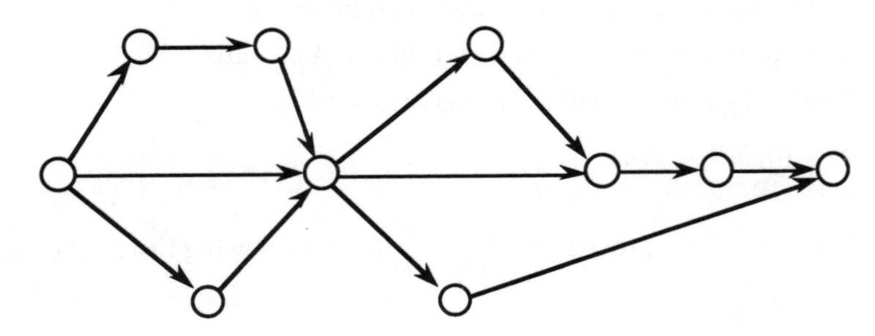

Figure 3.25. *Exercise 3-20*

3-20 Establish that the digraph Figure 3.25 is edge series-parallel by reducing it to a single directed edge. Also, give its decomposition tree.

3-21 Can two nonisomorphic, edge series-parallel digraphs exhibit isomorphic line graphs?

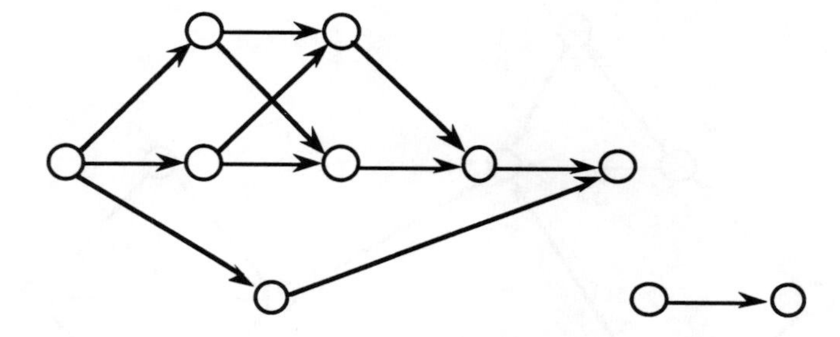

Figure 3.26. *Exercise 3-25*

3-22 Give a vertex series-parallel digraph that yields two nonisomorphic decomposition trees.

3-23 Can any acyclic digraph that is not vertex series-parallel be made so by adding arcs?

3-24 Any digraph can be made vertex series-parallel by removing arcs. Consider the problem of determining the least number to remove. Either give a fast algorithm for this problem or prove it to be $\mathcal{NP}$-Hard.

3-25 Solve the problem $1|prec|\Sigma w_i c_i$ on the instance below and for the digraph shown in Figure 3.26.

$$\tau = (5, 7, 6, 3, 8, 4, 2, 9, 4, 6)$$
$$w = (3, 2, 4, 5, 7, 3, 2, 5, 4, 1)$$

3-26 Show that the digraph in Figure 3.27 is not edge series-parallel by finding a subgraph homeomorphic to the forbidden subgraph of Theorem 3.20.

3-27 Form the line graph of the digraph in Exercise 3-26 and repeat the stated exercise for the resultant vertex series-parallel digraph.

3-28 Solve problem $1|prec|T_{\max}$ on the digraph given in Figure 3.28 and with durations and due-dates shown below:

$$\tau = (7, 3, 8, 2, 7, 9, 5, 8, 2, 1, 5, 4)$$
$$d = (9, 7, 14, 6, 9, 15, 12, 18, 17, 19, 24, 22)$$

3-29 Prove Theorem 3.25.

3-30 Obtain the decomposition tree of Figure 3.12 from that of Figure 3.14 by forming appropriate line graphs relative to the latter.

3-31 The *earliness* of a job, E_j, is given by $\max(0, d_j - c_j)$. Prove the minimizing total earliness is the same as minimizing total tardiness; *i.e.*, the two problems have the same complexity status.

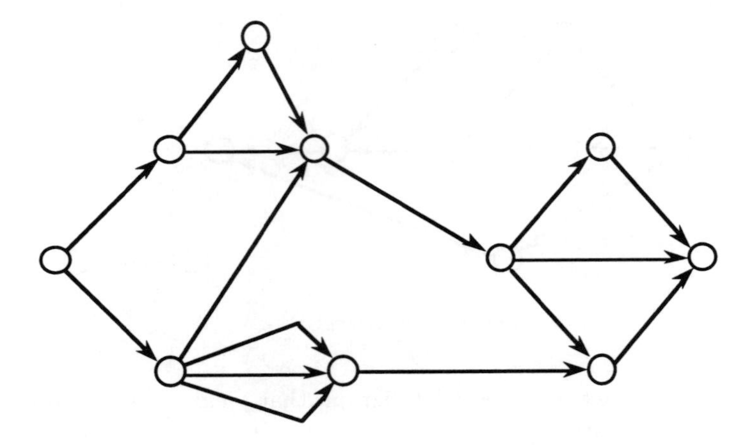

Figure 3.27. *Exercise 3-26*

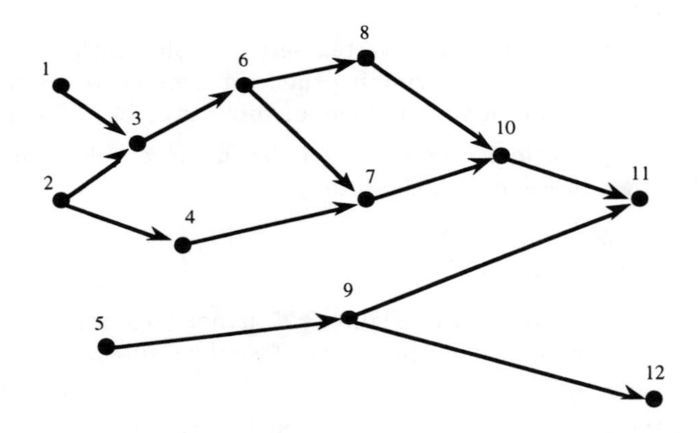

Figure 3.28. *Exercise 3-28*

3-32 Describe fast algorithms for problems: $1|r_i \geq 0,\ \tau_i = 1|\Sigma w_i T_i$, $1|r_i \geq 0,\ \tau_i = 1|\Sigma w_i u_i$, and $1|r_i \geq 0,\ prec|C_{\max}$.

3-33 Consider the problem $1||\Sigma c_i^t$ where $t > 1$. Is this easy or difficult? Explain.

3-34 In Lenstra *et al.* (1977) the reduction 3-PARTITION $\propto$ $1||\Sigma w_i T_i$ is given specifically as follows:

$$n = 4t - 1;$$
$$\tau_i = w_i = a_i,\ d_i = 0,\ 1 \leq i \leq 3t$$
$$\tau_i = 1,\ w_i = 2,\ d_i = (i - 3t)(b + 1),\ 3t + 1 \leq i \leq 4t - 1;$$

and threshold on $\Sigma w_i T_i$ as $\sum_{1 \le i \le j \le 3t} a_i a_j + \frac{1}{2}t(t-1)b$. Complete the details of the proof.

3-35 Propose an approximation strategy for $1||\Sigma w_i u_i$ and analyze its performance.

3-36 Propose some measures of performance that are not *regular*.

3-37 Suppose our instance of $1|prec|\Sigma w_i c_i$ is defined on $4k$ jobs where these are given by k modules each taking the form of the z-graph of Figure 3.10. Either propose a fast algorithm for this restricted case or show that it remains hard.

3-38 Give an explicit illustration of the construction in the proof of Theorem 3.23 by using K_4 as the input graph of problem k-CLIQUE.

3-39 Consider a restriction of $1|prec, \tau_i = 1|\Sigma T_i$ to the class of series-parallel digraphs. Does this make the problem easier or does it retain its hard status?

3-40 Consider a modification to problem $1||\Sigma w_i c_i$ such that if a job is placed in the kth sequential position, then its weight is kw_i. Resolve the complexity of this altered problem.

CHAPTER 4

PARALLEL-PROCESSOR PROBLEMS

In this chapter, we take up scheduling problems arising in an environment characterized by two or more processors and where these processors possess similar capability (but perhaps different performance qualities, *i.e.*, speed, etc.). That is, any of $m \geq 2$ processors are available for processing a given job and the aim is to find an assignment of all jobs to existing processors that makes optimal some predetermined measure. As before, we begin with problems having no dependencies.

4.1 Independent Jobs

4.1.1 Makespan Case

The classic version of a parallel-processor problem seeks an assignment of a set of n jobs to a set of m identical processors so as to minimize makespan $C_{\max}$. More formally: Given a finite set J of jobs, a nonnegative duration τ_i for each $i \in J$, a number $m \geq 2$ of processors, and a completion time threshold $D > 0$, does there exist a partition $\{J_1, J_2, \ldots, J_m\}$ of J such that

$$\max\left\{\sum_{i \in J_k} \tau_i ; 1 \leq k \leq m\right\} \leq D?$$

Cast in this more formal context, it is easy to determine the complexity status of the problem. We have

Theorem 4.1 *The (recognition version) problem $P||C_{\max}$ is $\mathcal{NP}$-Complete.*

Proof. The reduction is immediate from the problem PARTITION which we state below:

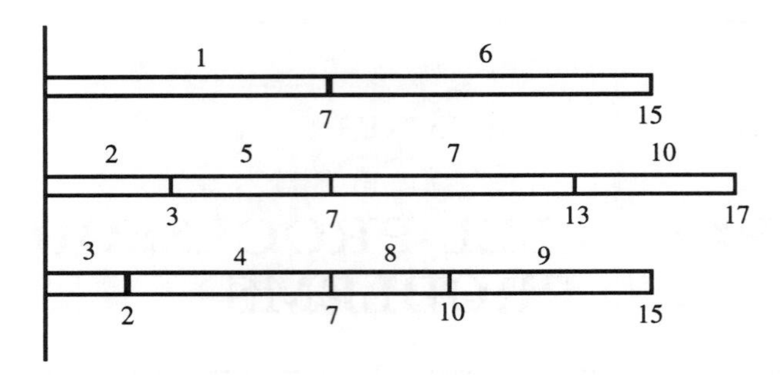

Figure 4.1. *Timing diagram of Example 4.1*

Given a finite set A with elements weighted as $w(a) \in \mathbb{Z}^+$ for all $a \in A$, is there a subset $A' \subseteq A$ such that

$$\sum_{a \in A'} w(a) = \sum_{a \in A \setminus A'} w(a)?$$

Inclusion in $\mathcal{NP}$ is also easy. □

The following (obvious) corollary demonstrates even further the degree to which this problem remains hard.

Corollary 4.2 *The (recognition version) problem* $P2||C_{\max}$ *is* $\mathcal{NP}$*- Complete.* □

We also note that the problem can be solved in pseudo-polynomial time for any fixed number of processors m but remains strongly $\mathcal{NP}$-Complete when m is free.

In the face of this difficulty, it should come as no surprise that the problem has been much studied in the context of approximation approaches. Following then, we shall examine the behavior of several of these approximations in some depth.

Common to numerous approximation procedures is the notion commonly referred to as *list processing*.

A_L: List Processing

> Create a list of jobs L and from this list, form a schedule as follows: whenever a processor becomes available, schedule the first available job from the list. □

Example 4.1 Let $m = 3$ and consider job duration times given by $\tau = (7, 3, 2, 5, 4, 8, 6, 3, 5, 4)$. Then for $L = \{1, 2, \ldots, 10\}$, the list-processing generated schedule would appear as shown in Figure 4.1. □

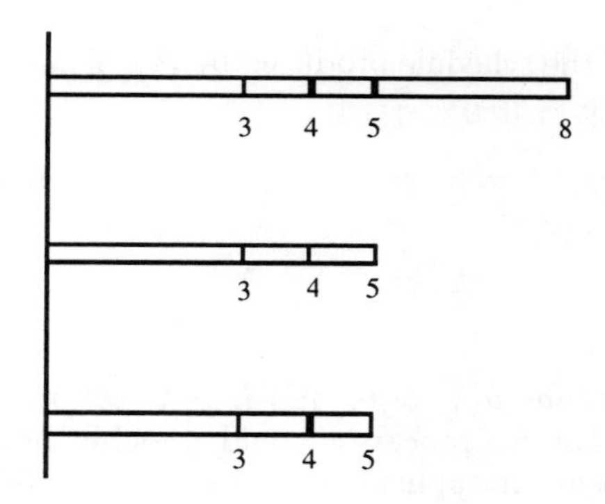

Figure 4.2. *Timing diagram of Example 4.2*

Of course, the statement of $\boldsymbol{A_L}$ leaves open the issue of how the list L might be formed. In practice one might expect to be guided by some sense of priority among jobs with the latter based on attributes such as job duration, due-dates, etc. Indeed, the nature of specific manifestations of $\boldsymbol{A_L}$ affect precisely constructions of L. We begin with an interesting (albeit dated) illustration motivated by Knuth and Kleitman (*cf.* Graham, 1969).

$\boldsymbol{A_{KK}}$: Knuth-Kleitman Procedure

For an integer $k \geq 0$, select the k longest jobs and arrange them in an optimal way on the m processors. Assign the remaining $n-k$ jobs arbitrarily. □

Note that the "arbitrary" assignment alluded to in $\boldsymbol{A_{KK}}$ is restricted in the obvious way; any assignment is required to at least adhere to the list-processing rule, *i.e.*, whenever a processor is available no available job is left waiting.

Example 4.2 Suppose 10 jobs have duration times given by $\tau = (3,3,3,3,1,1,1,1,1,1)$ and let $m = k = 3$. Then a correct application of $\boldsymbol{A_{KK}}$ produces the schedule shown in Figure 4.2 (where we have assumed a list to be formed as $\{1,2,3,5,6,7,8,9,10,4\}$). □

Of course, it is easy to see that the schedule created in the previous example is not optimal. The latter would partition on duration times as (3,1,1,1), (3,1,1,1), (3,3) with $\nu^* = 6$. Accordingly then, our interest would be in a performance bound for $\boldsymbol{A_{KK}}$.

Following Graham (1969), let

$$\nu_k \triangleq \text{value of the (optimal) } k\text{-job "preschedule."}$$

$\nu(k) \triangleq$ value of the schedule produced by $\boldsymbol{A_{KK}}$ when operating on the given instance.

Denoting, as before, ν^* to be the optimal schedule length, we have:

Theorem 4.3

$$\frac{\nu(k)}{\nu^*} \leq 1 + \frac{1 - \frac{1}{m}}{1 + \lfloor \frac{k}{m} \rfloor}$$

Proof. We may assume $\nu(k) > \nu_k$ and $n > k$. Let $\tau^* = \max_{k+1 \leq i \leq n}(\tau_i)$. Now, it is clear that no processor could possibly be idle prior to time $\nu(k) - \tau^*$. Otherwise, our application of $\boldsymbol{A_{KK}}$ under list-processing would not have been correct. Thus,

$$\sum_i \tau_i \geq (\nu(k) - \tau^*)m + \tau^*.$$

But

$$\nu^* \geq \frac{\Sigma \tau_i}{m} \geq (\nu(k) - \tau^*) + \frac{\tau^*}{m} = \nu(k) - \tau^* \left(1 - \frac{1}{m}\right).$$

Now, there are $k + 1$ jobs having duration no less than τ^* and some processor must work on at least $1 + \lfloor \frac{k}{m} \rfloor$ such jobs. Hence,

$$\nu^* \geq \left(1 + \lfloor \frac{k}{m} \rfloor\right) \tau^*$$

or

$$\tau^* \leq \frac{\nu^*}{1 + \lfloor \frac{k}{m} \rfloor}.$$

But

$$\nu(k) \leq \nu^* + \tau^* \left(1 - \frac{1}{m}\right);$$

so we have

$$\nu(k) \leq \nu^* + \nu^* \frac{(1 - \frac{1}{m})}{1 + \lfloor \frac{k}{m} \rfloor} = \nu^* \left(1 + \frac{1 - \frac{1}{m}}{1 + \lfloor \frac{k}{m} \rfloor}\right)$$

and the proof is complete. □

It is evident that the result of Theorem 4.3 provides a family of bounds, each depending on the choice of k. The following instances demonstrate realizability. For $k = 0$, we have $\nu(0)/\nu^* \leq 2 - \frac{1}{m}$. Hence, for a list of jobs ordered (for ease) on duration times as $L = \{1, 1, 2\}$ and for $m = 2$, we obtain the 3/2 result as shown in Figure 4.3. When $k = m$, our bound is $\frac{3}{2} - \frac{1}{2m}$ and the previous list $L = \{3, 3, 3, 1, 1, 1, 1, 1, 1, 3\}$ provides a realizable instance. For $k = 2m$ we have $\frac{\nu(2m)}{\nu^*} \leq \frac{4}{3} - \frac{1}{3m}$ and so forth. So,

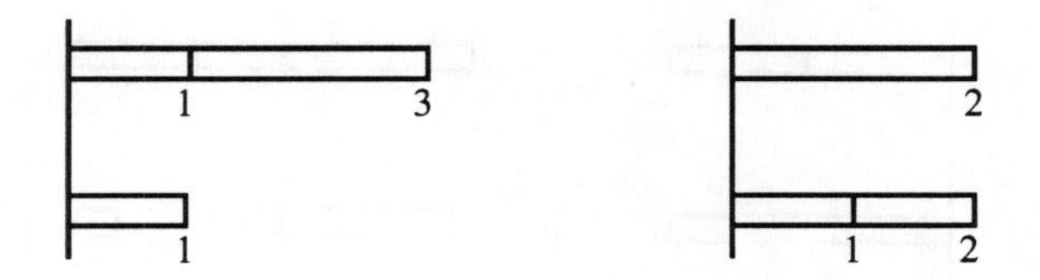

Figure 4.3. *Bad instance for A_{KK} with $k = 0$*

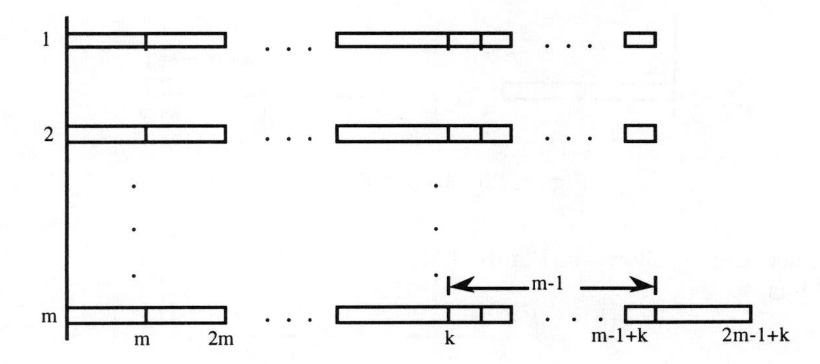

Figure 4.4. *Schedule generated by $L(k)$*

as k grows, the bound gets (as expected by definition) close to 1. But, this is not meaningful of course, since determining an optimal arrangement of the k prescheduled jobs is itself a hard problem. (We know how to do this when $k = 0$ or m as in the above illustrations but not for, say, the $2m$ case.) Still, the following result formalizes the sense of the quality of the bound.

Theorem 4.4 *The bound on $\boldsymbol{A_{KK}}$ is the best possible for $k \equiv 0(mod\, m)$.*

Proof. We need only give a class of instances that are realizable. To this end, let $k \equiv 0(mod\, m)$ and set $n = k + 1 + m(m-1)$ with

$$\tau_i = \begin{cases} m, & 1 \leq i \leq k+1 \\ 1, & k+2 \leq i \leq k+1+m(m-1). \end{cases}$$

Also, let us form our list as

$$L(k) = \{1, 2, \ldots, k, k+2, \ldots, k+1+m(m-1), k+1\}.$$

Then our schedule appears as in Figure 4.4. Here $\nu_k = \frac{km}{m}$ and $\nu(k) = k + 2m - 1$. The optimal schedule length results as

$$\nu^* = \frac{\Sigma \tau_i}{m} = \frac{m(k+1) + m(m-1)}{m} = k + m.$$

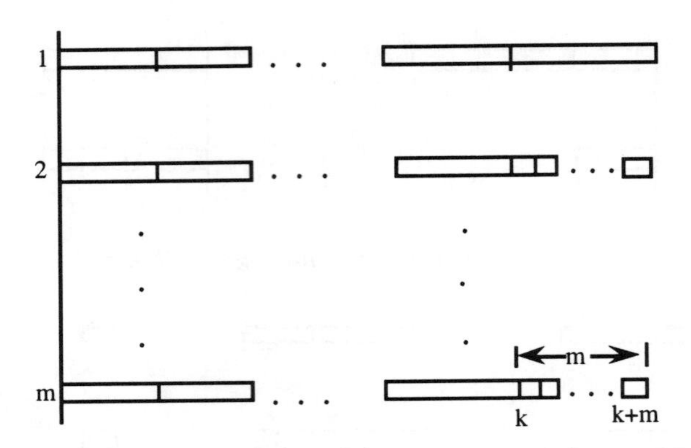

Figure 4.5. *An optimal schedule*

The schedule is shown in Figure 4.5.

Thus,

$$\frac{\nu(k)}{\nu^*} = \frac{k+2m-1}{k+m} = 1 + \frac{1-1/m}{1+\lfloor k/m \rfloor}$$

(recall $k \equiv 0(mod\, m)$). □

Now, suppose we employ the following alternative heuristic:

$\boldsymbol{A_{LPT}}$: **Longest Processing Time Heuristic**

Create L with jobs arranged in nonincreasing τ-order (longest processing time or LPT order). □

Example 4.3 Let our list be ordered by LPT and again, specified with actual duration times rather than job number as $L = \{7, 7, 6, 6, 5, 5, 4, 4, 4\}$. For $m = 4$, the schedule generated by $\boldsymbol{A_{LPT}}$ is shown in Figure 4.6 along with the accompanying optimal schedule. □

Interestingly, this LPT heuristic achieves the same bound as that for the $k = 2m$ case earlier. Letting the outcome of $\boldsymbol{A_{LPT}}$ be ν_{LPT}, we have:

Theorem 4.5

$$\frac{\nu_{LPT}}{\nu^*} \leq \frac{4}{3} - \frac{1}{3m}.$$

Proof. We may safely assume that $n > m$, since otherwise we would assign at most one job to each processor, which would clearly be optimal. Now, following Graham (1969), let us suppose the jobs are numbered in LPT order as $\tau_1 \geq \tau_2 \geq \ldots \geq \tau_n$. Then as in a previous proof, we have $\nu_{LPT} = t + \bar{\tau}$ where t is the start time of job $\bar{j}$ which finishes last and which has duration time $\bar{\tau}$. Moreover, we know then that no processor is idle prior to

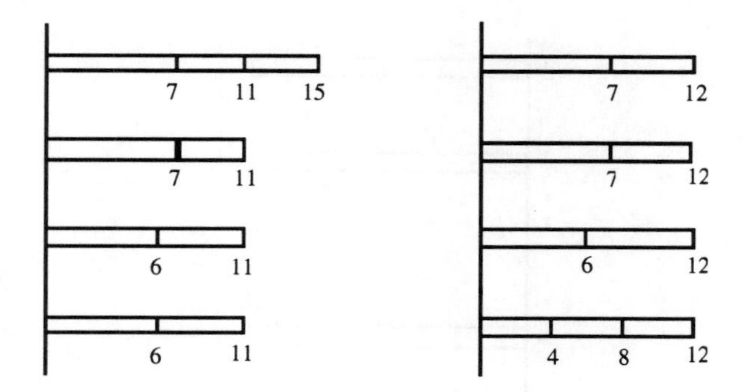

Figure 4.6. *An LPT schedule*

t and thus

$$\sum_{j \neq \bar{j}} \tau_j \geq mt$$

or

$$\nu_{LPT} \leq \frac{1}{m} \sum_{j \neq \bar{j}} \tau_j + \bar{\tau} = \frac{\Sigma_j \tau_j}{m} + \left(\frac{m-1}{m} \right) \bar{\tau}.$$

Now, let us assume that the stated bound is incorrect. That is, let us suppose that an instance exists for which

$$\frac{\nu_{LPT}}{\nu^*} > \frac{4}{3} - \frac{1}{3m}$$

and further, let this instance be minimum in the sense that no smaller instance exists which invalidates the bound. Accordingly, we may safely assume that $\bar{j} = n$ since, otherwise, we could simply create a truncated instance by removing all jobs with duration smaller than that of $\bar{j}$ and that has optimal value no greater than that on the original instance, but whose $\boldsymbol{A_{LPT}}$ generated length is the same as that for the original. But this smaller instance would also be a counterexample and the aforementioned minimality assumption is contradicted.

So, upon rewriting, we have

$$\frac{\nu_{LPT}}{\nu^*} \leq 1 + \left(\frac{m-1}{m} \right) \frac{\bar{\tau}}{\nu^*},$$

and by hypothesis

$$1 + \left(\frac{m-1}{m} \right) \frac{\bar{\tau}}{\nu^*} > \frac{4}{3} - \frac{1}{3m},$$

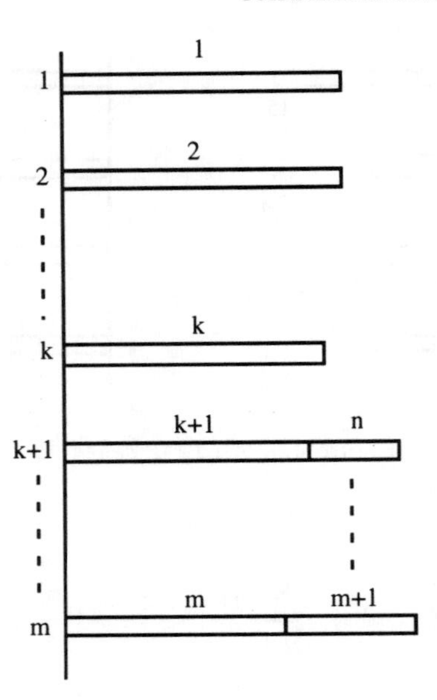

Figure 4.7. *LPT for* $k = 2m - n$

or

$$\left(\frac{m-1}{m}\right)\frac{\bar{\tau}}{\nu^*} > \frac{1}{3} - \frac{1}{3m} = \frac{m-1}{3m},$$

which leads to

$$\bar{\tau} > \frac{\nu^*}{3}.$$

That is, if the stated bound is false, then in an optimal solution, no processor can have in excess of two jobs assigned to it. However, Graham (1969) was able to show that under these conditions, $\boldsymbol{A_{LPT}}$ will produce a schedule like that indicated by the timing diagram in Figure 4.7 and that is optimal; *i.e.*, $\nu^* = \nu_{LPT}$. But this contradicts the assumption that $\nu_{LPT}/\nu^* > \frac{4}{3} - \frac{1}{3m}$ and we are done. □

While the bound of the theorem above is equivalent to the Knuth-Kleitman bound when $k = 2m$ one clearly cannot conclude that this implies that LPT is optimal for an instance defined on $2m$ jobs. Example 4.4 makes the case.

Example 4.4 Consider an instance with six jobs on three processors and with $\tau = (8, 5, 4, 3, 3, 3)$. It is easy to see that an optimal schedule length has value 9. On the other hand, the LPT schedule has length 10. Both

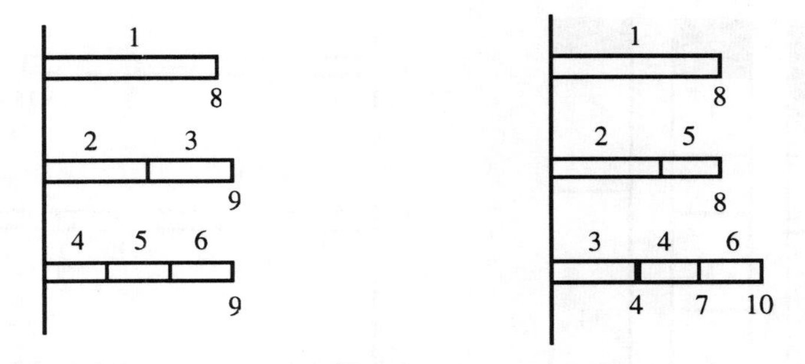

Figure 4.8. *Suboptimality of LPT with $n = 2m$*

schedules are shown in Figure 4.8. □

For some time the so-called 4/3 bound of Theorem 4.5 was the best known in terms of a performance guarantee. However, this position of prominence was relinquished when an alternative heuristic was offered by Coffman *et al.* (1978). Referred to as MULTIFIT, this approximation employs, in a clever way, a natural dual-like relationship between the problem $P||C_{\max}$ with the well-known (hard) BIN-PACKING problem.

Stated informally (but sufficient for our present purpose), BIN-PACKING seeks an admissible assignment or packing of a finite set of "chips," each with some positive weight, into the fewest number of finite capacity "bins." This dual-like relationship should be evident:

BIN-PACKING		$P\|\|C_{\max}$
bins	$\longleftrightarrow$	processors
capacity	$\longleftrightarrow$	$C_{\max}$ threshold
chips	$\longleftrightarrow$	jobs

Moreover, if there exists a packing into no more than m bins each with capacity C, then there exists a suitable schedule with makespan no greater than C.

Example 4.5 Suppose nine chips are packed into bins of capacity 23 as shown on the left in Figure 4.9. Chips are indicated by their weight for ease of depiction. The interpretation as a schedule is shown on the right. □

Put simplistically, the MULTIFIT heuristic asks for a minimum size bin capacity admitting an m-bin packing. It attempts to find this by utilizing a process of binary search over an interval $[\beta_L, \beta_U]$ which bounds initially the capacity and which guarantees an m-bin packing. Of course, this search is not exact since we are forced to use an approximation procedure to treat the bin-packing problem itself.

To this end, we consider the following popular heuristic for bin-packing.

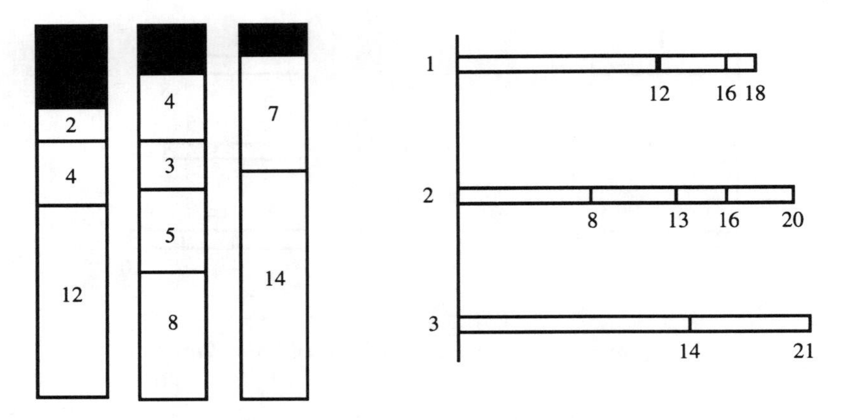

Figure 4.9. *Parallel-processor scheduling ↔ bin-packing*

Referred to as *first-fit, decreasing weight (FFD)* the procedure fills bins in a greedy fashion with largest weight chips selected first. Formally, we have:

A_{FFD} : First-Fit, Decreasing Weight Heuristic

Create a list L of chips arranged in nonincreasing weight-order. Select chips from L in this order, placing a given selection in the first available bin into which it will fit. □

Consider an illustration taken from Coffman *et al.* (1978).

Example 4.6 Let $C = 61$ and consider the list of chips given as $L = \{44, 24, 24, 22, 21, 17, 8, 8, 6, 6\}$. Applying A_{FFD} produces the four-bin packing in Figure 4.10. □

We now specify the MULTIFIT procedure more formally.

A_{MF} : MULTIFIT Heuristic

Step 0: *Initialization.* Let T be the set of jobs and fix upper and lower bounds relative to T and m, as $\beta_U[T, m]$ and $\beta_L[T, m]$, respectively. Let $\beta_1(0) \leftarrow \beta_U$ and $\beta_2(0) \leftarrow \beta_L$. Choose a number of iterations t and an iteration counter $i \leftarrow 1$.

Step 1: *Capacity Change.* If $i > t$, stop. Otherwise, set

$$C \leftarrow (\beta_2(i-1) + \beta_1(i-1))/2.$$

Step 2: *Upper Bound.* If the number of bins required by A_{FFD} operating on T with capacity C, given as FFD$[T, C]$, is no greater than m, set $\beta_1(i) \leftarrow C, \beta_2(i) \leftarrow \beta_2(i-1)$, update $i \leftarrow i+1$, and go to 1.

Step 3: *Lower Bound.* If FFD$[T, C] > m$, set $\beta_2(i) \leftarrow C, \beta_1(i) \leftarrow \beta_1(i-1)$, update $i \leftarrow i+1$ and go to 1. □

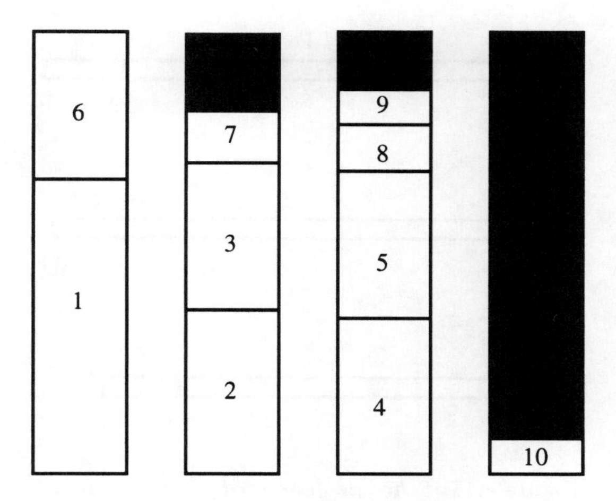

Figure 4.10. A_{FFD} *application*

Needed are values of $\beta_L[T,m]$ and $\beta_U[T,m]$. Accordingly, we may specify these as follows:

$$\beta_L[T,m] = \max\{\Sigma\tau_j/m, \max_j(\tau_j)\}$$

$$\beta_U[T,m] = \max\{2\Sigma\tau_j/m, \max_j(\tau_j)\}$$

The correctness of $\beta_L[T,m]$ is obvious so let us examine the validity of $\beta_U[T,m]$. Assume $C \geq \beta_U[T,m]$ and that FFD$[T,C] > m$ with T arranged in nonincreasing τ-order. Let j^* be the first chip/job to be placed into bin $m+1$ using $\boldsymbol{A_{FFD}}$. But $j^* \geq m+1$. So, if $\tau_j^* > \frac{C}{2}$, then by the ordering in T, $\tau_i > \frac{C}{2}$ for $1 \leq i \leq m+1$. Thus,

$$\sum_{j=1}^{n} \tau_j > \frac{mC}{2} \geq \frac{\beta_U[T,m]}{2} \geq \sum_{j=1}^{n} \tau_j,$$

which is a contradiction. Alternately, suppose $\tau_j^* \leq \frac{C}{2}$. Then since j^* would not fit into bins $1, 2, \ldots, m$ each must have had items whose total weight exceeded $\frac{C}{2}$. This means that $\Sigma\tau_j > \frac{mC}{2}$ and we are led to the same contradiction as before.

Following, we demonstrate the $\boldsymbol{A_{MF}}$ procedure.

Example 4.7 Let $m = 3$ and $t = 6$ with duration times for seven jobs given as $(59, 47, 38, 22, 13, 12, 11)$. Applying the algorithm, we would start by calculating $\beta_L = 67.3$ and $\beta_U = 134.6$. The computation is summarized in the table below.

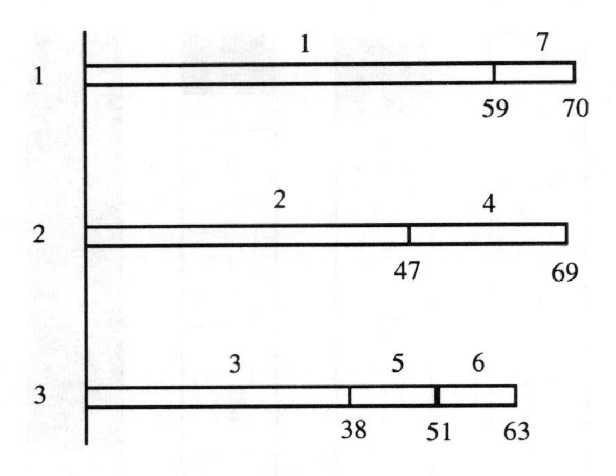

Figure 4.11. *Schedule generated for Example 4.7*

$i = 1:$	$C = 100.9, \text{FFD}[T, C] \leq 3, \beta_1(1) = 100.9, \beta_2(1) = 67.3$
$i = 2:$	$C = 84.1, \text{FFD}[T, C] \leq 3, \beta_1(2) = 84.1, \beta_2(2) = 67.3$
$i = 3:$	$C = 75.7, \text{FFD}[T, C] \leq 3, \beta_1(3) = 75.7, \beta_2(3) = 67.3$
$i = 4:$	$C = 71.5, \text{FFD}[T, C] \leq 3, \beta_1(4) = 71.5, \beta_2(4) = 67.3$
$i = 5:$	$C = 69.4, \text{FFD}[T, C] > 3, \beta_1(5) = 71.5, \beta_2(5) = 69.4$
$i = 6:$	$C = 70.5, \text{FFD}[T, C] \leq 3, \beta_1(6) = 70.5, \beta_2(6) = 69.4$
$i = 7:$	Stop.

The final packing yields the schedule shown in Figure 4.11. The reader may wish to check that the LPT schedule has length 71. □

Let us now examine the performance of $\boldsymbol{A_{MF}}$. Before proceeding in a formal way, however, suppose we look at a somewhat unsettling outcome.

Example 4.8 Consider the list of chip weights given before as $L = \{44, 24, 24, 22, 21, 17, 8, 8, 6, 6\}$. In the earlier example we saw that $\text{FFD}[L, 61] = 4$. Suppose now we use $C = 60$. Then applying FFD the packing in Figure 4.12 results. Here, $\text{FFD}[L, 60] = 3$. □

Thus, we have that under the FFD heuristic, $C_1 < C_2$ and $\text{FFD}[T, C_1] \leq m \not\Rightarrow \text{FFD}[T, C_2] \leq m$ for $m \geq 3$. Clearly, this anomaly with a bin-packing heuristic suggests similar problems when it is employed in $\boldsymbol{A_{MF}}$.

So, we know that binary search is not guaranteed to yield the smallest C such that $\text{FFD}[T, C] \leq m$. However, suppose that for a value $\bar{C}$ we knew that $\text{FFD}[T, \bar{C}] \leq m$ and moreover, that $\text{FFD}[T, C'] \leq m$ for all $C' \geq \bar{C}$

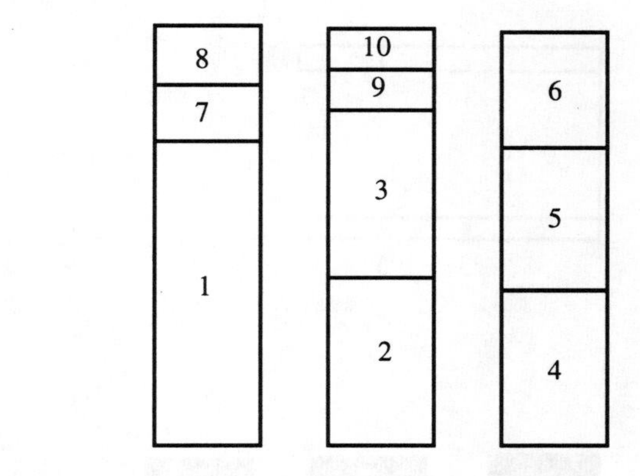

Figure 4.12. A_{FFD} *anomaly*

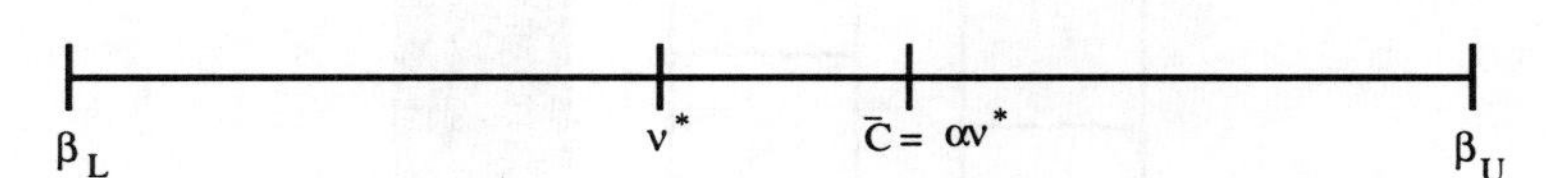

Figure 4.13. *Anomaly-proof target capacity* $\bar{C}$

(*i.e.*, no anomalies occur such as that in Example 4.8). Then the situation (with binary search) is greatly improved. The effect of this in terms of the interval $[\beta_L, \beta_U]$ is shown in Figure 4.13. So, although less ambitious in its intent, our binary search strategy will now seek a capacity $C \leq \alpha\nu^*$.

Formally, let r_m be a *least expansion factor* satisfying the following: for given r_m, then for any T and $r \geq r_m$, FFD$[T, r\nu^*] \leq m$. Key (and most complicated) here is the determination of values for r_m. What we know can be shown in the following table:

	$m = 2$	$m = 3$	$4 \leq m \leq 7$	$m \geq 8$
upper bound on r_m	$\frac{8}{7}$	$\frac{15}{13}$	$\frac{20}{17}$	1.220
lower bound on r_m	$\frac{8}{7}$	$\frac{15}{13}$	$\frac{20}{17}$	$\frac{20}{17}$

Note that the lower bounds (in contrast to the upper bounds) are easy to produce. All we need to do is construct suitable instances. For example, for $m = 2$ we can let $C = 7$ with $T = \{3, 3, 2, 2, 2, 2\}$. Here, $\nu^* = 7$ and FFD$[T, 7] = 3$ so $r_2 \geq \frac{8}{7}$ as shown in Figure 4.14. For $m = 3$, set $C = 13$ and $T = \{7, 5, 4, 4, 4, 3, 3, 3, 3, 3\}$. Here, $\nu^* = 13$ and FFD$[T, 13] = 4$ yielding $r_3 \geq 15/13$ as demonstrated in Figure 4.15.

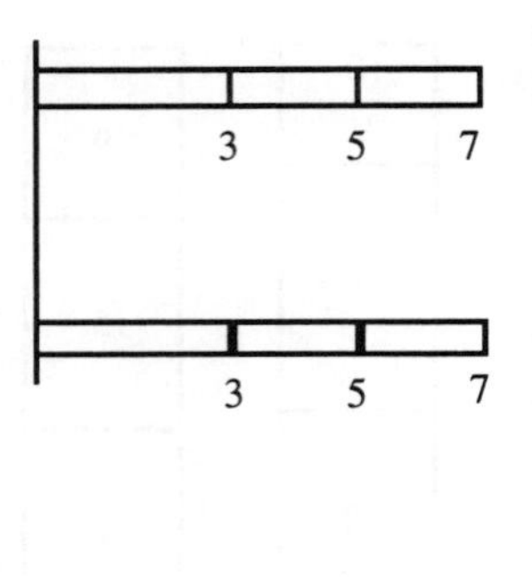

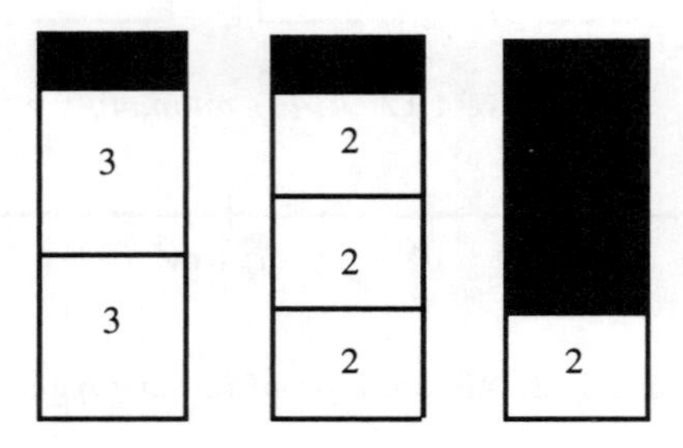

Figure 4.14. *Bound on* r_2

And, finally, let $m \geq 4$, with $C = 17$ and set $T = \{\underbrace{17, 17, \ldots, 17,}_{1,2,\ldots,m-4}$ $9, 7, 6, 5, 5, 4, 4, 4, 4, 4, 4, 4, 4, 4\}$. This produces $\nu^* = 17$ with FFD[T, 17] $= m + 1$ so $r_m \geq \frac{20}{17}$. We demonstrate this in Figure 4.16.

Essentially, the values in the previous table yield bound values for $\boldsymbol{A_{MF}}$. Formally, we have the following:

Theorem 4.6 *Let the solution value produced by* $\boldsymbol{A_{MF}}$ *when applied to an instance on* m *processors with* t *iterations specified be* ν_m^t *and let the optimal value be* ν^*. *Then for all* $m \geq 2$ *and* $t \geq 0$

$$\frac{\nu_m^t}{\nu^*} \leq r_m + \left(\frac{1}{2}\right)^t.$$

Proof. Assume the result is not true. That is, suppose $\nu_m^t/\nu^* > r_m + (1/2)^t$ for some m and t. Then there exists some job set T such that $\boldsymbol{A_{MF}}$ produces a schedule having makespan strictly greater than $(r_m + (1/2)^t)\nu^*$. Since $\nu_m^t \geq \beta_2(t)$ we have $\beta_2(t) \geq (r_m + (1/2)^t)\nu^*$. Now, consider the final value of $\beta_1(t)$. From the search process, $\beta_2(t) - \beta_1(t) = (1/2)^t(\beta_U - \beta_L) \leq \left(\frac{1}{2}\right)^t \nu^*$. This follows since $\beta_U \leq 2\beta_L$ and $\beta_L \leq \nu^*$, *i.e.*, $\left(\frac{1}{2}\right)^t (\beta_U -$

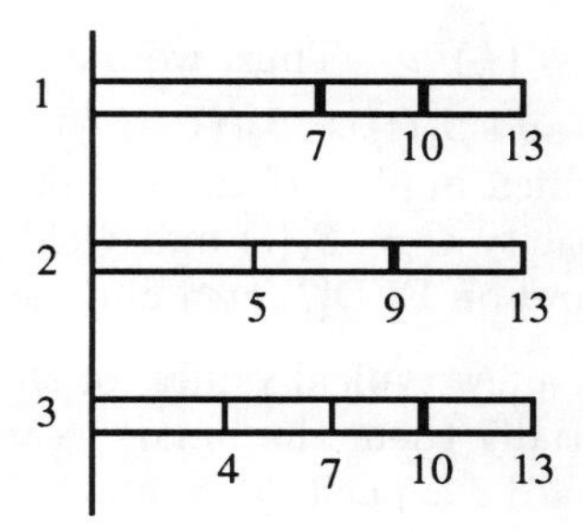

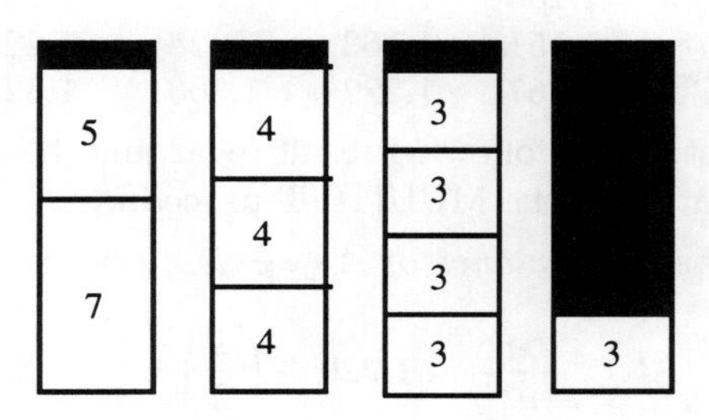

Figure 4.15. *Bound on* r_3

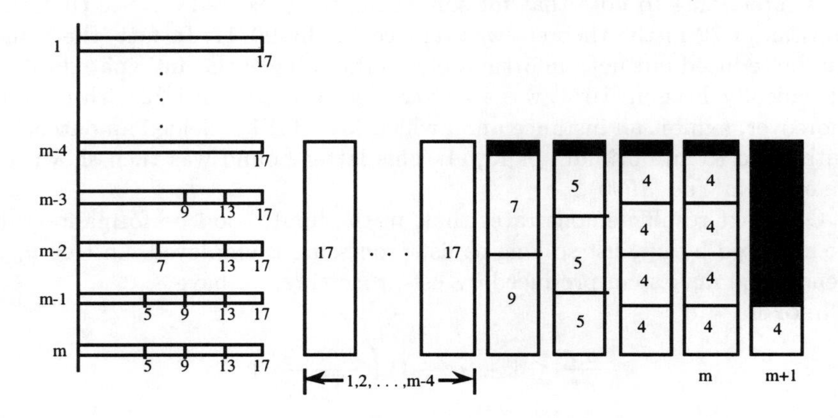

Figure 4.16. *Bound in* r_m, $m \geq 4$

$\beta_L) \leq \left(\frac{1}{2}\right)^t (2\beta_U - \beta_L) \leq \left(\frac{1}{2}\right)^t \nu^*$. Thus, we have that $\beta_1(t) \geq r_m\nu^*$ since $\beta_2(t) \geq r_m\nu^* + \left(\frac{1}{2}\right)^t \nu^*$ and $\beta_1(t) \geq \beta_2(t) - \left(\frac{1}{2}\right)^t \nu^*$. However, $r_m > 1$ for $m \geq 2$ so $\beta_1(t) > \beta_L$ which implies that FFD must have been performed at some point with capacity $C = \beta_1(t)$ and yielded FFD$[T, C] > m$. But this contradicts the bound on FFD$[T, r\nu^*]$ and we are finished. □

The table below gives a few typical values for the performance of $\boldsymbol{A_{MF}}$ as m and t vary. Essentially then, the binary search employed by $\boldsymbol{A_{MF}}$ moves progressively toward a capacity $C \leq r_m\nu^*$. The overall running time of the heuristic is $O(n \log n + tmn)$.

t	$m = 2$	$m = 3$	$m = 10$	$m = 50$
5	1.174	1.185	1.251	1.251
6	1.158	1.169	1.236	1.236
7	1.151	1.162	1.228	1.228
LPT	1.167	1.222	1.300	1.327

Finally, we can state the following result regarding the guaranteed quality of solutions generated by the MULTIFIT procedure:

Theorem 4.7 *The performance of* $\boldsymbol{A_{MF}}$ *is given by*

$$\frac{\nu_{MF}}{\nu^*} \leq 1.220 + \left(\frac{1}{2}\right)^t.$$

□

It is interesting to note that for some time there existed a sense that the constant 1.22 in the theorem was replaceable by 20/17. In fact, the value can be reduced but not, unfortunately, to the value of the anticipated ratio. Specifically, Friesen (1984) was able to replace the value of 1.22 with 1.2 and moreover, exhibit an instance upon which MULTIFIT yielded an outcome, with ratio to an optimum of 13/11. This latter bound was then shown to be exact in Yue (1990).

Our next result demonstrates that, in the limit, good performance will be achieved by *any* list so long as list-processing is employed. Letting ν_{LP} denote the makespan produced by list-processing, we have

Theorem 4.8

$$\frac{\nu_{LP}}{\nu^*} \leq 1 + (m-1)\left(\frac{\max(\tau_i)}{\sum_i \tau_i}\right).$$

Proof. Define τ^* to be the maximum duration time. Then as in the proof of Theorem 4.3, no processor is idle prior to time $\nu_{LP} - \tau^*$. So, since some processor completes at ν_{LP}, we have

$$\sum_i \tau_i \geq \nu_{LP} + (m-1)(\nu_{LP} - \tau^*).$$

But since

$$\nu^* \geq \frac{\sum \tau_i}{m} \geq \frac{1}{m}(\nu_{LP} + (m-1)(\nu_{LP} - \tau^*)) \geq \frac{1}{m}(m\nu_{LP} - \tau^*(m-1))$$

or

$$1 \geq \frac{1}{\nu^* m}(\nu_{LP} m - \tau^*(m-1)),$$

then,

$$\frac{\nu_{LP}}{\nu^*} \leq 1 + \frac{\tau^*(m-1)}{\nu^* m}.$$

But $\nu^* \geq \frac{\Sigma \tau_i}{m}$ and so we have

$$\frac{\nu_{LP}}{\nu^*} \leq 1 + \frac{(\max_i \tau_i)(m-1)}{\Sigma \tau_i}.$$

□

We leave it to the reader to show that the following instance establishes realizability for the bound stated above:

$$\begin{aligned} n &= m(m-1)+1, \tau_i = 1 \text{ for } 1 \leq i \leq n-1, \tau_n = m \\ L &= \{1, 2, \ldots, n\} \end{aligned}$$

The corresponding $2 - 1/m$ result should be familiar.

4.1.2 Total Completion Time Problems

Suppose now that our measure is total completion time Σc_i and in addition, that the environment allows for processors that are not identical. In this regard, we extend our duration time parameter to denote processor-dependent values. That is, let τ_{ij} indicate the duration time of job i on processor j and let τ be the $n \times m$ matrix of all such duration times. For example, for times given by τ below, suppose we process job 1 on processor 1 with jobs 2 and 3 on the second processor in the order $\{3, 2\}$. From the timing diagram we see that the total completion time of the resultant schedule, shown in Figure 4.17, is 13.

$$\tau = \begin{bmatrix} 4 & 3 \\ 2 & 5 \\ 6 & 2 \end{bmatrix}.$$

Interestingly, this problem, $R||\Sigma c_i$, possesses a polynomial time solution following Bruno *et al.* (1974), which also employs a bipartite matching strategy. To see this (and to appreciate how the result does not extend to the weighted case), suppose job i is scheduled on processor j in the last position. Then clearly it contributes τ_{ij} to total completion time. Similarly, if

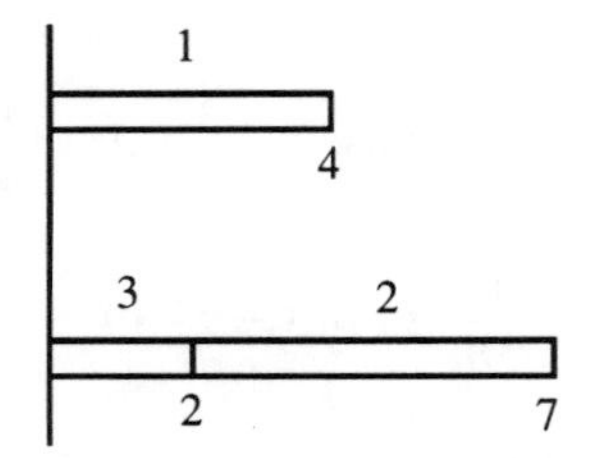

Figure 4.17. *Schedule for instance of* $R||\Sigma c_j$

it is next-to-last, it contributes $2\tau_{ij}$ and so forth, contributing $k\tau_{ij}$ if scheduled in the position k from the last (it contributes to its own completion in addition to all those jobs it "pushes" later in the schedule).

Now, suppose we form a bipartite graph $G = (X, Y, E)$ with $X = \{1, 2, \ldots, n\}$ corresponding to the jobs and Y given by vertices labeled as $\{11, 12, \ldots, 1m, 21, 22, \ldots, 2m, \ldots, n1, n2, \ldots, nm\}$ where a pair (uv) corresponds to the uth position from the end on processor v [*e.g.* (22) represents the next-to-last sequential position on the second processor]. The edges correspond to the options relating jobs to scheduling positions. These edge weights are given by matrix $\bar{\tau}$ defined as follows:

$$\bar{\tau} = \begin{bmatrix} [1\tau_{ij}] \\ [2\tau_{ij}] \\ \vdots \\ [n\tau_{ij}] \end{bmatrix}.$$

It is not hard to see that a minimum weight matching saturating every vertex in X solves the problem. We illustrate the strategy below:

Example 4.9 For the three-job illustration above our construction would appear as shown in Figure 4.18. □

An obvious question is whether or not this matching formulation yields any further efficiencies when used in specialized settings. For example, if it is applied to $P||\Sigma c_j$, are there shortcuts implied? Happily, there are in that when employed in the identical processor case, the outcome is equivalent to simply applying the SPT scheduling strategy, generalized to handle multiple processors. The reader is asked to formalize this in Exercise 4-4.

Similarly, when a *uniform* processor environment is considered, the total completion time problem is also particularly easy to solve. Understood is that in this case, each processor, while related, is distinguished by a possibly different "speed" that scales fixed processing times on the respective processors accordingly. Handling this in a manner like that just alluded to in the identical processor case is the subject of Exercise 4-41.

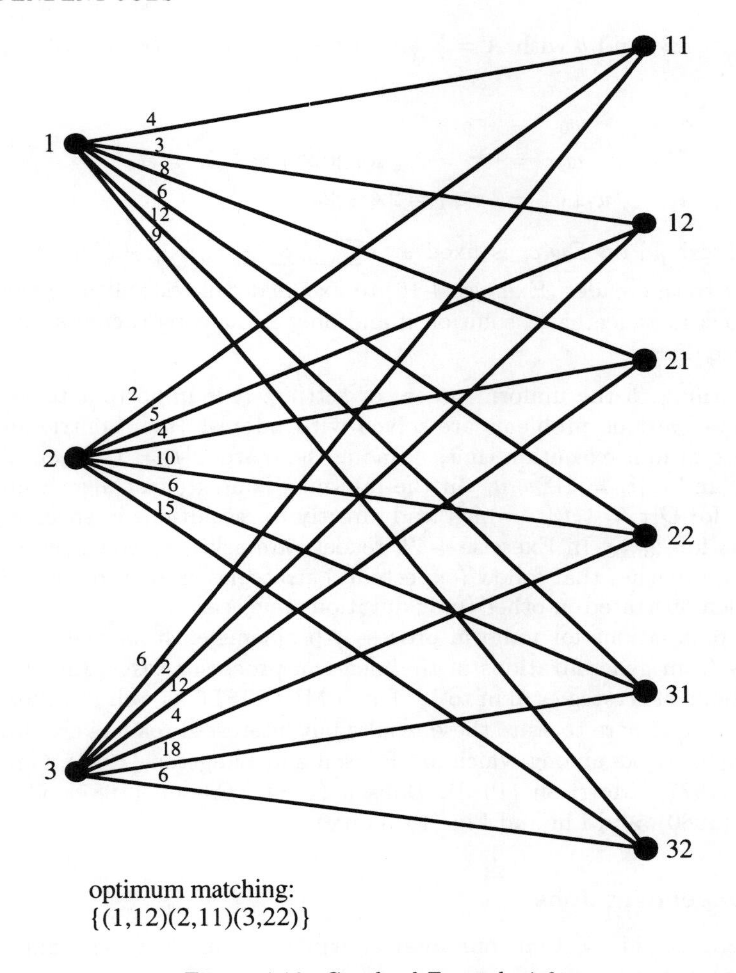

Figure 4.18. *Graph of Example 4.9*

Unfortunately, the matching procedure cannot be extended to the case where jobs possess weights. This would require not only knowledge of a job's position in a sequence but also specific knowledge of which jobs precede it. Expectedly, we have

Theorem 4.9 *The (recognition version) problem* $P2||\Sigma w_i c_i$ *is* $\mathcal{NP}$*-Complete.*

Proof. We will provide only the reduction. Accordingly, after Bruno *et al.* (1974) let us create, from an instance of KNAPSACK defined on integers

$a_1, a_2, \ldots, a_t$, and b with $A = \sum_i a_i$, a $t+2$ job instance of $P2||\Sigma w_i c_i$ as follows:

$$\begin{aligned} w_0 &= \tau_0 = 2b - 1 \\ w_i &= \tau_i = 2a_i \text{ for jobs } i = 1, 2, \ldots, t \\ w_{t+1} &= \tau_{t+1} = 2A - 2b - 1. \end{aligned}$$

The threshold on $\Sigma w_i c_i$ is fixed at $\left(\sum_{i=0}^{t+1} \sum_{j=0}^{i} w_i w_j\right) - (2A-1)^2$. We leave it to the reader (Exercise 4-16) to fix the details establishing that the knapsack instance has a solution if and only if the constructed scheduling instance does. □

Returning to the uniform processor setting, it is important to at least note that various problems are solved with interest (*i.e.*, nontrivially) in the case of unit-execution times. Among these are $Q|\tau_i = 1|\Sigma w_i c_i$, $Q|\tau_i = 1|\Sigma T_i$, and $Q|\tau_i = 1|\Sigma w_i u_i$. In the min-max context, fast algorithms are known for $Q|\tau_i = 1|f_{\max}$ which lead directly to resolutions in specific cases such as for $L_{\max}$. In Exercise 4-42, readers are asked to consider some of these approaches that follow (expectedly) strategies employed and that we have demonstrated in other (unit duration time) cases.

Approximations for uniform processor problems exist as well, following largely from generalizations in the identical processor cases. Included are extensions of strategies akin to LPT and MULTIFIT as well as others. We take no space here to state these results but interested readers are directed to various works among which are Friesen and Langston (1983), Gonzalez *et al.* (1977), Morrison (1988), Dobson (1984), Friesen (1987), Cho and Sahni (1980), and Liu and Liu (1974a,b,c).

4.2 Dependent Jobs

Throughout this section, our interest will be confined to the makespan measure of performance.

4.2.1 General Precedence

In general, problems tend to become hard very quickly when $\prec \neq \phi$ and particularly so in the case of parallel processor scheduling.

Theorem 4.10 *The (recognition version) $P|prec, \tau_i = 1|C_{\max}$ is $\mathcal{NP}$-Complete.*

Proof. As shown in Lenstra and Rinnooy Kan (1978), the reduction is from the $\mathcal{NP}$-Complete problem of deciding if a graph $G = (V, E)$ possesses a clique of size at least k. Accordingly, from an instance of the latter [defined on $G = (V, E)$ and with parameter k] let us construct an instance of the scheduling problem.

Our job set is given by

$$T = V \cup E \cup B \cup C \cup D,$$

where

$$\begin{aligned} B &= \{b_1, b_2, \ldots, b_x\} \\ C &= \{c_1, c_2, \ldots, c_y\} \\ D &= \{d_1, d_2, \ldots, d_z\} \end{aligned}$$

with $|B|$, $|C|$, and $|D|$ satisfying:

(i) $m = k + |B| = |V| - k + \binom{k}{2} + |C| = |E| - \binom{k}{2} + |D|$

(ii) $\min(|B|, |C|, |D|) = 1$.

The precedence constraints $\prec$ are constituted as follows:

$$(b_i, c_j), 1 \leq i \leq x, 1 \leq j \leq y,$$

$$(c_j, d_k), 1 \leq j \leq y, 1 \leq k \leq z,$$

and

$$(v_i, e_j), \text{ if } v_i \text{ is incident to } e_j \text{ in } G.$$

Let us set a threshold value on $C_{\max}$ of 3. Now, observe that $|T| = 3m$ and thus no admissible schedule can have $C_{\max} < 3$ (this follows also from the length-3 chains $\{b_i, c_j, d_k\}$). In addition, no idle time can exist in any feasible schedule.

Now, for the first direction, assume G possesses a k-clique C. Then it is easy to see how a suitable schedule can be constructed:

- Schedule the b-jobs in the first period for all processors.
- Schedule the c-jobs in the second period.
- Schedule the d-jobs in the third period.
- Schedule the vertex-jobs for $v \in C$ in the first period (there are k of these).
- Schedule the other vertex-jobs in period 2. There are $|V| - k$ of these.
- Schedule the edge jobs relative to edges induced by C in period 2. There are $\binom{k}{2}$ of these.
- Schedule the remaining $|E| - \binom{k}{2}$ edge-jobs in period 3.

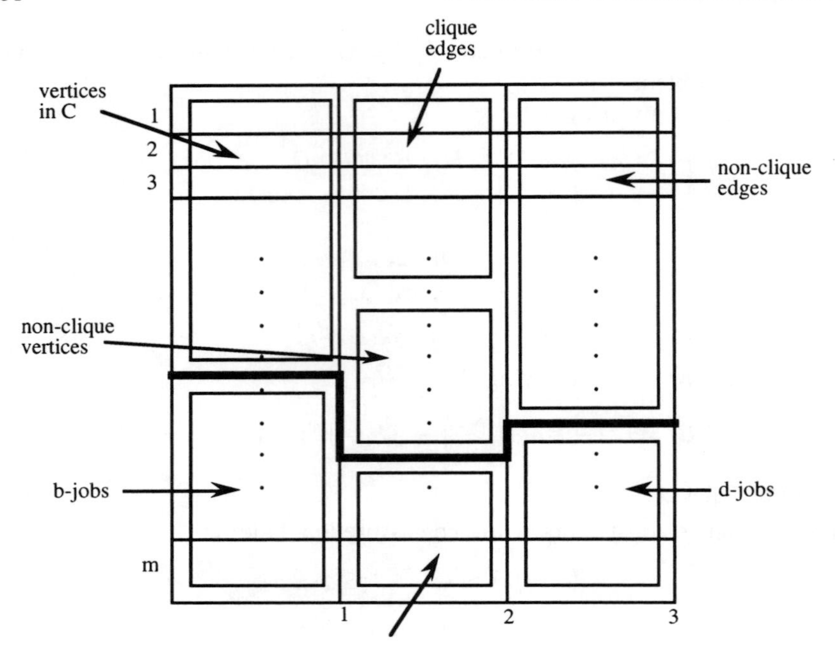

Figure 4.19. *Construction of Theorem 4.10*

This construction is easy to see from the diagram of Figure 4.19. Clearly, we have produced a feasible schedule that fits perfectly into the $3m$ available positions.

Conversely, let us suppose a feasible schedule exists having $C_{\max} = 3$. Clearly, the b-, c-, and d-jobs have to be scheduled in the 1st, 2nd, and 3rd periods respectively. The diagram of Figure 4.20 illustrates.

Now, by hypothesis, k slots are available in period 1, $|V| - k + \binom{k}{2}$ are available in period 2, and $|E| - \binom{k}{2}$ are available in period 3. Further, these remaining positions are the only candidates into which the other jobs must have been scheduled. In the 3rd period, the corresponding slots have to be used by edge jobs (G has no isolated vertices) because they are preceded by vertex jobs. The other $\binom{k}{2}$ edge jobs have to go into the 2nd period for the same reason. Now, the vertex jobs can go into the 1st or 2nd periods. But the vertex-jobs in period 1 *must* include the endpoints of the

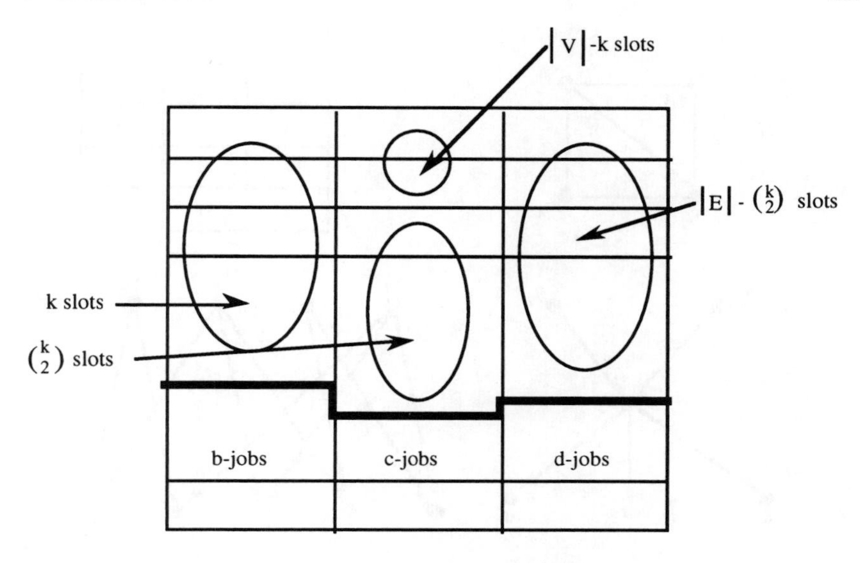

Figure 4.20. *A suitable schedule with* $C_{\max} = 3$

$\binom{k}{2}$ edge-job edges in period 2. However,the only way that k vertices can include the endpoints of all $\binom{k}{2}$ edges is by forming a k-clique. Hence, the existing feasible schedule implies the corresponding existence of a set of vertices R given by the k vertex-jobs in period 1. Setting $C \leftarrow R$ yields a suitable clique in G. Membership in $\mathcal{NP}$ is easy to see and the proof is complete. □

Example 4.10 Let G be given as shown in Figure 4.21 with vertices and edges labeled as indicated and let $k = 3$. Then we have $m = 3 + |B| = 4 + |C| = 2 + |D|$ and we may set $|B| = 2$, $|C| = 1$, and $|D| = 3$ so that $m = 5$. The precedence structure follows as shown in Figure 4.21 along with a feasible schedule (note the presence of 3-cliques in G). □

The following corollary is now obvious:

Corollary 4.11 *Determining if a schedule of length* $3\,(C_{\max} \leq 3)$ *exists for* $P|prec, \tau_i = 1|C_{\max}$ *is* $\mathcal{NP}$*-Complete.* □

Observe that if we replace the makespan threshold of 3 by 2 in the statement of the corollary, the resulting problem is trivial. On the other hand, Tovey has shown that if $\prec$ contains no chains in excess of length 2, $P|prec, \tau_i = 1|C_{\max} \leq 3$ is still hard. We have

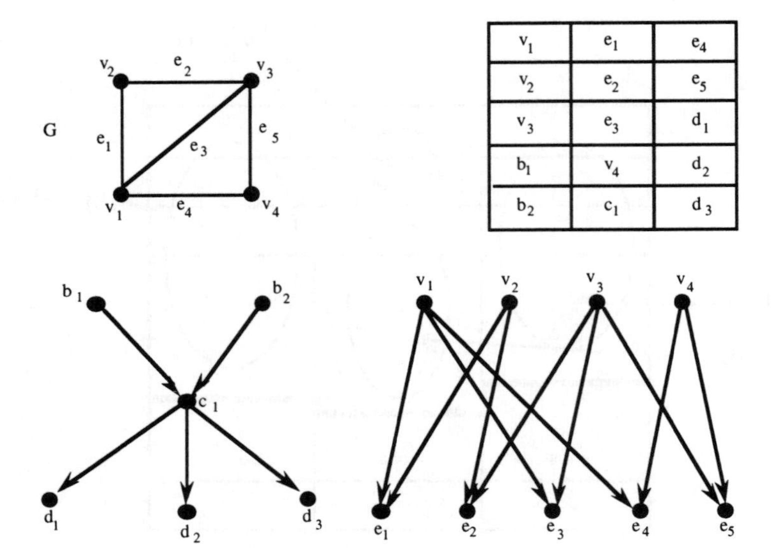

Figure 4.21. *Sample reduction from Theorem 4.10*

Theorem 4.12 *The (recognition version) problem $P|prec, \tau_i = 1|C_{\max}$ with constant threshold value of 3 is hard even if $\prec$ has no chains of length 3 or greater.*

Proof. We can obtain a reduction from the $\mathcal{NP}$-Complete problem MIN SET UNION which takes as an instance a collection C of subsets of some ground set of elements and integers a and b, asking if there are b members of C whose union is a set of size no greater than a. Accordingly, our scheduling instance is defined on a precedence structure formed from a bipartite graph having bipartition $\{X, Y\}$ with X representing elements in the ground set and Y, subsets in C. Arcs (i, j) are simply the oriented edges between X and Y created by membership of elements in X with the respective subsets captured by Y. We fix $m = a$ and form X, Y, and b satisfying the following:

$$|X| + |Y| = 3m, \ b = 2m - |X|.$$

Observe that we may "pad" the scheduling instance, as needed, and particularly with additional members of C (thus Y) where each additional subset contains all elements of the (possibly augmented) ground set X. It is easy to see that a three-period schedule exists exactly when a suitable subcollection of sets exists. □

Note that the complexity status of problem $P|prec, \tau_i = 1|C_{\max}$ remains open when the number of processors is fixed with $m \geq 3$. Also interesting is that when approximations are employed, we should also temper our expectations as well:

Corollary 4.13 *Unless $\mathcal{P} = \mathcal{NP}$, there can exist no polynomial-time approximation algorithm for $P|prec, \tau_i = 1|C_{\max}$ having performance bound strictly less than 4/3.*

Proof. If such an algorithm existed, it would solve the stated makespan problem of Corollary 4.11, thus establishing the equivalence of $\mathcal{P}$ and $\mathcal{NP}$. □

4.2.2 Special Precedence Structures

There are some (interesting) solvable unit duration-time cases. One is given by a classic result of Hu (1961) which takes $\prec$ as an arborescence with vertex out-degree ≤ 1.

A_H : Hu's Algorithm

Step 0: Compute the length of a longest path from each vertex. Call these values ℓ_i.

Step 1: Create a list L of jobs arranged in nonincreasing ℓ-order. Perform list-processing on L. □

Example 4.11 Consider a set of jobs constrained as shown by the digraph of Figure 4.22. Values for ℓ_i are given by each vertex, the list L is formed, and the resultant schedule is created as shown. □

Although not complicated, the argument that A_H is correct is more interesting than one might expect. There are at least two "short" proofs. Following we employ the one by Lawler and Lenstra (1982). Another is due to McHugh (1984).

Theorem 4.14 *Algorithm A_H solves $P|tree, \tau_i = 1|C_{\max}$.*

Proof. We will show that the length of a schedule produced by a correct application of A_H is not improvable by another feasible schedule. To this end, let us apply the algorithm and let the resulting completion time $C_{\max}$ be ν^*. Now, select any smaller value $\nu < \nu^*$. A logical selection arises by examining various lower bounds on $C_{\max}$. Among these, we have

$$C_{\max} \geq \max_{1 \leq i \leq h+1} \left\{ (i-1) + \left\lceil \frac{|V_i|}{m} \right\rceil \right\},$$

where h is the height of the precedence arborescence and V_i is the set of vertices in the structure at or above level i. (We ask the reader to prove the validity of this bound as an exercise.)

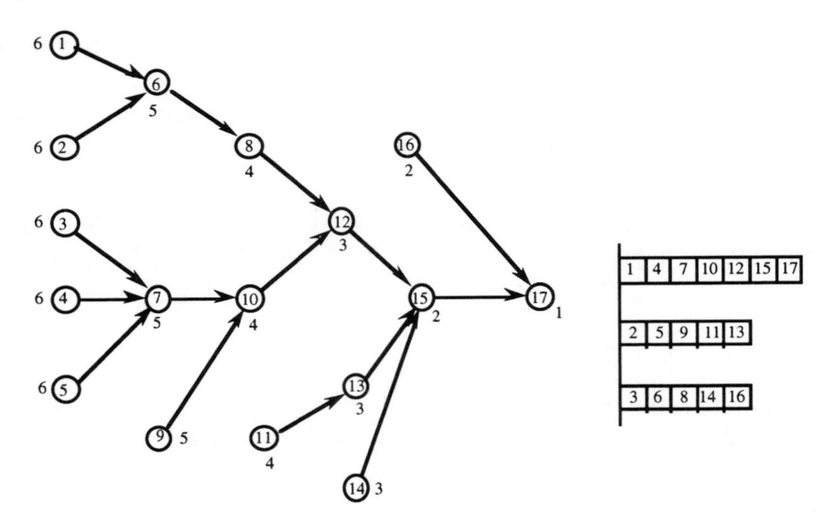

Figure 4.22. *Digraph for Example 4.11*

Now, let us define a label for each vertex/job determined as the difference between ν and the respective job's level. Clearly, each job has a label one less than its direct successor. Then upon application of $\boldsymbol{A_H}$, a schedule of length strictly greater than ν resulted which implies that in some (unit) time interval j, $1 \leq j \leq \nu^*$, a job is scheduled having a label less than j (recall that $\boldsymbol{A_H}$ will schedule jobs with these new, smaller labels first). Let us denote by s the smallest such period and let a job with label $\ell < s$ be scheduled in period s. We want to show that in every period prior to s, all processors are busy (i.e., there is no idle time). Suppose otherwise. Then there is a period $s' < s$ with fewer than m jobs scheduled. But $s' \neq s - 1$ since the job labeled ℓ waited until period s, it must have been blocked by a predecessor job scheduled in s' where the latter would be labeled $\ell - 1 < s'$, contradicting the given choice of s. So, $s' < s - 1$.

However, this means that a period (*i.e.*, s') would have greater idle time than its immediate successor period (*i.e.*, $s' + 1$) which is inconsistent with the precedence (arborescence) structure. Thus, we have that for each period $s' < s$, m jobs are being worked on and all of these have labels less than s. Moreover, this means that during s at least one job having label value less than s is actually unavoidable, implying that no schedule of length $\nu < \nu^*$ is possible. This completes the proof. □

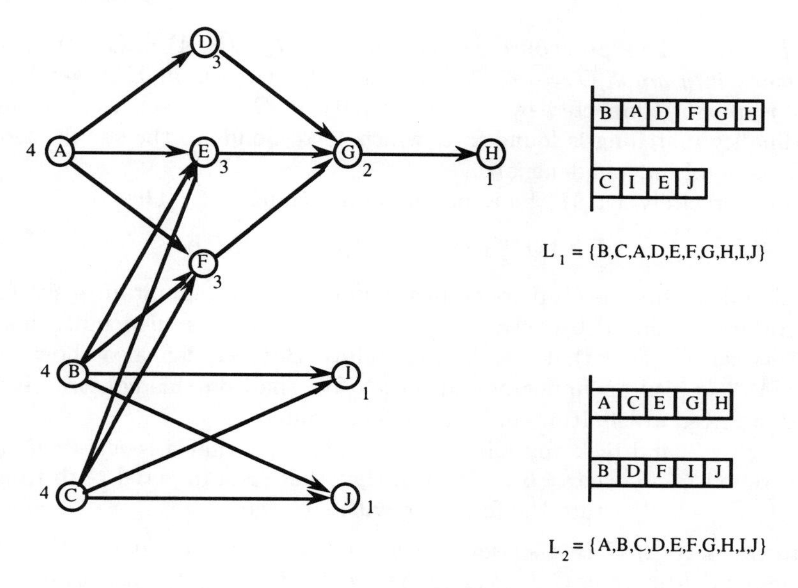

Figure 4.23. *Digraph on which A_H fails*

Hu's algorithm will also work when $\prec$ is a forest with each component having the stated arborescence property. It will also work with so-called *antiforests*, *i.e.*, arborescences having vertex in-degree ≤ 1. These cases are obvious since, as suggested before, in the first we need only connect the disconnected components to a single terminal vertex while in the second, a reversal of orientation creates a structure satisfying the conditions in Theorem 4.14.

On the other hand, algorithm $\boldsymbol{A_H}$ can be shown to fail on more general precedence structures. Consider the following illustration from Graham (1978).

Example 4.12 Consider the structure pictured in Figure 4.23. Values ℓ_i are given by respective vertices and list L_1 is shown along with the schedule it produces. The latter is not optimal as the second schedule, formed by L_2, indicates. □

4.2.3 The Two-Processor Case

Particularly interesting is the case of $P2|prec, \tau_i = 1|C_{\max}$. Following we give three polynomial-time procedures each with their own ingenious attributes. We begin with what is apparently the original offering in Fujii *et al.* (1969). This procedure constructs from $G = (V, A) = (T, \prec)$ an *incomparability graph* $\bar{G} = (V, E)$ where an edge (i, j) is in E iff vertices i and j are not connected by a directed path in $G = (T, \prec)$. A maximum cardinality matching is found in $\bar{G}$ which corresponds to the simultaneous processing of independent jobs.

More precisely, let M^* be a maximum matching in $\bar{G}$. Then

$$\nu^* \geq |M^*| + (n - 2|M^*|) = n - |M^*|,$$

which follows since $|M^*|$ pairs of jobs can be processed simultaneously with the others required, because of precedence, to be scheduled individually (otherwise we deny that M^* is maximum). But, we can also show that $n - |M^*|$ is also an upper bound on ν^*. We shall do this constructively. First we need a definition and a technical lemma.

Let $S \subseteq T$ and pick any job $j \in S$. We shall say that j is *maximal* in S if there exists no other job $i \in S$ such that there is a directed path from i to j in $(T, \prec)$. We state the following without proof.

Lemma 4.15 *If S is a subset of jobs and M is a matching of S in the incomparability graph $\bar{G} = (V, E)$, then at least one of the following must hold:*

i) *there exists a maximal job in S that is not matched under M,*

ii) *there exists a pair of jobs matched under M where both jobs are maximal in S,*

iii) *there exists two job-pairs (i, j) and (k, h) in M such that i and k are maximal in S and (j, h) is an edge in $\bar{G}$.* □

This lemma can be applied in an algorithmic context. We state such a procedure as follows:

A_{FKN} : The Two-Processor Algorithm of Fujii, Kasami, and Ninomiya

Begin with $S = T$ and apply Lemma 4.15 recursively to find sets ξ_i, $1 \leq i \leq n - |M^*|$. The final schedule is formed as $\{\xi_1, \xi_2, \ldots, \xi_{n-|M^*|}\}$. If (iii) is used, set the respective $\xi_j = \{i, k\}$ and alter the relevant matching M as $M \cup (j, h) \backslash \{(i, j), (k, h)\}$. □

Example 4.13 Consider the graph in Figure 4.24. The application of A_{FKN} is summarized in the table shown. The resultant schedule is also given. □

Any prospects that the matching based algorithm A_{FKN} can be extended would appear to be limited as the following result in Lenstra and Rinnooy Kan (1978) suggests:

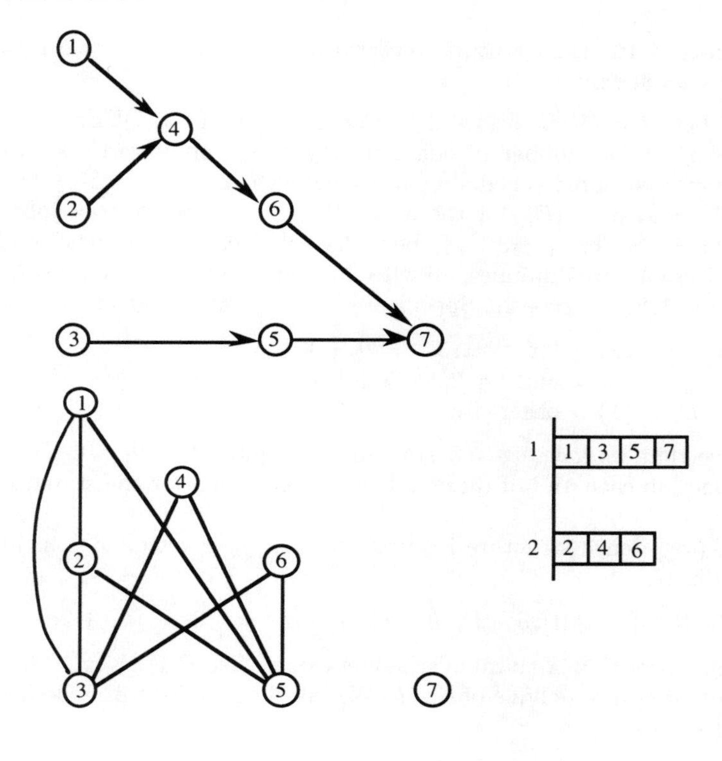

$M^* = \{(1,2),(3,6),(4,5)\}$

$S_1 = \{1,2,\ldots,7\}$, $M_1 = M^*$ Apply (ii) to (1,2). Set $\xi_1 = \{1,2\}$, $S_2 = \{3,4,5,6,7\}$, $M_2 = \{(3,6),(4,5)\}$.
Apply (iii) to (3,6) and (4,5), *i.e.*, 3 and 4 are maximal in S and $(5,6) \in E$. Set $\xi_2 = \{3,4\}$, $S_3 = \{5,6,7\}$, $M_3 = \{(5,6)\}$.
Apply (ii) to (5,6). Set $\xi_3 = \{5,6\}$, $M_4 = \phi$, $S_4 = \{7\}$.
Apply (i) and set $\xi_4 = \{7\}$.

Figure 4.24. *Graph of Example 4.13*

Theorem 4.16 *The (recognition version) problem $P2|prec, \tau_i \in \{1,2\}|C_{\max}$ is $\mathcal{NP}$-Complete.*

Proof. Let $G = (V, E)$ and k define an instance of CLIQUE. Also, let us denote by ℓ the number of edges in any clique on k vertices. Now, we construct a suitable scheduling instance with $u = 4v + 3e$ jobs where $v = |V|$ and $e = |E|$. Of these, let $i_1, i_2, \ldots, i_v$ be "vertex jobs" with duration 1 and let $j_1, j_2, \ldots, j_e$ be "edge jobs" each with duration 2. The remaining jobs are dummies, all with duration 1, given by pairs $[h, t]$ where $h \epsilon D_t$, $t = 1, 2, \ldots, 2v + 2e$. Specifically, the sets D_t are given as follows:

$$\begin{aligned} D_t = \{1,2\} \quad & \text{for } t = 1, 3, \ldots, 2k-1 \\ & \text{and } t = 2k + 2\ell + 1, 2k + 2\ell + 3, \ldots, 2v + 2\ell - 1, \\ D_t = \{1\} \quad & \text{otherwise.} \end{aligned}$$

Observe that we have $v + e$ vertex and edge jobs, $2v + 2e$ sets D_t with at least one job each and of these, v have a second job for the stated total of n.

The precedence structure is given as: $i_x \prec j_y$ if vertex x is incident to edge y in G,

$$[g, t] \prec [h, t+1], g \in D_t, h \in D_{t+1}, \quad \text{for } t = 1, 2, \ldots, 2v + 2e - 1.$$

We thus have that a suitable schedule exists with $C_{\max} \leq y = 2v + 2e$ if and only if G has a clique of size k. We ask the reader to provide details in Exercise 4.18. □

A second algorithm for this two-processor problem is due to Coffman and Graham (1972). The procedure is based on a lexicographic ordering of jobs constructed on labels that are assigned as a function of a job's successors.

Let $\ell = (n_1, n_2, \ldots, n_k)$ and $\ell' = (n'_1, n'_2, \ldots, n'_{k'})$ be sequences of integers where $k, k' \geq 1$. Then recall that ℓ is said to be lexicographically smaller than (or less than) ℓ' if

- **i)** there exists j, $1 \leq j \leq k$, such that for every i, $1 \leq i < j$ we have $n_i = n'_i$ but $n_j < n'_j$, or
- **ii)** $k < k'$ and $n_i = n'_i$ for all i, $1 \leq i \leq k$.

For example, $\ell = (6, 5, 2, 2, 1)$ is lexicographically smaller than $\ell' = (6, 5, 3)$. Similarly $\ell = (6, 5)$ is smaller than $\ell' = (6, 5, 4)$ where often we write these as $\ell \overset{L}{<} \ell'$. We now state the algorithm:

A_{CG} : The Two-Processor Algorithm of Coffman and Graham

Step 0: Assign labels $1, 2, \ldots, s$ to jobs (vertices of $G = (T, \prec)$) which have no successors. Now, assume labels $1, 2, \ldots, j-1$ have been assigned and let S be the set of unlabeled jobs having no unlabeled

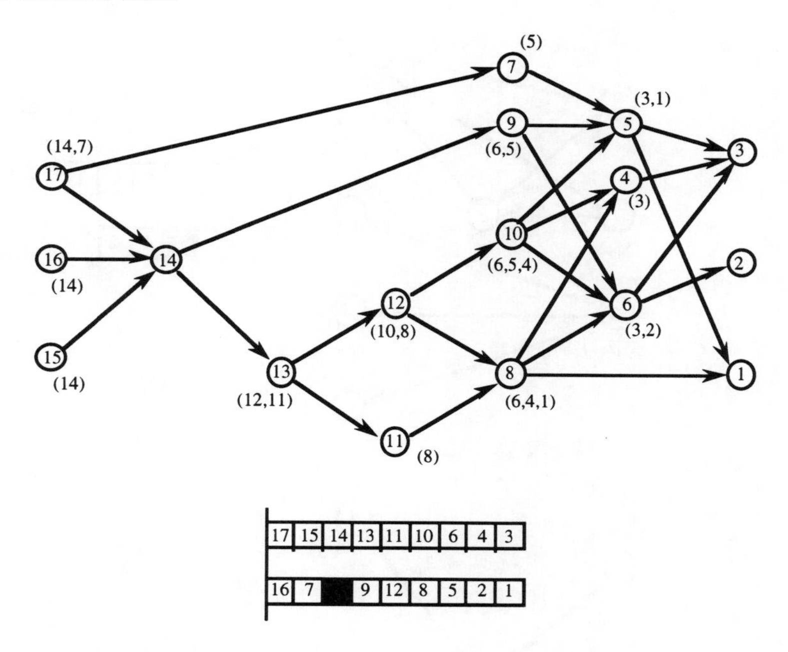

Figure 4.25. *Instance of Example 4.14*

successors. For each job in S form a sequence ℓ arranged in decreasing order of its immediate successors' labels. Let $x \in S$ be such that

$$\ell(x) \overset{L}{\leq} \ell(x') \text{ for all } x' \in S$$

and label vertex x by j.

Step 1: When all jobs have been labeled, perform list-processing on L with jobs/vertices arranged in decreasing label order. □

Example 4.14 The key notion underlying $\boldsymbol{A_{CG}}$ is that jobs that head long chains or have many successors possess higher labels and are thus placed earlier in L. We apply the strategy to the instance in Figure 4.25 [this as well as the instances of the next two examples is borrowed from Coffman (1976)] where the sequences $\ell(x)$ are specified next to each vertex. □

Note that in applying $\boldsymbol{A_{CG}}$, we assume that $G = (T, \prec)$ is *transitively reduced, i.e.*, $(i,j), (j,k) \in \prec \Rightarrow (i,k) \notin \prec$. That we make such an assumption is for reasons far more important than convenience. To make the point, consider the next illustration.

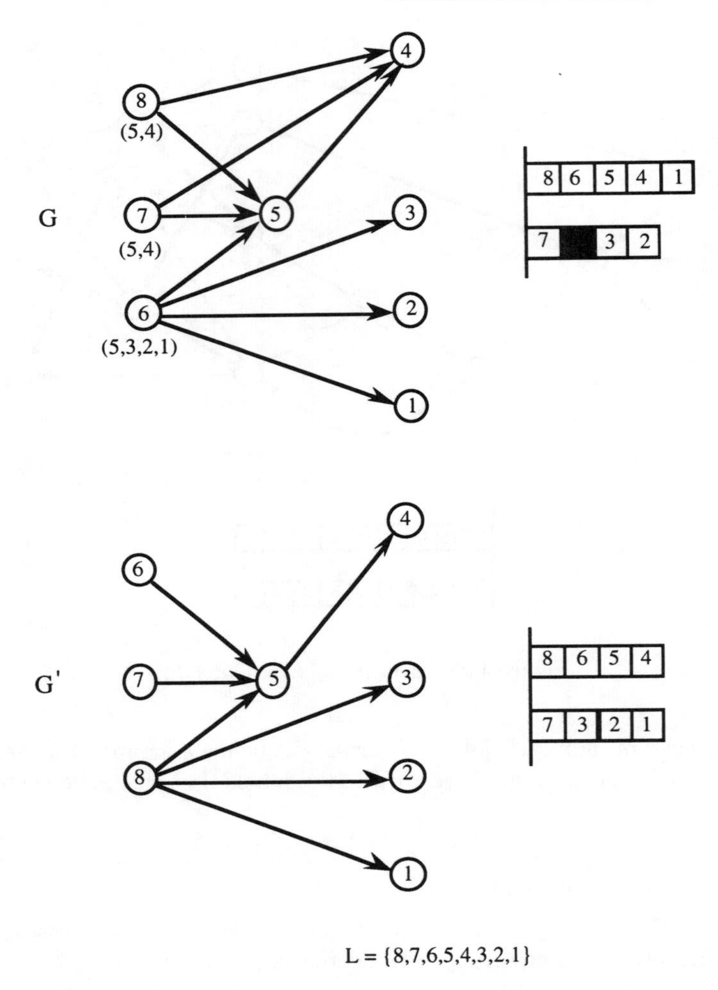

Figure 4.26. *Non-transitively reduced graph*

Example 4.15 Suppose G appears as indicated in Figure 4.26. Applying A_{CG} produces the schedule shown with makespan value 5. But, this is not optimal and the offending culprit is the presence of non-transitively reduced arcs. Forming G' on the other hand (which is so reduced) allows a correct application. □

The A_{CG} procedure is also not apparently extendable to $m \geq 3$.

Example 4.16 We can demonstrate the failure when $m = 3$ by the in-

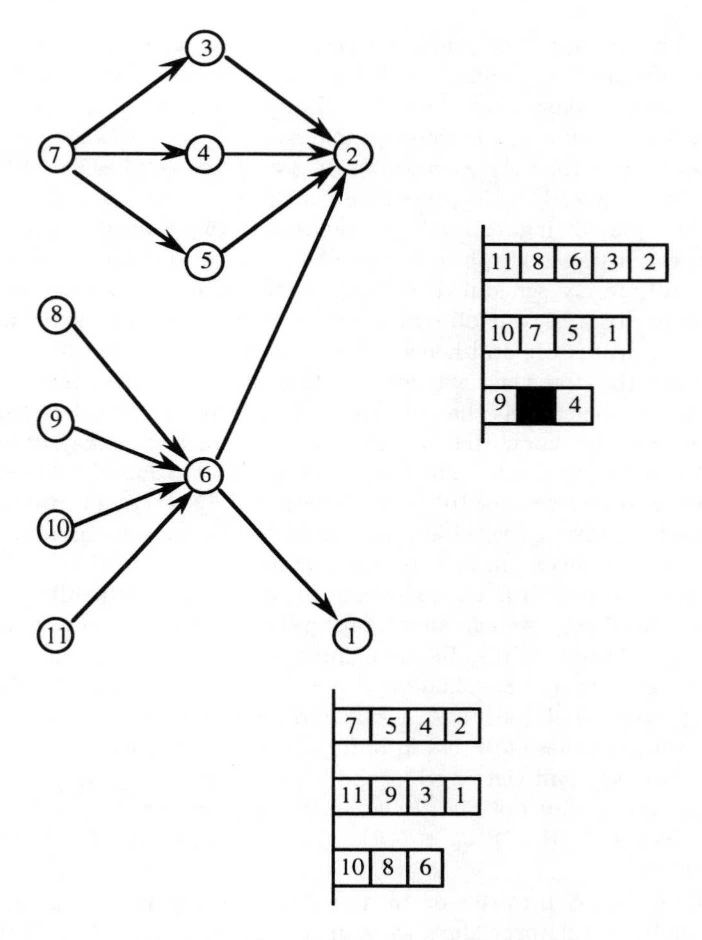

Figure 4.27. *A_{CG} failure when $m = 3$*

stance shown in Figure 4.27. The first schedule is produced by a valid application of $\boldsymbol{A_{CG}}$. However, this is not optimal as the schedule at the bottom indicates. □

The proof that $\boldsymbol{A_{CG}}$ is valid is longer than space permits here. Following, we state the desired result and sketch briefly the basis for its argument. Accordingly, we have:

Theorem 4.17 *Algorithm $\boldsymbol{A_{CG}}$ will correctly solve $P2|prec, \tau_i = 1|C_{\max}$.* □

In spirit, the proof for this theorem is not unlike that underlying the argument for the Fujii *et al.* matching-based strategy given previously. That is, we determine a minimum length for the makespan of any schedule and then proceed to show that algorithm $\boldsymbol{A_{CG}}$ produces an admissible outcome with this value and is thus optimal.

Also evident is that $\boldsymbol{A_{CG}}$ captures, in a way, the emphasis on the level-by-level strategy of the Hu procedure which, as we saw, is correct for arborscences; jobs at higher levels get higher priority than do jobs at lower levels in the instance graph, are thus placed earlier in the list, and therefore are ultimately scheduled earlier. In fact, if each level possessed an even number of jobs, our objective would be accomplished since we have exactly two processors and hence all jobs at one level would have to complete before those at the next level could begin. Of course, this is clearly not enough in general, as the instance in Example 4.12 revealed. But from the latter, we also know that, at least, we need to have a way of examining and comparing a job's successors, where in addition this comparison would have to be sensitive to the priorities of the successors. Happily, the lexicographic ordering-based application in the Coffman-Graham strategy establishes that this is, in fact, all that we need.

The essential notion is the following: From a correct schedule produced by algorithm $\boldsymbol{A_{CG}}$, we can show constructively that it is always possible to identify a family of disjoint, nonempty subsets of jobs $J_t, J_{t-1}, \ldots, J_0$ each having some odd cardinality $|J_i| = 2k_i - 1$ and where all jobs in J_i precede (under $\prec$) all jobs in J_{i-1}. Clearly, no schedule could do any better than to schedule these job sets in order, assuming at least $\sum_{0 \leq i \leq t} k_i$ time units accordingly. But the length of the schedule formed by $\boldsymbol{A_{CG}}$ is exactly this value and is thus optimal. Relative to the graph of Example 4.14, our sets are $J_3 = \{15, 16, 17\}$, $J_2 = \{14\}$, $J_1 = \{4, 5, 6, 8, 9, 10, 11, 12, 13\}$, and $J_0 = \{3\}$.

Our final approach to this problem is due to Garey and Johnson (1975). Interestingly, in this procedure we want to maintain or create all the arcs implied by precedence in $G = (T, \prec)$ (these were precisely the ones eliminated in the transitive reduction and we saw in Chapter 3 that a graph formed in this way is called the transitive closure of G). Loosely, this approach takes as input a bound on the completion time of all jobs and then aims to find a schedule that finishes by this time, or, alternately, concludes that one does not exist. Operationally, the procedure computes a set of artificial due-dates that would have to be met if this assumed completion time threshold is to be met by some feasible schedule. The key result is that for a given threshold, say D, if there is *any* (feasible) schedule that completes by D, then performing list-processing on L formed by nondecreasing due-date order will produce such a schedule.

In fact, this third $P2|prec, \tau_i = 1|C_{\max}$ algorithm actually arises as a special application of a more general setting. In the latter, each job j is identified with a nonnegative, integer start time and due-date specified say, by the pair (s_j, d_j) and the objective is one of deciding if a schedule exists where each job is processed in the respective interval $[s_j, d_j]$. Trivially, a fast strategy for dealing with this problem is immediately transferable to the makespan case by the usual binary search tactic. However, here we will adopt the language of this more general setting in order to make directly accessible the key results that underlie the fundamental ideas of the common scheduling strategy.

Following Garey and Johnson (1977), for any job i and for integers s and d satisfying $s_i \leq s \leq d_i \leq d$, let $N[i, s, d]$ be defined as the set of jobs $\{j \neq i | d_j \leq d$ and either $i \prec j$ or $s_j \geq s\}$. That is, $N[i, s, d]$ consists of those jobs that are either successors of job i and/or are restricted to start no earlier than i. Conceptually, these are jobs that are *constrained* to be assigned to the two processors no earlier than when job i is assigned. Indeed, this rather obvious notion coupled with the following lemma should help in clarifying why the ensuing strategy works.

Lemma 4.18 *For any job i and integers s, d that satisfy $s_i \leq s \leq d_i \leq d$, if $|N[i, s, d]| \geq 2(d-s)$, then job i must be finished by time $d - \lceil |N[i, s, d]|/2 \rceil$ in* any *admissible schedule meeting all job due-dates.* □

From this lemma, it is easy to see that when the stated conditions are satisfied, another valid constraint on the latest completion time of a job i results; if $\lceil |N[i, s, d]| \rceil \geq 2(d - s)$ and if $d - \lceil |N[i, s, d]|/2 \rceil < d_i$, then we may change d_i to $d - \lceil |N[i, s, d]|/2 \rceil$ and in so doing, not eliminate any admissible schedule that meets all due-dates. More formally, we may conclude that an admissible schedule satisfies all of the original due-dates if and only if it satisfies the modified due-dates resulting from the indicated alteration to d_j. Moreover, these alterations can be performed over and over until no further changes are possible *or* until some d_i is computed having value strictly less than $s_i + 1$. In the latter case, it follows that no admissible schedule exists (*i.e.*, one meeting all of the original due-dates). Naturally, some care needs to be exercised in order that this due-date modification process be accomplished efficiently. As it turns out, this can be dealt with handily (see Garey and Johnson, 1977).

Now, let us assume that for every job i we have that $d_i \geq s_i + 1$ and for every pair of integers s, d satisfying $s_i \leq s \leq d_i \leq d$, if $|N[i, s, d]| \geq 2(d - s)$, then $d_i \leq d - \lceil |N[i, s, d]|/2 \rceil$. Due-dates satisfying these conditions are referred to as *internally consistent* and accordingly yield the following properties:

(i) $i \prec j \Rightarrow d_i < d_j$, and

(ii) $s \leq d \Rightarrow |\{i | s \leq s_i$ and $d_i \leq d\}| \leq 2(d - s)$.

The importance of these "consistency" properties is revealed by the following theorem which is the key result in producing an algorithm. Accordingly, we have:

Theorem 4.19 *Let L be a list of jobs arranged in nondecreasing modified due-date order. Then if the due-dates are internally consistent, the outcome produced by operating on L in list-processing manner will yield a schedule where all jobs meet their respective due-dates.* □

We are now in a position to state the algorithm. Recall that we are dealing with the case where every job possesses the start-time, due-date pair (s_i, d_i). We will describe the makespan minimization modification shortly.

A_{GJ} : The Two-Processor Algorithm of Garey and Johnson

Step 1: Using the result of Lemma 4.18, successively modify due-dates until either all d_i are internally consistent or some $s_i \geq d_i$ in which case we may halt, reporting correctly that no admissible schedule exists.

Step 2: Create a list L with jobs arranged in nondecreasing order of the modified due-dates found in the previous step. Perform list-processing on L. □

Now, in order to solve an instance of $P2|prec, \tau_i = 1|C_{\max}$, we simply specify an integer threshold value D and set artificial due-dates $d_i = D$ for all jobs i. (We can take all s_i to be zero.) Application of A_{GJ} then decides the existence of a feasible schedule relative to fixed D and the process can be repeated in an iterative manner by affecting a binary search over a suitable interval, stopping finally with the smallest D admitting feasibility. Since jobs have unit duration times, an obvious upper bound on this interval is n. We can demonstrate A_{GJ} with the following example.

Example 4.17 In fact, for ease of exposition, we can implement the Step 1 requirements in the following way. For those jobs that are unsucceeded, let their due-dates be given by the threshold value D. Now, for each job i whose successor(s) have been assigned due-dates and for each such due-date, say, d', determine $N[i, 0, d']$. Then set d_i as the minimum of $d' - \lceil |N[i, 0, d']|/2 \rceil$ taken over the values of d'. We leave it to the reader to make the argument that this strategy suffices for Step 1 in A_{GJ}.

Accordingly, let us apply the procedure to the instance of Example 4.15. If we set D at 4, the due-dates result as indicated on the graph of Figure 4.28. For example, $|N[x, 0, 4]| = 2$ and $|N[x, 0, 3]| = 1$ so $d_x = \min(4-1, 3-1) = 2$ as shown. On the other hand, $|N[z, 0, 4]| = 5$ and $|N[z, 0, 3]| = 1$ yielding $d_z = \min(4-3, 3-1) = 1$. The list L is formed relative to these due-dates and the optimal schedule is produced, *i.e.*, see the second schedule associated with the earlier example. □

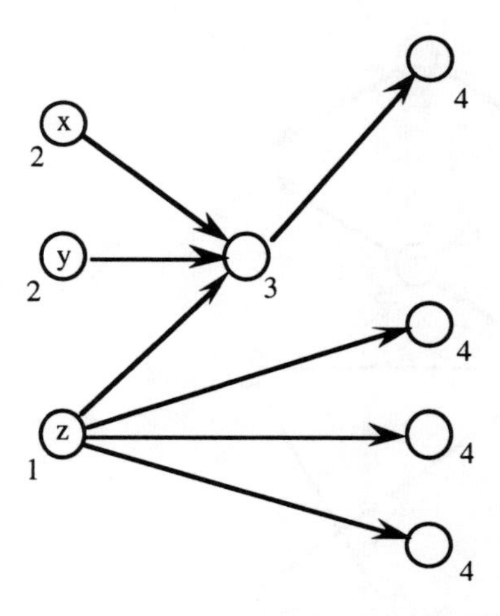

Figure 4.28. *Application of A_{GJ}*

4.2.4 Approximations

Returning to $P|prec|C_{\max}$, it is clear that approximations are in order. Let us recall the procedure given earlier and that was optimal in certain restricted cases. Stated now for general purposes, we have:

A_{LP} : Longest Path Heuristic

Step 0: Compute the length of a longest path ℓ_i from each vertex i in $G = (T, \prec)$.

Step 1: Create L with jobs arranged in nonincreasing ℓ-order. Perform list-processing on L. □

Example 4.18 Let $m = 2$ and consider the graph of Figure 4.29 and which was used in the bicycle assembly illustration of Chapter 1. Values ℓ_i are given next to each vertex. L is formed as stated and the list-processor generated schedule is shown. □

The schedule produced in the example above is clearly optimal. Interestingly, however, if we reapply the algorithm to the same list but with the instance altered so that now $m = 3$, the schedule shown in Figure 4.30 is produced.

Strangely, we have increased the number of processors but taken longer to perform the same amount of work! These sorts of anomalies were first described in Graham's classic paper (1969), and have since been the subject of substantial study and refinement. We shall return to this later, but first we will establish a performance bound on the heuristic A_{LP}.

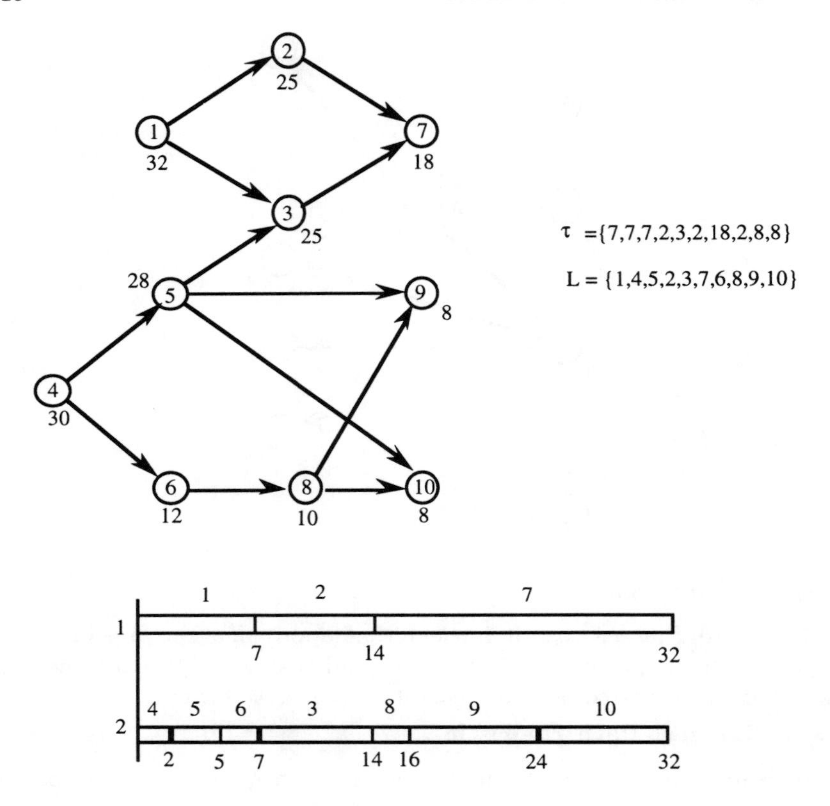

Figure 4.29. *Bicycle assembly graph*

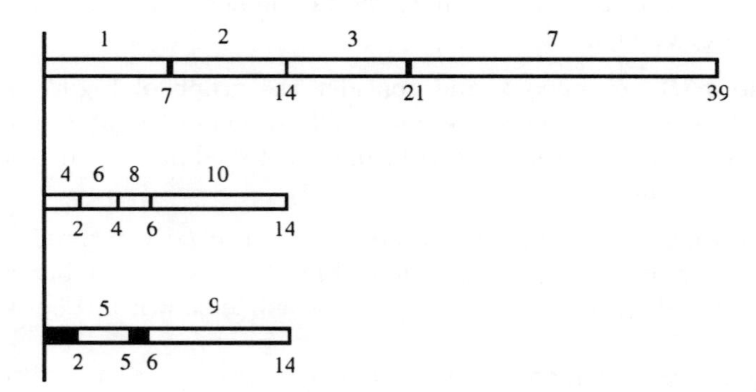

Figure 4.30. *Effect of increasing m from 3 to 2*

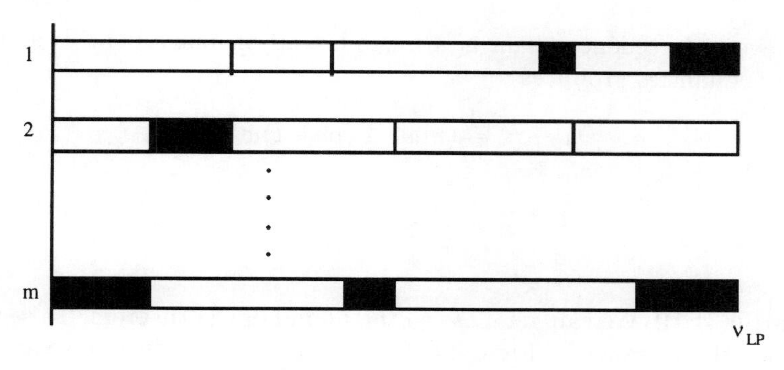

Figure 4.31. *List-processing generated schedule*

Let us define ν_{LP} to be the schedule length produced by $\boldsymbol{A_{LP}}$. As before, ν^* denotes the optimal value. We have

Theorem 4.20

$$\frac{\nu_{LP}}{\nu^*} \leq 2 - \frac{1}{m}.$$

Proof. Consider the timing diagram for any schedule generated by $\boldsymbol{A_{LP}}$. The depiction in Figure 4.31 is typical where we denote processing times by the nonshaded portions.

Suppose now we define ϕ to be an "idle-time" slot and let I be the set of all of these. Then, indexing the latter as ϕ_j, we have

$$\nu_{LP} = \frac{1}{m}\left[\sum_{j\in T} \tau_i + \sum_{\phi_j \in I} \tau_{\phi j}\right].$$

Now, let C be a precedence chain having the property that whenever a processor is idle, some other processor is working on some job in C. Thus

$$\sum_{\phi_j \in I} \tau_{\phi_j} \leq (m-1)\sum_{j\in C} \tau_i,$$

which would be satisfied as equality in the worst case for an n-job precedence chain defining $G = (T, \prec)$. Also, trivially

$$\nu^* \geq \sum_{j\in T} \tau_j / m$$

and

$$\nu^* \geq \sum_{j\in C} \tau_i,$$

since every precedence chain is a lower bound on makespan. Combining these inequalities produces

$$\nu_{LP} \leq \frac{1}{m}(mw^* + (m-1)w^*)$$

or

$$\frac{\nu_{LP}}{\nu^*} \leq 2 - \frac{1}{m}.$$

□

Example 4.19 We can get close to the bound of the theorem. To see this consider the instance of Figure 4.32 with $m = 4$ and duration values as shown. The relevant list produces the value $\nu_{LP} = 23$ with $\nu^* = 14$. This approaches the derived ratio of 7/4 (a more sophisticated instance can be constructed, but that the bound can be approached should be evident with the instance shown). □

The observant reader will also notice that nowhere in the proof of Theorem 4.20 was the prescribed longest path ordering used. On the other hand, the instance in the above example employed such a list. Clearly then, in the worst-case sense a randomly constituted list is as good as one "more cleverly" formed by the stated ℓ-ordering. Indeed, with the longest path ordering, the bound is improved (asymptotically) only to $2 - \frac{2}{m+1}$ even if $G = (T, \prec)$ takes the form of an arborescence. We can demonstrate this in Figure 4.33 (*cf.*, Coffman, 1976). Note further that with unit duration times (and arbitrary precedence structure) this performance remains essentially unchanged as well (*cf.* Exercise 4-15).

Let us now return to the rather counterintuitive phenomenon demonstrated earlier. There we saw that increasing the number of processors was no assurance that makespan would not worsen–in fact, we created an instance where it did precisely that. But, besides the novelty of such an apparent anomaly, there is a practical basis around which we might like to examine the existence/behavior of these outcomes. If the number of processors is subject to change, *e.g.*, breakdowns and repairs, we would presumably be interested in a *robust* schedule that gives good performance for varying m. But unfortunately, even optimality is no guarantee of robustness. For example, it can be shown that operating on $G = (T, \prec)$ in Figure 4.34 with $L = \{1, 2, 3, \ldots\}$ is optimal for $m + 1$ processors but as bad as any possible list schedule for m processors. Specifically, the application of list-processing on $L = \{1, 2, 3, \ldots, m\}$ with m processors and unit duration times, would produce the schedule shown in Figure 4.35 having a completion time of $2m - 1$. It should also be easy to see that if another processor is added, this same list would produce an optimal schedule with completion time m.

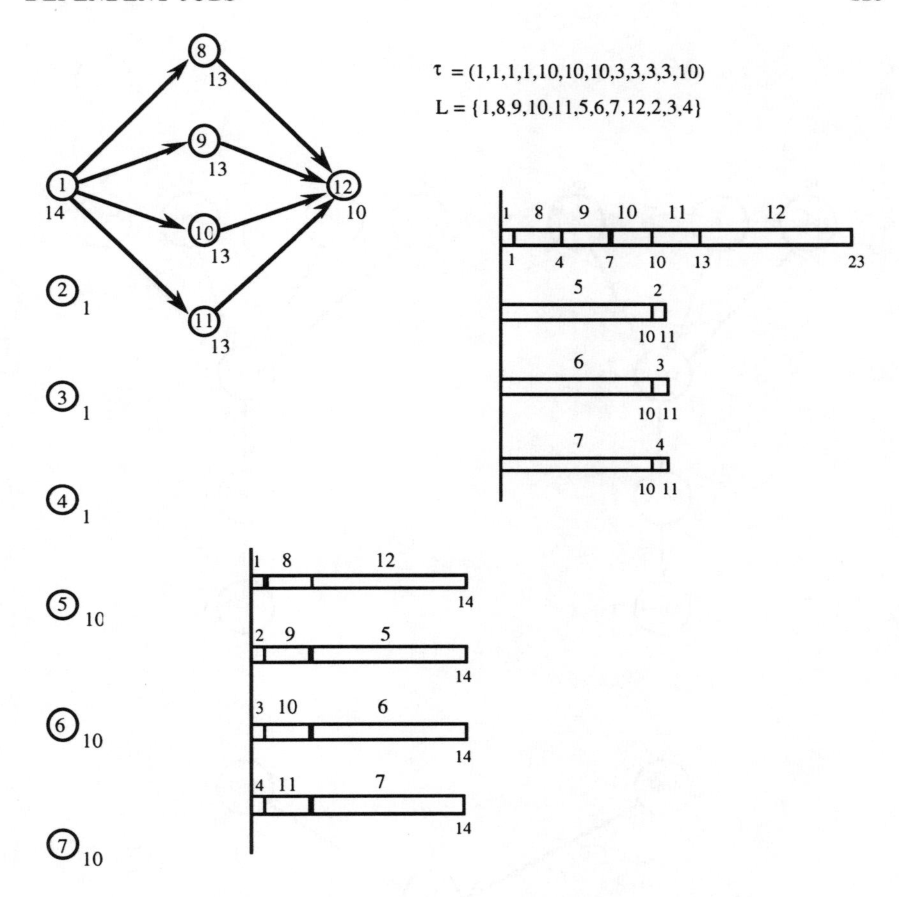

Figure 4.32. *Worst-case sense of Theorem 4.20 bound*

On the other hand, the schedule shown in Figure 4.36 is optimal for m processors. That is, the given list performs optimally for the $m + 1$ processor case but yields the *worst possible* schedule when the number of processors is m. Clearly, this list schedule is not robust. The next example helps by making these options explicit.

Example 4.20 Suppose we consider an instance with $m = 5$ processors and $n = 25$ jobs constrained as indicated in Figure 4.37. The list $L = \{1, 2, \ldots, 25\}$ would produce the schedule on the left below while the one on the right is optimal. But for this particular instance (*i.e.*, with $n = m^2$ and jobs constrained as shown), it is easy to give a list that is

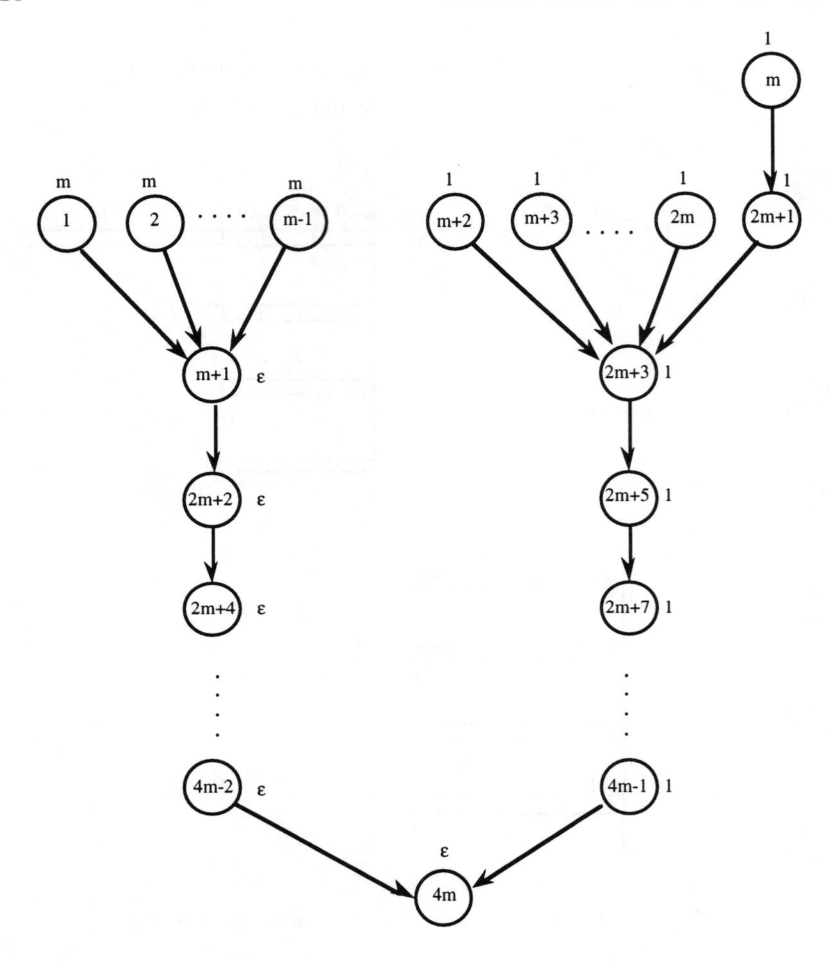

Figure 4.33. *Worst-case arborescences*

robust in the sense that it produces schedules that are optimal for m as well as $m+1$ processors. For example, in the $m = 5$ case shown, let $L = \{1, 2, 3, 4, 6, 5, 7, 8, 9, 12, 10, 11, 13, 14, 18, 15, 16, 17, 19, 24, 20, 21, 22, 23, 25\}$. We leave verification of the aforementioned robustness to the reader. □

In Example 4.20, it was easy to see that there exists a list schedule that is robust in the sense described yielding an optimal schedule for both m and $m+1$ processors. Unfortunately, even this may not always be possible as shown first by Graham (*cf.* Coffman, 1976).

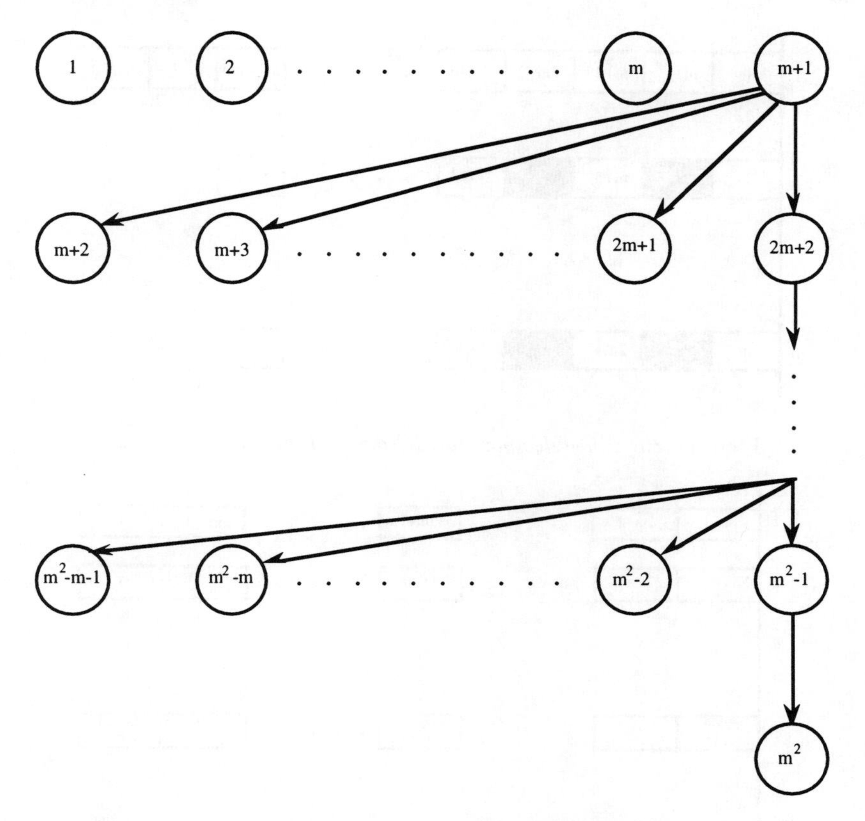

Figure 4.34. *Insufficiency of optimality and robust schedules*

The instances formed by Graham were quite large but the point was established where for a particular 30-job structure, it was shown that no single list existed that led to an optimal schedule on both two and three processors. Since then, this phenomenon has also been demonstrated by Tovey but on much smaller instances. Our coverage follows the latter.

As described in Tovey (1990), for any $m > 2$, let us construct an instance on $3(m+1)$ jobs constrained as defined by the bipartite digraph $G = (X, Y, \prec)$ where $X = \{v_1, v_2, \ldots, v_{2m+1}\}, Y = \{u_1, u_2, \ldots, u_{m+2}\}$, and with precedence specified as follows:

$$\{(v_i, u_j) | 1 \leq i \leq m, \text{ for all } j\},$$
$$\{(v_i, u_j) | m+1 \leq i \leq 2m, \text{ for all } j \neq 1\}, \quad \text{and}$$
$$\{(v_{2m+1}, u_j) | \text{ for all } j \neq 2, 3\}.$$

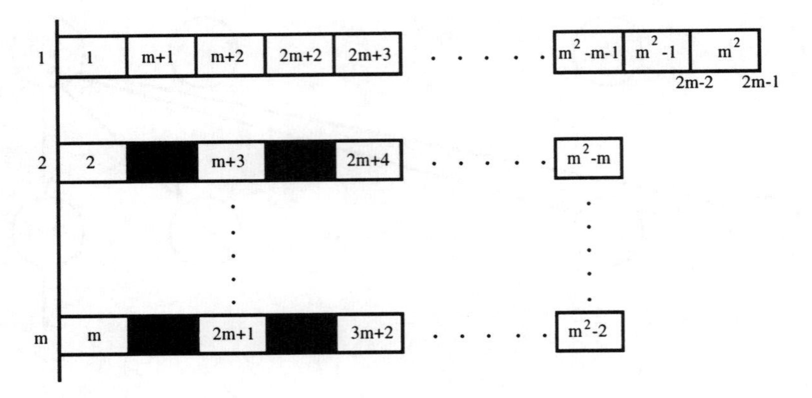

Figure 4.35. *Schedules generated from instance of Figure 4.34*

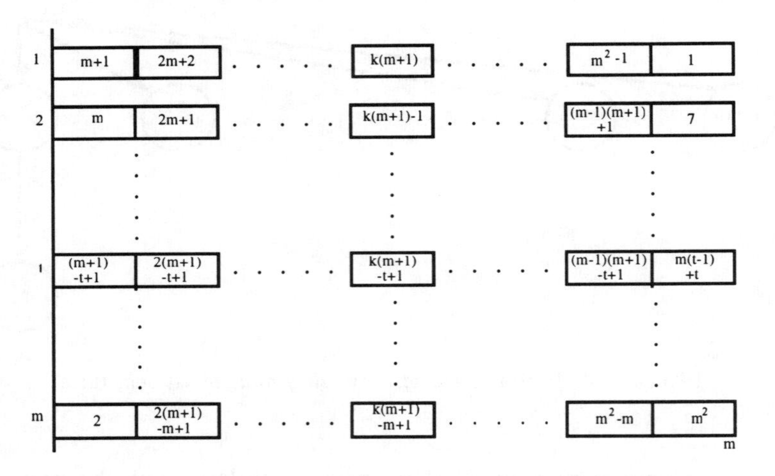

Figure 4.36. *Optimal/suboptimal* $m + 1\backslash m$ *processor schedules*

The case for $m = 3$ is shown in Figure 4.38.

Now, for $m+1$ processors, every optimal schedule must place jobs $v_1, v_2, \ldots, v_m$, and v_{2m+1} in the first period. This is so since any alternative would force an idle slot during the second period and extend makespan beyond the optimal value of m. But for the case of m processors, it is only optimal to schedule the jobs $v_1, v_2, \ldots, v_{2m}$ during the first two periods. We illustrate with the following example.

Example 4.21 Consider the 12-job instance for the $m = 3$ case in Figure 4.38. The optimal policy just specified for $m + 1$ processors results in the

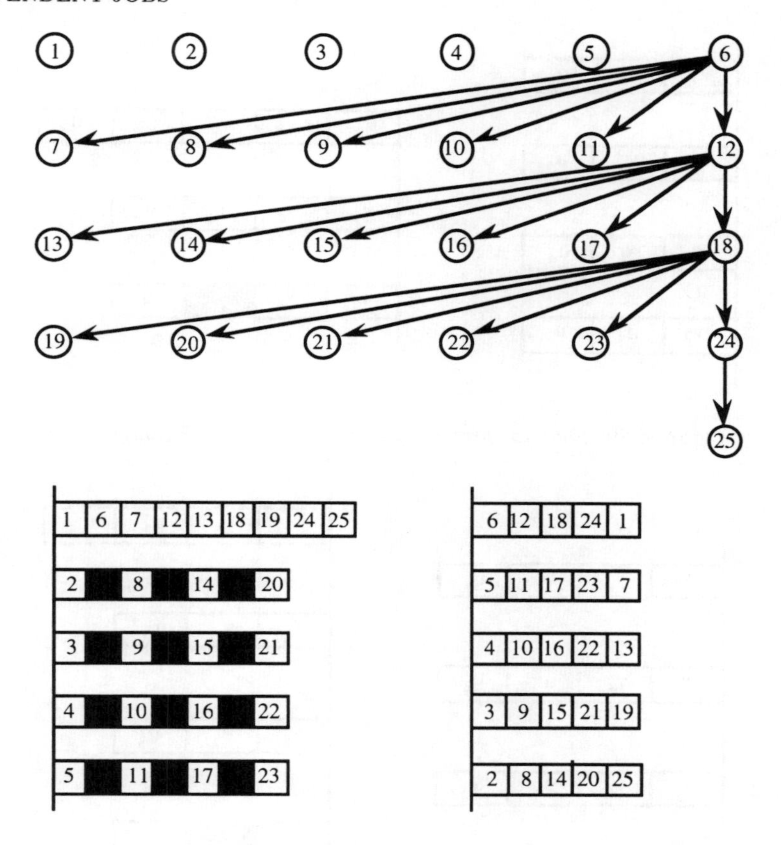

Figure 4.37. *Schedules of Example 4.20*

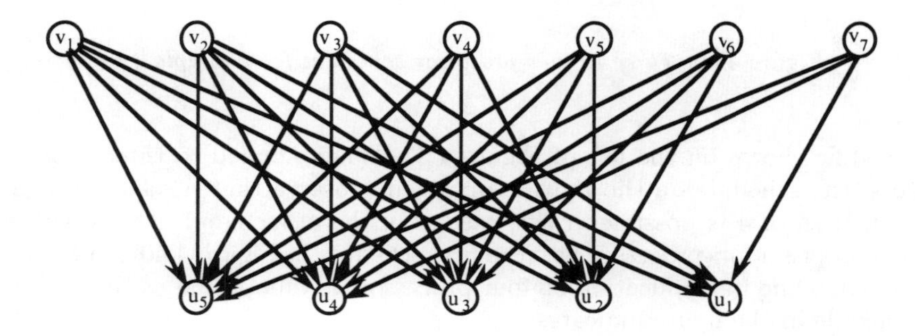

Figure 4.38. *Tovey's construction*

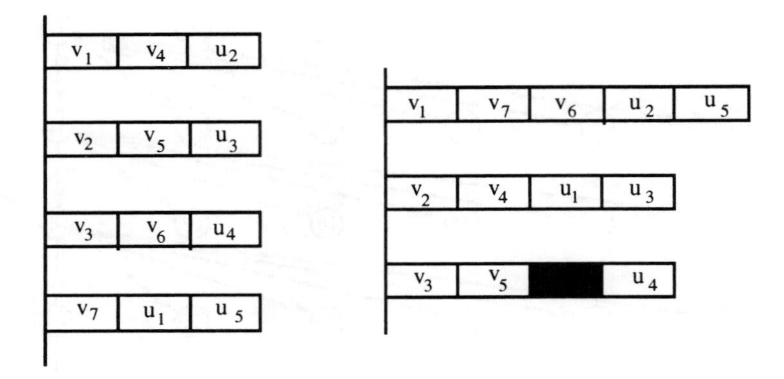

Figure 4.39. *Four to three-processor schedules for Example 4.21*

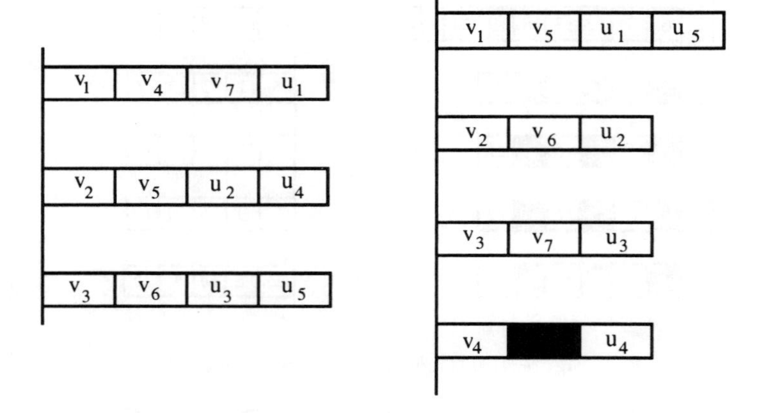

Figure 4.40. *Three to four-processor schedules for Example 4.21*

schedule shown on the left of Figure 4.39. When applied to three processors, the schedule on the right is obtained (observe that a schedule with makespan of 4 is possible in this case). On the other hand, if we operate on an optimal m-processor list, the first schedule in Figure 4.40 can be obtained, while its application on four processors is suboptimal as the second schedule in the figure indicates.

So, we have from the example that the optimal list for an $m = 3$ instance is not optimal for the $m \leftarrow m + 1 = 4$ instance and misses by the ratio 4/3 accordingly. Similarly, the $(m + 1)$-optimal strategy on the four-processor instance missed optimality on the three-processor instance by a ratio of

5/4. In fact, Tovey (1990) conjectures that these may be the largest possible ratios for nonoptimality of the most robust schedules when processors change from m to $m+1(m+1$ to $m)$. □

As we finish, it is worth noting that in the work leading to the construction for the troublesome instance above, Tovey also produced a nice side result that fixed the complexity status of finding schedules of length no greater than 3 for the problem $P|prec, \tau_i = 1|C_{\max}$ where no precedence chain has length greater than 2. The reader will recall that this was given by Theorem 4.12 earlier.

4.3 Exercises

4-1 Propose a pseudo-polynomial time procedure for $P2||C_{\max}$. Apply it to an instance with $\tau = (7,4,3,2,1,1,5,4,6,3)$.

4-2 Apply the MULTIFIT procedure to the following instance of $P||C_{\max}$.

$$m = 3, \tau = (5,6,9,3,2,1,4,7,5,4,6), t = 5$$

4-3 Apply the Bruno *et al.* algorithm for $R||\Sigma c_i$ to the instance below:

$$\tau = \begin{bmatrix} 4 & 5 \\ 3 & 7 \\ 1 & 3 \\ 6 & 2 \end{bmatrix}$$

4-4 Suppose we apply the Bruno *et al.* matching strategy to the identical processor case, *i.e.*, $P||\Sigma c_i$. Are there any refinements? Explain.

4-5 Give a fast algorithm for $P|prec, \tau_i = 1|C_{\max} \leq 2$.

4-6 Give the explicit proof of Corollary 4.13.

4-7 Give an instance for which no list L exists that produces, by list-processing, an optimal schedule.

4-8 Give an instance with $\prec$ an arborescence but with $\tau_i \epsilon \{1,2\}$ where Hu's algorithm ($\boldsymbol{A_H}$) fails.

4-9 Solve the instance of $P2|prec, \tau_i = 1|C_{\max}$ in Figure 4.41 by using algorithm $\boldsymbol{A_{FKN}}$.

4-10 Apply (correctly) algorithms $\boldsymbol{A_{CG}}$ and $\boldsymbol{A_{GJ}}$ to the instance of the previous exercise.

4-11 Propose a fast algorithm for finding the transitive reduction of a digraph.

4-12 Apply the approximation $\boldsymbol{A_{LP}}$ to the instance of $P|prec|C_{\max}$ given in Figure 4.42 and with $m = 3$.

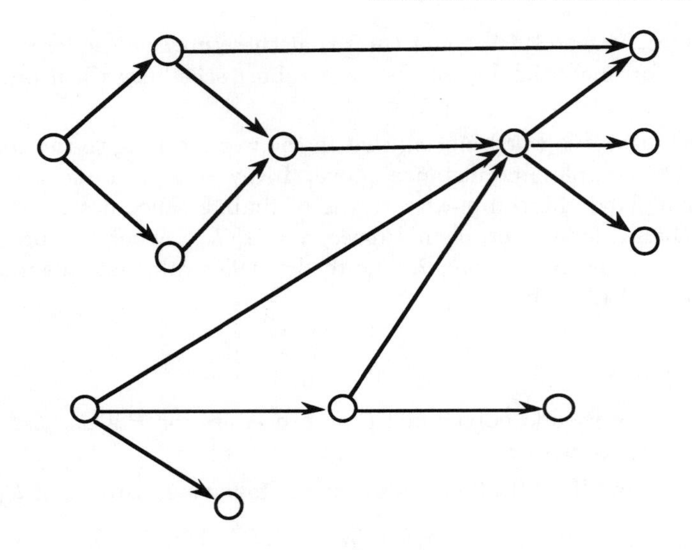

Figure 4.41. *Exercise 4-9*

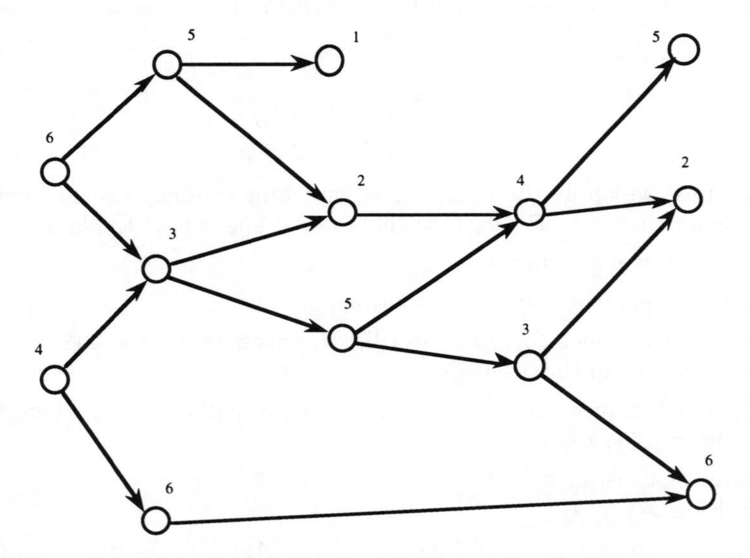

Figure 4.42. *Exercise 4-12*

4-13 Create an instance of $P|prec|C_{\max}$ with duration times τ and another instance that is identical but with duration times τ' where $\tau_j' = \tau_i - 1$ everywhere and such that algorithm $\boldsymbol{A_{LP}}$ operates on $P|prec, \tau|C_{\max}$ yielding a schedule with value ξ but when applied to $P|prec, \tau'|C_{\max}$ yields one with value $\xi' > \xi$.

4-14 Repeat the previous exercise by creating a pair of instances of $P|prec|$ C_{max} that are identical except that $\prec' \subset \prec$ but upon which $\boldsymbol{A_{LP}}$ yields $\xi' > \xi$.

4-15 Suppose $\boldsymbol{A_{LP}}$ is applied to $P|prec$, $\tau_i = 1|C_{\max}$. Give as tight a bound as you can on its performance accordingly.

4-16 Provide the details for the proof of Theorem 4.9.

4-17 Provide the details for the proof of Theorem 4.12.

4-18 Provide the details for the proof of Theorem 4.16.

4-19 Prove that there can be no constant performance bound for the bin packing problem with value strictly less than 3/2 unless $\mathcal{P} = \mathcal{NP}$.

4-20 Describe a MULTIFIT-like strategy for $R||C_{\max}$.

4-21 Given the obvious dual-like relationship between BIN PACKING and $P||C_{\max}$, how does the result of Exercise 4-19 square with a result like that of Theorem 4.5?

4-22 Give an instance of BIN PACKING for which a correct application of the FFD heuristic fails badly.

4-23 Give a polynomial-time solution for BIN PACKING assuming all chip weights are defined over the interval $(C/3, C]$.

4-24 If algorithm $\boldsymbol{A_{CG}}$ is applied on instances with $m \geq 3(\tau_i = 1)$ we still approach a bound of 2 (*i.e.*, $2 - 2/m$). Substantiate this on the graph in Figure 4.43 (*cf.* Lam and Sethi, 1977).

4-25 Verify the validity of the bound on $C_{\max}$ given in the proof of Theorem 4.14.

4-26 Consider $P2|prec|C_{\max}$ and let us replace instances accordingly by $P2|prec$, $\tau_i = 1|C_{\max}$ where a job in the former with duration τ_i is broken into τ_i unit execution time jobs, arranged in series. Upon application of any of the two-processor algorithms (*e.g.*, $\boldsymbol{A_{FKN}}$, $\boldsymbol{A_{CG}}$, $\boldsymbol{A_{GJ}}$) to the latter, will the outcome solve $P2|prec, pmtn|C_{\max}$? Explain.

4-27 Consider problem $P||C_{\max}$. Generate some sample instances upon which $\boldsymbol{A_L}$ performs very badly.

4-28 Repeat Exercise 4-27 for $P|prec|C_{\max}$.

4-29 Apply $\boldsymbol{A_{KK}}$ to the instance below:

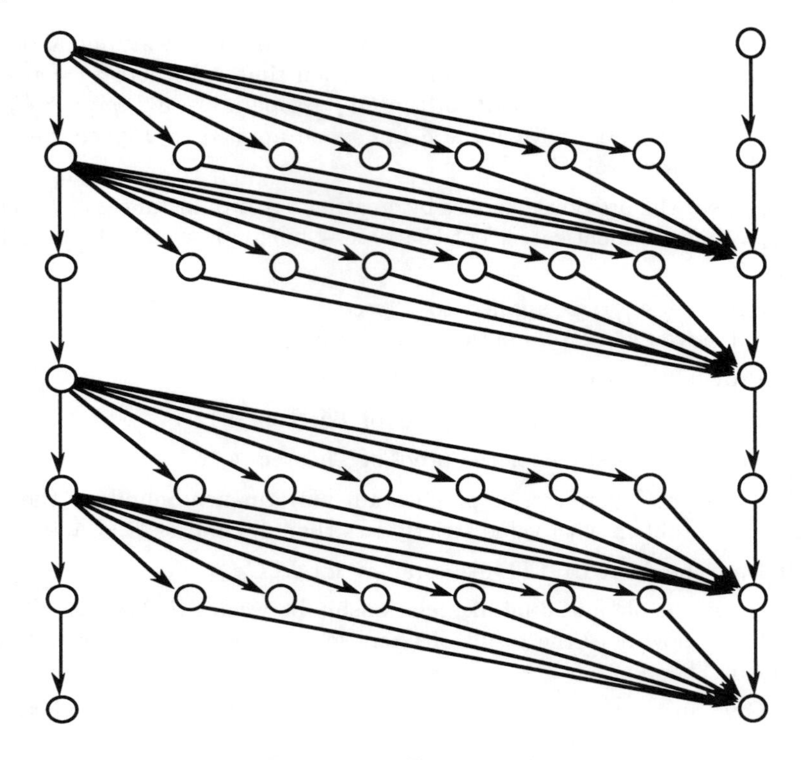

Figure 4.43. *Exercise 4-24*

$m = 5$, $\tau = (4,7,8,12,3,7,9,17,4,2,8,7,15,9,8,11)$, and $k = 7$.

4-30 Find another instance with $n = 2m$ (*cf.* Example 4.4) and where $\boldsymbol{A_{LPT}}$ is not optimal.

4-31 Produce an anomaly like that exposed in Examples 4.6 and 4.8.

4-32 Consider $R||\Sigma w_i c_i$ and suppose the Bruno *et al.* matching approach used for the unweighted case is applied here as an approximation. Give an instance that forces this strategy to produce a very bad solution.

4-33 Repeat Exercise 4-32 with the restricted case that all weights are drawn from $\{1, 2, \ldots, k\}$ for fixed $k \geq 2$.

4-34 Reproduce the sense of the proof of Theorem 4.10 *vis-a-vis* another illustration like that in Example 4.10.

4-35 Apply $\boldsymbol{A_H}$ to the instance of $P|tree, \tau_i = 1|C_{\max}$ given in Figure 4.44.

4-36 Repeat Exercise 4-35 on the structure in Figure 4.45.

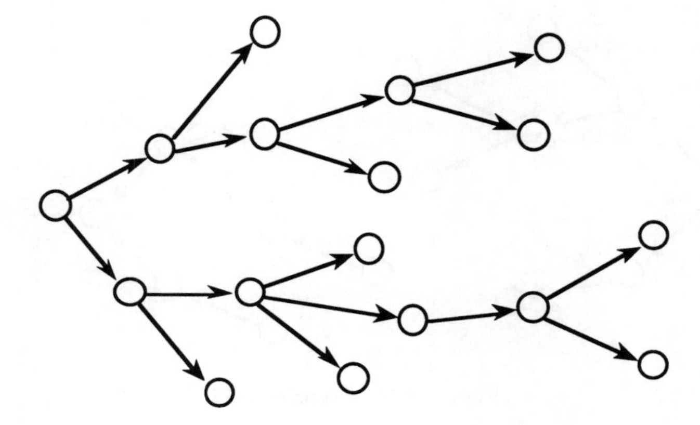

Figure 4.44. *Exercise 4-35*

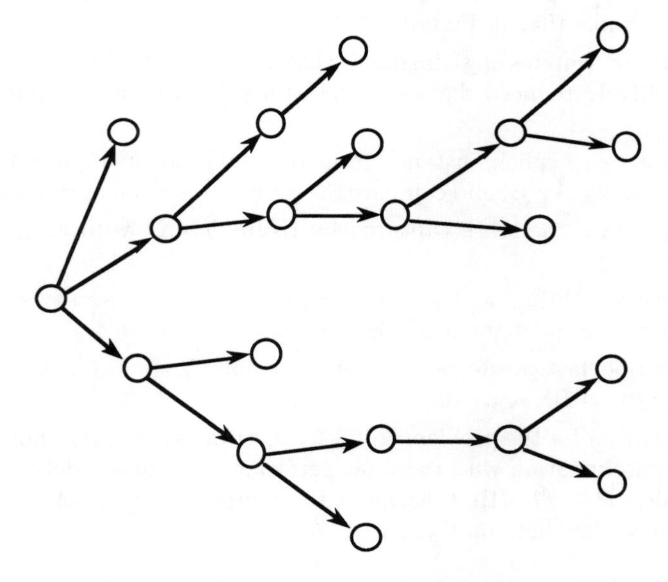

Figure 4.45. *Exercise 4-36*

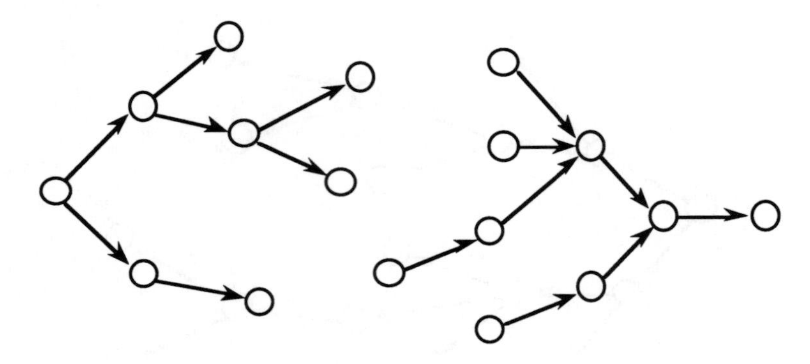

Figure 4.46. *Exercise 4-37*

4-37 Will $\boldsymbol{A_H}$ "work" for instances of $P|tree,\ \tau_i = 1|C_{\max}$ defined on structures like that in Figure 4.46?

4-38 Give an (interesting) instance that shows, as in Figure 4.26, that a transitively reduced digraph is necessary for a correct application of $\boldsymbol{A_{CG}}$.

4-39 Create an explicit instance from the structure in Figure 4.33 with $m = 5$. Apply $\boldsymbol{A_{LP}}$, and compare the outcome with an optimal solution.

4-40 Reproduce the sense captured by Example 4.21 with an $m = 4$ instance.

4-41 Consider $Q||\Sigma c_i$ and suppose each processor possesses a "speed" given by s_i. Repeat the analysis called for in Exercise 4-4.

4-42 Describe fast strategies for the problems $Q|\tau_i = 1|\Sigma w_i c_i$, $Q|\tau_i = 1|\Sigma T_i$, $Q|\tau_i = 1|\Sigma w_i u_i$, and $Q|\tau_i = 1|L_{\max}$.

4-43 Show that for the problem $R||C_{\max}$, there does not exist a polynomial-time approximation with constant performance bound strictly less than 3/2 unless $\mathcal{P}=\mathcal{NP}$. (Hint: Examine the complexity status of the problem for a fixed threshold on $C_{\max}$.)

CHAPTER 5

FLOW SHOPS, JOB SHOPS, AND OPEN SHOPS

In this chapter, we continue with the multiple-processor assumption but now introduce the notion that jobs possess dedicated processor assignments. Accordingly, let us begin with a classic problem.

5.1 Flow Shops

A much studied model, particularly in early publications of scheduling problems, is the so-called *flow shop*. A plausible explanation for this proliferation stems from the problem's interesting mix of complication with structure where the latter produces some particularly interesting results which remain among the most fundamental in the theory of scheduling.

In the traditional flow shop, each job is taken to consist of at most $m \geq 2$ operations (*i.e.*, $m(j) \leq m$ for all j). These operations are defined by processors that perform work on a given job. In particular, we denote a flow shop operation by the pair (jm_ℓ) signifying the operation of a job j on the processor in the ℓth order position. We assume $\prec$ to be defined by relations $jm_\ell \prec jm_{\ell+1}$ for each j and $\ell = 1, 2, \ldots, m-1$. Moreover, we take $m_\ell = \ell$ and so for each job the order of processing is the same. Note that we introduce the subscripted index m_ℓ now even though it is redundant for flow shops; for more general models it is meaningful since the so-called processor orderings in the latter are allowed to vary across jobs. The directed graph representation of a flow shop is shown by $(T, \prec)$ in Figure 5.1. Observe that T is relativized in the obvious way to denote the set of operations.

The fundamental flow shop problem $F||C_{\max}$ is easy to state: Given an appropriate instance, determine a permutation of the jobs on each processor that creates a schedule having minimum makespan. By the structure of $\prec$, all such permutations are feasible. A schedule is shown in Figure 5.2 for a so-called 4-job, 3-machine, or 4×3 instance, the makespan of which is 30. Note that, for ease, we shall hereafter assume that every job requires

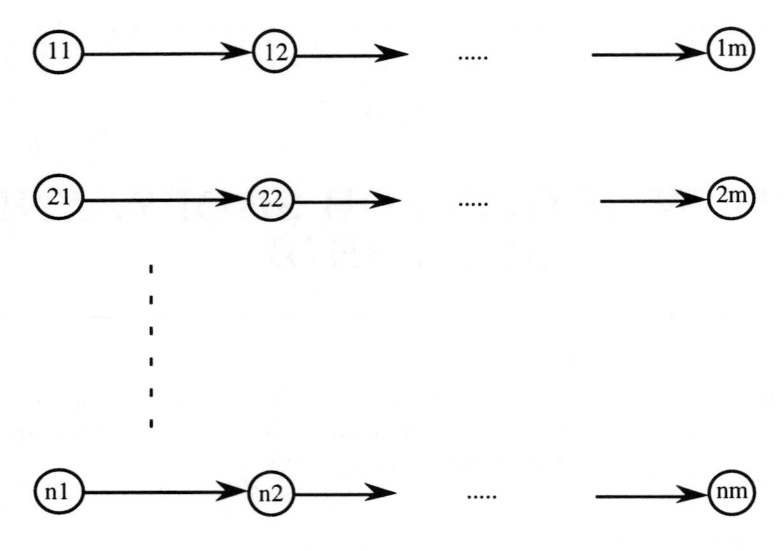

Figure 5.1. *Flow shop structure*

exactly m operations, one on each processor.

5.1.1 Permutation Schedules

The following pair of related properties is easy to see. We have:

$\boldsymbol{P_1}$: For $F||C_{\max}$, there exists an optimal solution having the same processing permutation on the *first* two processors.

To see $\boldsymbol{P_1}$, consider any solution where different orders exist on the first two processors. Then there must be a pair of adjacent jobs, say a and b, on the first processor permutation that appear in reverse order in the permutation on the second. But these two jobs can be reversed on the first processor without increasing the start time (and thus the completion time) of any job on the second processor. Inductively, we can repeat this pairwise switching process until the permutation on the first processor is made to agree with the (original) order on the second.

It is not surprising that a similar result exists for the last two processors.

$\boldsymbol{P_2}$: For $F||C_{\max}$, there exists an optimal solution having the same order on the *last* two processors.

If the permutations do not agree on the last two processors, we can change the order on the last one so that it agrees with the order on processor $m-1$. Again, the argument is based on an easy pairwise switch on processor m creating ultimately an ordering identical to that on $m-1$ and that yields a

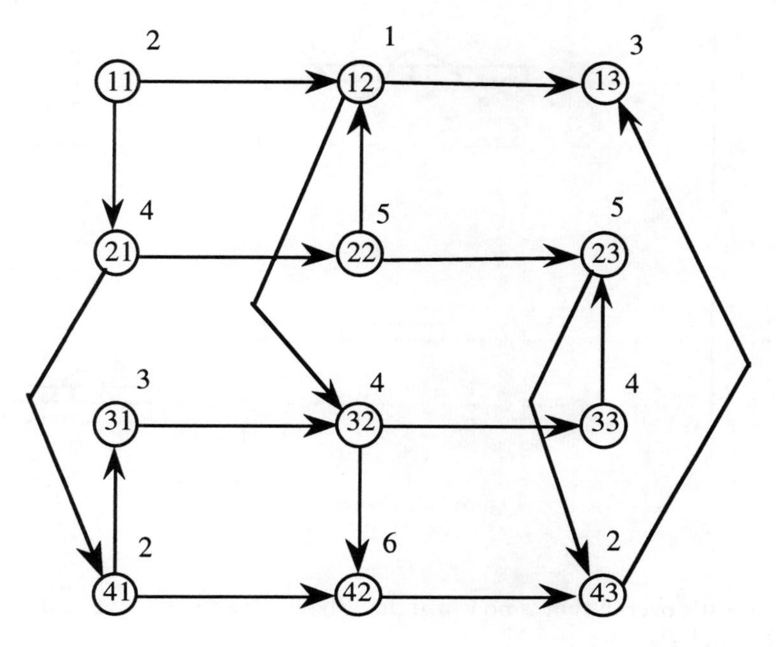

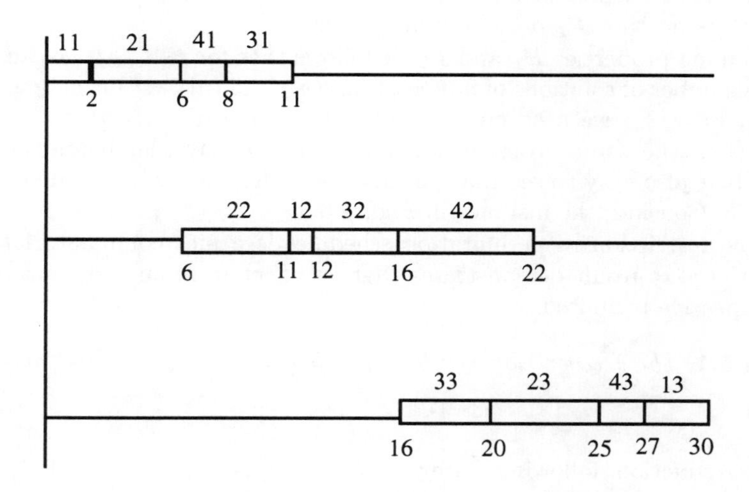

Figure 5.2. *Flow shop solution/schedule*

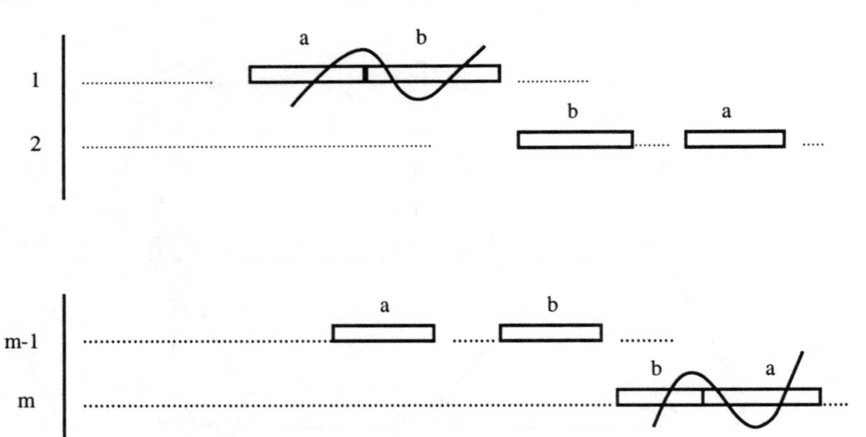

Figure 5.3. *Properties P_1 and P_2*

schedule overall that is no worse than the original one. Figure 5.3 illustrates $\boldsymbol{P_1}$ and $\boldsymbol{P_2}$.

As an aside, we note that property $\boldsymbol{P_2}$ does not hold under the total completion time measure, Σc_i. The instance in Figure 5.4 makes the point. Observe, however, that $\boldsymbol{P_1}$ does hold in this case.

By combining properties $\boldsymbol{P_1}$ and $\boldsymbol{P_2}$, it follows that for $F||C_{\max}$ we can reduce the number of solutions of interest from $(n!)^m$ to $(n!)^{m-2}$ for $m \geq 3$. In fact, for $m = 2, 3$ we need only consider the family of $n!$ *permutation schedules* (*i.e.*, the same order on all processors) in any search for the optimum. It is also easy to see that permutation schedules will not suffice when $m > 3$. Consider the instance in Figure 5.5.

While the restriction to permutation schedules is a nice combinatorial outcome, the next result demonstrates that its effect from any computational perspective is limited.

Theorem 5.1 *The (recognition version) problem $F3||C_{\max}$ is $\mathcal{NP}$-Complete.*

Proof. The reduction, following Garey *et al.* (1976), is from the 3-PARTITION problem: Given positive integers n, B, and a set of integers $A = \{a_1, a_2, \ldots, a_{3n}\}$ with $\sum_{i=1}^{3n} a_i = nB$ and $B/4 < a_i < B/2$ for all i, does there exist a partition of A as $\{A_1, A_2, \ldots, A_n\}$ such that $|A_j| = 3$ for $1 \leq j \leq n$ and where $\sum_{a \in A_j} a = B$? Specifically, the mapping creates a

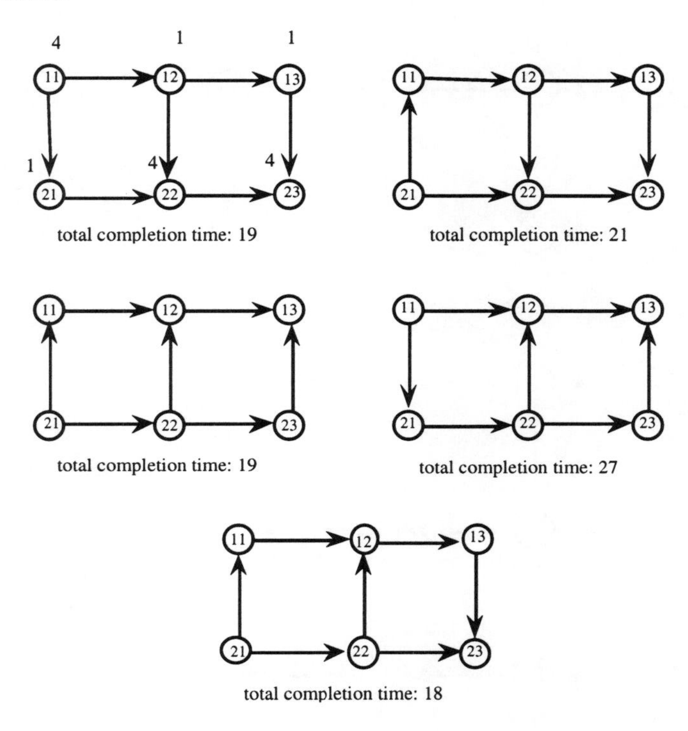

Figure 5.4. *P_2 does not hold for Σc_i*

$4n + 1$ job instance having duration times shown as follows:

$$\begin{array}{c} 1 \\ 2 \\ \vdots \\ n-1 \\ n \\ n+1 \\ n+2 \\ \vdots \\ 4n+1 \end{array} \begin{bmatrix} 0 & B & 2B \\ 2B & B & 2B \\ \vdots & \vdots & \vdots \\ 2B & B & 2B \\ 2B & B & 0 \\ 0 & a_1 & 0 \\ 0 & a_2 & 0 \\ \vdots & \vdots & \vdots \\ 0 & a_{3n} & 0 \end{bmatrix}.$$

Then it can be shown that a suitable three-processor flow shop schedule exists with makespan no greater than $(2n + 1)B$ exactly when a desired partition of A exists. □

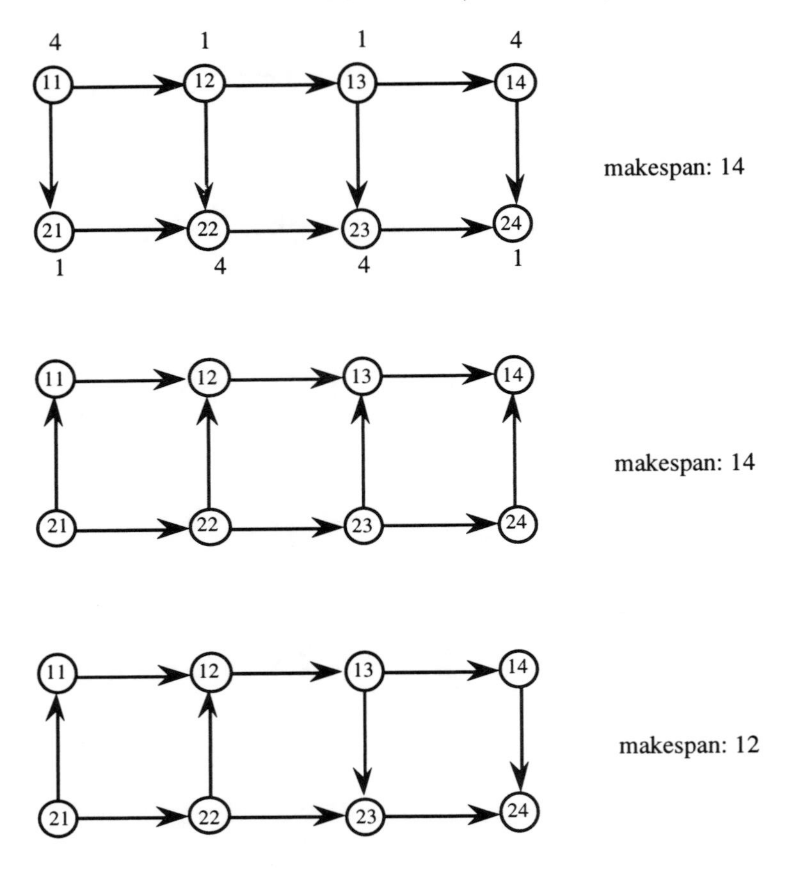

Figure 5.5. *Permutation solutions are not sufficient with* $m \geq 4$

The threshold value of $(2n+1)B$ in Theorem 5.1 is easy to justify. No matter how ordered on the first processor, the jobs labeled $1, 2, \ldots, n$ complete at time $2nB$ and since every such job has a duration time of B for its second operation, the earliest that all of the latter can complete is thus $2nB + B$. To achieve the stated threshold, the $3n$ jobs labeled $n+1$ through $4n+1$ then have to be scheduled during idle time gaps on the second processor.

Interestingly, and much more complicated, is the proof that the mean completion time problem $F2||\Sigma c_i$ is also difficult. Details of this result can be found in Garey *et al.* (1976) as well.

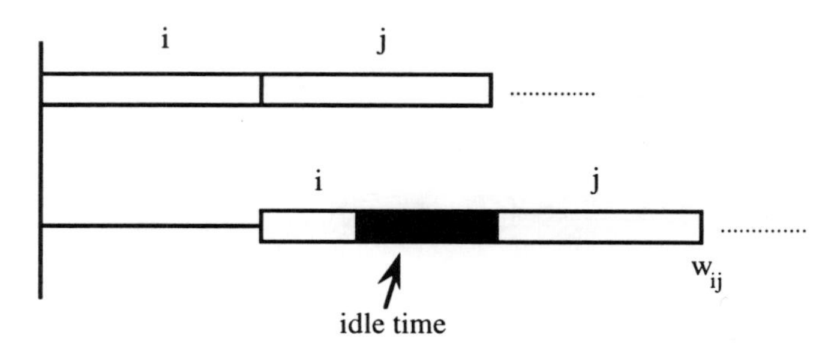

Figure 5.6. *Jobs i and j are initial jobs*

5.1.2 Solvable Flow Shop Cases

Arguably the most classic outcome in scheduling theory pertains to the solution of the two-processor case $F2||C_{\max}$. The basis for the result stems from the following property due to Johnson (1954).

Theorem 5.2 *Let S be a two-processor flow shop (permutation) schedule in which job i is processed before and adjacent to job j. If*

$$\min(\tau_{j1}, \tau_{i2}) \leq \min(\tau_{i1}, \tau_{j2})$$

then the schedule created by interchanging i and j will have completion time no longer than that of S.

Proof. Consider first that i and j are the initial two jobs to be processed in the schedule as indicated in Figure 5.6. Here, we have $w_{ij} = \tau_{i1} + \tau_{j2} + \max(\tau_{j1}, \tau_{i2})$. Now, let us assume that

$$\min(\tau_{j1}, \tau_{i2}) \leq \min(\tau_{j2}, \tau_{i1})$$

or equivalently,

$$\max(-\tau_{j1}, -\tau_{i2}) \geq \max(-\tau_{j2}, -\tau_{i1}).$$

Upon adding the constant $\tau_{i1} + \tau_{i2} + \tau_{j1} + \tau_{j2}$, we obtain

$$\max(\tau_{i1} + \tau_{i2} + \tau_{j2}, \tau_{i1} + \tau_{j1} + \tau_{j2}) \geq \max(\tau_{i1} + \tau_{i2} + \tau_{jl}, \tau_{i2} + \tau_{j1} + \tau_{j2})$$

or upon rearrangement

$$\tau_{i1} + \tau_{j2} + \max(\tau_{i2}, \tau_{j1}) \geq \tau_{i2} + \tau_{j1} + \max(\tau_{i1}, \tau_{j2}).$$

Thus, $w_{ij} \geq w_{ji}$ and the completion time claim follows.

For the other possibility, suppose i and j are not first in the processing order. Then, we can break the situation into two cases as shown in Figure 5.7.

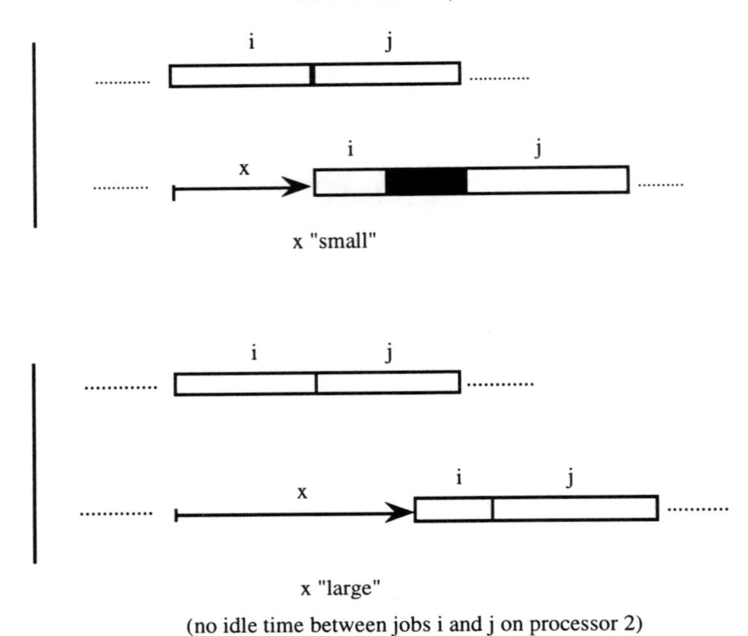

Figure 5.7. *Jobs i and j are not initial jobs*

If x is small, in the sense shown, we have $c_{j2} = s_{i1} + w_{ij}$. On the other hand, if x is large, the relationship is $c_{j2} = s_{i1} + x + \tau_{i2} + \tau_{j2}$ where the last three terms follows since there is no idle time on the second processor. So, if i and j are interchanged, we have

$$c_{i2} = s_{ji} + \max(w_{ji}, x + \tau_{j2} + \tau_{i2}),$$

and as before, switching the two does not lengthen the schedule. □

We are thus led to the renown *Johnson's Rule:*

***P*$_3$:** For $F2||C_{\max}$, a schedule is optimal if for every pair of jobs i and j where i is processed prior to j, the following inequality holds:

$$\min(\tau_{i1}, \tau_{j2}) \leq \min(\tau_{i2}, \tau_{j1}).$$

It is easy to show that Johnson's rule is not necessary. Nonetheless, the issue before us is to efficiently construct a schedule that satisfies property ***P*$_3$** and is thus optimal.

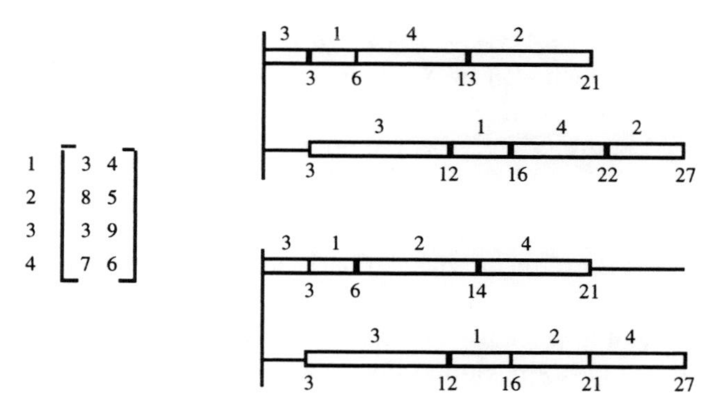

Figure 5.8. *Example 5.1*

Consider the following procedure:

A_{J_2}: Two-Processor Flow Shop Algorithm

Step 1: Let the jobs be labeled as $1, 2, \ldots, n$ and construct sets X and Y as

$$\begin{aligned} X &= \{j|\tau_{j1} < \tau_{j2}\} \\ Y &= \{j|\tau_{j1} \geq \tau_{j2}\} \end{aligned}$$

Step 2: Arrange jobs in X in nondecreasing τ_{j1}-order. Arrange jobs in Y in nonincreasing τ_{j2}-order. Call these ordered sets $\hat{X}$ and $\hat{Y}$.

Step 3: Concatenate $\hat{X}$ and $\hat{Y}$ forming the permutation $\{\hat{X}\hat{Y}\}$ which is the processing order for both processors. □

We can demonstate $\boldsymbol{A_{J_2}}$ with the following example.

Example 5.1 Consider a four-job instance having duration times given by the matrix in Figure 5.8. Sets $\hat{X}$ and $\hat{Y}$ are constructed as $\{3, 1\}$ and $\{4, 2\}$ respectively, yielding the total ordering $\{3, 1, 4, 2\}$ which results in a schedule with makespan 27 as shown. An alternative optimal schedule is shown as well. Observe that the second schedule establishes that Johnson's rule is not necessary [*i.e.*, $\min(\tau_{21}, \tau_{42}) \not< \min(\tau_{41}, \tau_{22})$]. □

Clearly, algorithm $\boldsymbol{A_{J_2}}$ is polynomial in the number of jobs n. More interesting is the establishment of its validity. We have:

Theorem 5.3 *Algorithm* $\boldsymbol{A_{J_2}}$ *will create a solution for the flow shop problem* $F2||C_{\max}$ *which satisfies Johnson's rule and is thus optimal.*

Proof. We can establish the result by considering three cases. We may also assume without loss of generality that in each, job i is processed prior to job j.

i) $i, j \in \hat{X}$. Here $\tau_{i1} \leq \tau_{j1} < \tau_{j2}$ since first, i and j are ordered in $\hat{X}$ and second, by inclusion in X. Hence, $\min(\tau_{i1}, \tau_{j2}) = \tau_{i1}$. But $\tau_{i1} < \tau_{i2}$ since $i \in X$ and again, by the ordering in $\hat{X}$ we have $\tau_{i1} \leq \tau_{j1}$. Thus, $\min(\tau_{i1}, \tau_{j2}) \leq \min(\tau_{i2}, \tau_{j1})$ as required.

ii) $i, j \in \hat{Y}$

iii) $i \in \hat{X}, j \in \hat{Y}$

The last two cases are dealt with similarly and the proof is complete. □

We take space at this point to examine another easy but potentially useful property of flow shops. Recall that $\prec$ dictates that for each job j we have $j1 \prec j2 \prec \ldots \prec jm$. Suppose we construct the *reversal* or *inverse* of $\prec$ denoted by $\prec_r$ and defined as (for given j) $jm \prec jm-1 \prec \ldots \prec j1$. We have:

***P*$_4$:** If S is any permutation schedule for $F||C_{\max}$ under $\prec$, then the reversal of the permutation for S yields an admissible permutation schedule under $\prec_r$ of the same makespan value.

This property is immediate by purely geometric means. Consider the timing diagrams in Figure 5.9. The first symbolizes a schedule S under $\prec$. If we rotate the diagram on an imaginary axis taken as the first processor, we produce the second diagram. With this, let us rotate on an imaginary axis fixed at the makespan value ν. We produce accordingly, the flow shop schedule shown in the third timing diagram where the initial processor is m and the last is processor 1. From the second rotation, we have reversed the permutation given by S forming say S^r and most pointedly, we have preserved the makespan value ν.

Example 5.2 Consider the 5×2 instance on the left below. Applying the two-processor algorithm yields the ordering $\{4, 3, 1, 2, 5\}$. The reversal instance is on the right. Here, we have $\hat{X} = \{5, 2, 1\}$ and $\hat{Y} = \{3, 4\}$ and upon concatenating, we obtain the total order $\{5, 2, 1, 3, 4\}$ which is indeed the reversal of the previous sequence.

$$\begin{array}{c} 1 \\ 2 \\ 3 \\ 4 \\ 5 \end{array} \begin{bmatrix} 6 & 4 \\ 5 & 3 \\ 7 & 8 \\ 2 & 5 \\ 3 & 1 \end{bmatrix} \qquad \begin{array}{c} 1 \\ 2 \\ 3 \\ 4 \\ 5 \end{array} \begin{bmatrix} 4 & 6 \\ 3 & 5 \\ 8 & 7 \\ 5 & 2 \\ 1 & 3 \end{bmatrix}$$

□

Regardless (and unfortunately), the efficacy with which the two-processor case is generally handled does not extend to three or more processors as Theorem 5.1 would substantiate. There are, however, some interesting special cases worth considering. We can capture these with the following results with proofs following those in Rinnooy Kan (1976):

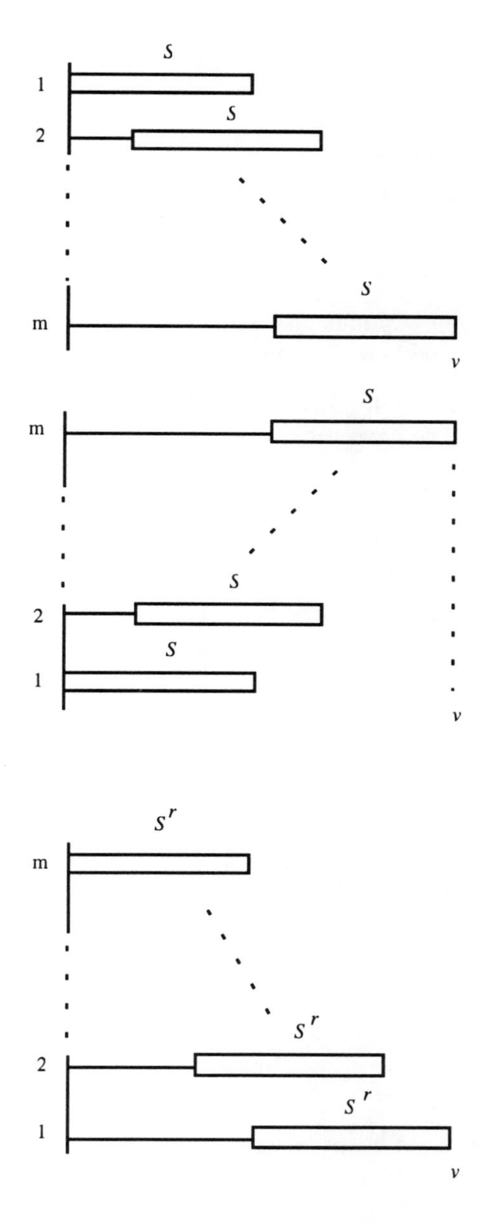

Figure 5.9. *Demonstration of property P_4*

Theorem 5.4 *For the problem $F3||C_{\max}$, if either*

i) $\tau_{i2} \leq \tau_{j1}$ *for all pairs i and j, or*

ii) $\tau_{i2} \leq \tau_{j3}$ *for all pairs i and j,*

then the problem solves by the same (optimal) ordering as $F2||C_{\max}$ where the latter is defined by an instance given by

$$\tau'_{i1} = \tau_{i1} + \tau_{i2} \text{ and } \tau'_{i2} = \tau_{i2} + \tau_{i3}$$

[If (i) above holds we say that processor 2 is dominated by processor 1 and if (ii) holds, then processor 3 dominates processor 2.]

Proof. Consider condition (i). Accordingly, it can be shown that for the problem $F3||C_{\max}$, we have

$$s_{[i]3} = \max(s_{[i-1]3} + \tau_{[i-1]3}, s_{[i]1} + \tau_{[i]1} + \tau_{[i]2}),$$

where $[i]$ denotes the job in the ith position of a sequence. Now, for $F2||C_{\max}$ constructed per the statement of the theorem, it can also be shown that

$$s'_{[i]1} = s_{[i]1} + \sum_{k=1}^{i-1} \tau_{[k]2}$$

and

$$s'_{[i]2} = s_{[i]3} + \sum_{k=1}^{i-1} \tau_{[k]2}.$$

Then,

$$\begin{aligned} c'_{[n]2} &= s'_{[n]2} + \tau_{[n]2} + \tau_{[n]3} \\ &= s_{[n]3} + \sum_{k=1}^{n-1} \tau_{[k]2} + \tau_{[n]2} + \tau_{[n]3} \\ &= s_{[n]3} + \sum_{k=1}^{n} \tau_{[n]2} + \tau_{[n]3} \\ &= c_{[n]3} + \text{ constant} \end{aligned}$$

and we have the desired invariance result. Case (ii) follows by applying the result for (i) to the reversal instance. This completes the proof. □

Example 5.3 The second processor is dominated by the first in the five-job instance shown below. Creating the corresponding artificial, two-processor instance results in the duration matrix on the right. Then applying our algorithm to the latter yields the ordering $\{4, 1, 2, 5, 3\}$, which is repeated on all three of the "original" processors.

$$\begin{array}{c} 1\\2\\3\\4\\5 \end{array}\begin{bmatrix} 5 & 3 & 4\\ 6 & 2 & 5\\ 7 & 1 & 2\\ 5 & 4 & 7\\ 4 & 3 & 3 \end{bmatrix} \qquad \begin{array}{c} 1\\2\\3\\4\\5 \end{array}\begin{bmatrix} 8 & 7\\ 8 & 7\\ 8 & 3\\ 9 & 11\\ 7 & 6 \end{bmatrix}$$

□

Another solvable case arises when, in essence, processor 2 is dominating. We have:

Theorem 5.5 *If in an instance of $F3||C_{\max}$, we have*

i) $\tau_{i1} \leq \tau_{j2}$ *for all pairs i and j, or*

ii) $\tau_{i3} \leq \tau_{j2}$ *for all pairs i and j*

then the problem can be solved by treating n instances of $F2||C_{\max}$ (each on $n-1$ jobs) corresponding to a choice of the first/last job respectively (whichever case applies).

Proof. Consider case (i). Here, we have

$$\begin{aligned} s_{[i]2} &= \max\left(s_{[i]1} + \tau_{[i]1}, s_{[i-1]2} + \tau_{[i-1]2}\right) \\ &= s_{[i-1]2} + \tau_{[i-1]2} \end{aligned}$$

and

$$s_{[i]3} = \max\left(s_{[i]2} + \tau_{[i]2}, s_{[i-1]3} + \tau_{[i-1]3}\right)$$

But these relationships define a two-processor problem on processors 2 and 3 with makespan values accordingly depending on $\tau_{[i]1}$. Minimizing over (polynomially many) choices for an initial job solves the problem. Again, we treat case (ii) by applying the same notion to the reversal. □

Example 5.4 Essentially, the result of Theorem 5.5 follows because under the parametric conditions specified, the first and/or last processor is inconsequential. The data in Figure 5.10 capture the first case. Examining each of three two-processor instances we find that the one in which job 2 is processed first is best. The other two jobs are processed as $\{3, 1\}$ and so the total ordering (on all processors) is $\{2, 3, 1\}$ with makespan 19. □

A particularly interesting, albeit somewhat esoteric, result regarding special cases is the following:

Theorem 5.6 *For problem $F3||C_{\max}$, suppose Π_{12} is the set of permutations produced by algorithm $\boldsymbol{A_{J_2}}$ for an instance of $F2||C_{\max}$ defined*

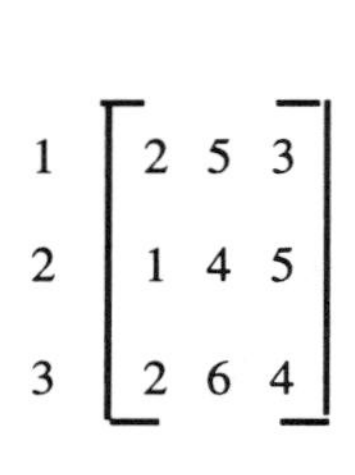

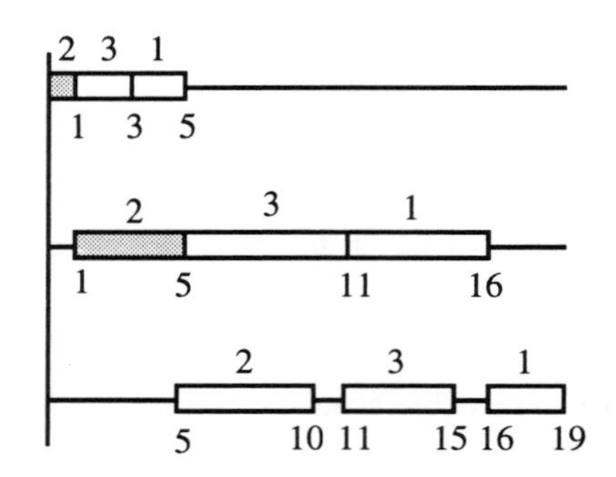

Figure 5.10. *Example 5.4*

by the first two machines in the three-processor problem. Let Π_{23} *be similarly defined relative to the last pair of processors. Then if* $\Pi_{12} \cap \Pi_{23}$ *is not empty, there is at least one such common permutation that solves the three-processor problem.* □

We leave the proof of this result in Johnson (1954) but demonstrate its sense with the following illustration.

Example 5.5 Consider the data for a 3×3 instance below. An optimal (from $\boldsymbol{A_{J_2}}$) permutation in Π_{12} which, in fact, is also in Π_{23}, is $\{3, 2, 1\}$ which solves the three-processor problem. Note that $\{1, 2, 3\} \in \Pi_{12} \cap \Pi_{23}$ but is not optimal, having $C_{\max} = 113$.

$$\begin{matrix} 1 \\ 2 \\ 3 \end{matrix} \begin{bmatrix} 50 & 1 & 3 \\ 6 & 1 & 6 \\ 3 & 1 & 50 \end{bmatrix}$$

□

In a sense, the instance of the previous example is less than interesting because processor 2 is dominated and the results of Theorem 5.4 apply. In Exercise 5-29, we ask the reader to construct an alternative.

Note that if we weaken the conditions of Theorem 5.6 by defining Π_{12} and Π_{23} as sets of optimal permutations, then there may be a nonempty intersection $\Pi_{12} \cap \Pi_{23}$ with a member not obtainable by applying $\boldsymbol{A_{J_2}}$ in the stated sense but that is optimal for the corresponding $F3||C_{\max}$ instance. Consider the following example.

Example 5.6 Relative to the 3×3 instance below, it can be shown that permutations $\{1, 2, 3\}, \{1, 3, 2\}, \{2, 1, 3\}$, and $\{2, 3, 1\}$ are optimal for the first two processors while $\{2, 3, 1\}$ and $\{3, 2, 1\}$ are optimal for the $F2||C_{\max}$ interpretation with the last two. However, the only common permutation, which is $\{2, 3, 1\}$, does not satisfy Johnson's rule (thus, is not obtainable by

$\boldsymbol{A_{J_2}}$, yet when applied to all three processors solves the relevant $F3||C_{\max}$ instance. We will leave the actual verification of these outcomes to the reader.

$$\begin{array}{c} 1 \\ 2 \\ 3 \end{array} \begin{bmatrix} 10 & 20 & 10 \\ 10 & 15 & 15 \\ 11 & 10 & 15 \end{bmatrix}$$

□

5.1.3 General Flow Shops

Void of special cases, the flow shop problem with $m \geq 3$ processors has to be treated in some enumerative fashion. Following we examine a basic approach. Accordingly, we shall assume our search to be confined to the family of permutation schedules. Recall that this is sufficient when $m = 3$ but for four or more processors, any claim of global optimality is lost. Now let us create (or at least imagine) a tree with leaves corresponding to all permutations of the integers $1, 2, \ldots, n$. We can build this tree in an obvious way as shown in Figure 5.11. Every level L of the tree corresponds to a sequential position and each interior node in the tree can be viewed as a partial schedule. In this regard, each non-root node is uniquely defined by its parent.

Now, let us denote a partial solution (partial permutation) by S_p and assume its parent node is identified by S'_p. Thus for $S_p = \{i_{[1]}, i_{[2]}, \ldots, i_{[L]}\}$ we have $S'_p = \{i'_{[1]} = i_{[1]}, i'_{[2]} = i_{[2]}, \ldots, i'_{[L-1]} = i_{[L-1]}\}$. To be precise, we should distinguish between arbitrary partial schedules in the tree. However, the ensuing process is highly structured and the context in which partial schedules are manipulated should be clear.

So, since every possible permutation solution is (trivially) accounted for in the tree shown, the only interesting aspect is the process by which we locate an optimal one. But in the branch-and-bound context, this matter rests essentially with how we implement our search, and most critically, how we establish a proof that at stopping we have a correct outcome. To this end, we develop the following function.

For some S_p and processor k define a lower bound on its completion as

$$B_k^L(S_p) = c_k^L(S_p) + \sum_{j \notin S_p} \tau_{jk} + \min_{j \notin S_p} \left(\sum_{\bar{m}=k+1}^{m} \tau_{j\bar{m}} \right),$$

where

$$c_k^L(S_p) = \max\left(c_{k-1}^L(S_p), c_k^{L-1}(S'_p)\right) + \tau_{[L]k}$$

and

$$c_0^L(S_p) = c_k^0(S_p) = 0.$$

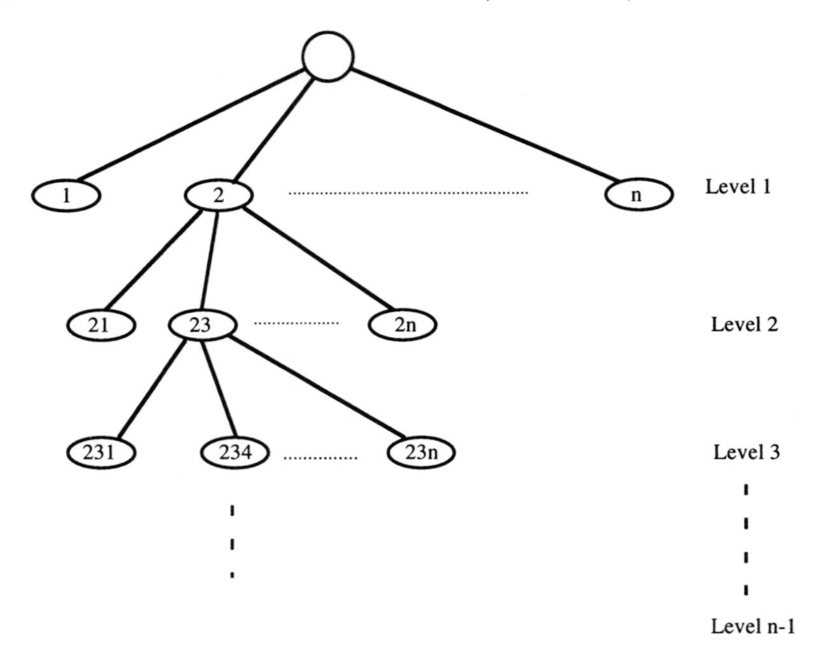

Figure 5.11. *Primative branch-and-bound tree for flow shops*

This bound is very simple, and in addition if each $B_k^L(S_p)$ is a valid lower bound then it follows that a comparison over the processors also yields a lower bound. That is, the value

$$B^L(S_p) = \max_{1 \leq k \leq m} \left(B_k^L(S_p)\right)$$

is also valid. Now, consider the timing diagram in Figure 5.12 and the computation of each $B_k^L(S_p)$. Given some S_p and its measured (actual) completion time on processor k, it is immediate that the remaining time (to complete jobs not in S_p) must extend at least by the amount of work remaining. (When these jobs are actually scheduled in an admissible way, idle time may be introduced.) Moreover, given any completion of S_p and hence every choice for a final job, then minimizing over the remaining processor times for each choice provides a bound on overall completion (recall that we are dealing with permutation schedules).

Example 5.7 Consider below a four-job instance of $F3||C_{\max}$ with the full tree shown in Figure 5.13 and with bound values indicated. We leave

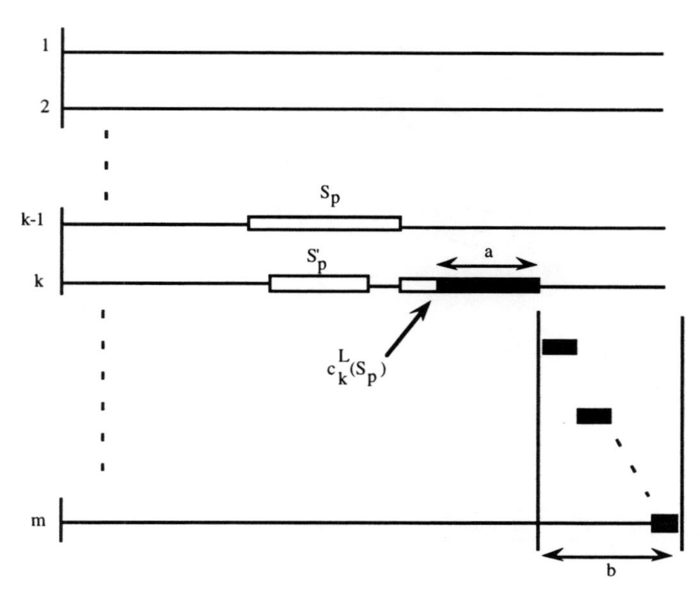

Figure 5.12. *Concept of* $B_k^L(S_p)$

it to the reader as an exercise to complete the specific calculations.

$$\begin{matrix} 1 \\ 2 \\ 3 \\ 4 \end{matrix} \begin{bmatrix} 3 & 4 & 2 \\ 8 & 5 & 4 \\ 3 & 9 & 6 \\ 7 & 6 & 2 \end{bmatrix}$$

□

Although only heuristic, it may be possible to realize some savings by treating a flow shop's reversal. A rule-of-thumb in this regard arises if the sum of processing times is, in some sense, much larger on the first processor than on the last. This follows since makespan values are taken from last processor completion times and these values are best estimated (in the lower bound sense) when less idle time results on the last processor. The latter tends to result when start times on the last processor are determined mostly by completion times of predecessors in a permutation rather than by predecessors in the processor-ordering.

Example 5.8 Suppose duration times for a four-job instance of $F3||C_{\max}$ are given by the matrix below. If this data are used, the first level bound values result as 38 for all partial sequences $(\{1\}, \{2\}, \{3\}, \{4\})$. But since the sum of first processor times exceeds that on the third, solving the reversal

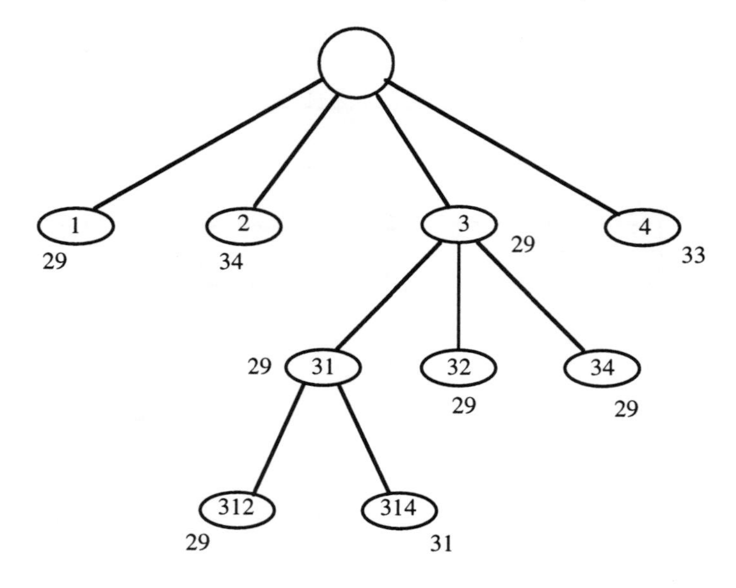

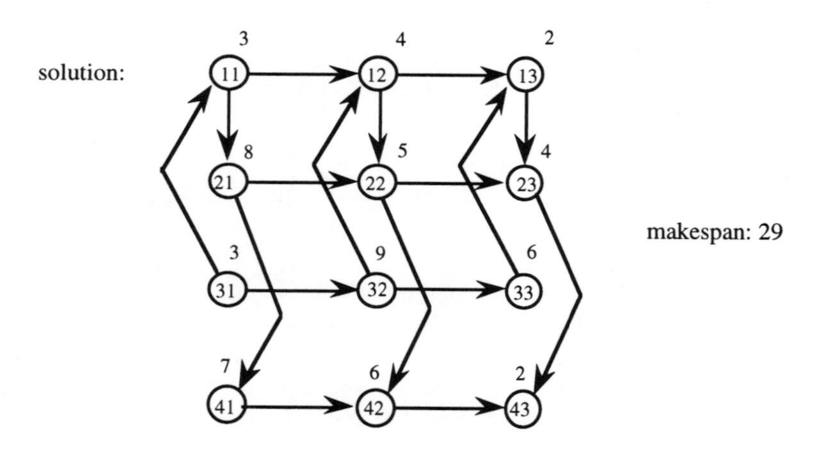

Figure 5.13. *Tree for Example 5.7*

might be advantageous computationally. Indeed, treating the reversal, the first level bound values are now 40, 38, 38, and 41.

$$\begin{matrix} 1 \\ 2 \\ 3 \\ 4 \end{matrix} \begin{bmatrix} 8 & 5 & 2 \\ 9 & 4 & 1 \\ 7 & 2 & 3 \\ 9 & 6 & 2 \end{bmatrix}$$

□

This brings to a close our coverage of the flow shop to which we add a note. Our treatment in terms of the results presented is admittedly rather pedestrian; it is not at great variance with what has been known, indeed with what has been exhibited in the (textbook) flow shop literature for at least twenty-five years. Now, while this is not to suggest that there have been no new flow shop results, we are not shy in intimating that it does suggest a dearth of results akin to those that have emerged in roughly the same period of time for the problem settings described in the previous two chapters—both in terms of elegance as well as relevance. Assuming that our sense reflects a true state of affairs, one justification may be that the highly structured nature of the traditional flow shop coupled with its rapid relegation to an intractable status have resulted in too little room for researchers to maneuver and as a consequence, the setting has long been mined of interesting results. But also incriminating is the unfortunate but growing view that there are simply not many *real* contexts that reflect the standard deterministic flow shop configuration and its concomitant assumptions. An interesting exposition regarding this position can be found in Dudek *et al.* (1992).

5.2 Job Shops

An immediate generalization of the flow shop problem can be created by relaxing the so-called "flow-structure" captured by $\prec$. Rather than requiring each job to progress through the processing stage in an identical fashion, we now allow jobs to have different ordering requirements. In this context, we also allow job operations to involve repetitious processing. That is, m_ℓ need not be different from $m_{\ell'}$ for a given job. We will, however, organize our coverage in terms of the structure of $\prec$. First, we adopt the classic context where $((im_\ell), (jm_{\ell'})) \in \prec$ if and only if $i = j$. That is, precedence exists only among operations of the same job. The second category relaxes this by allowing inter-job precedence.

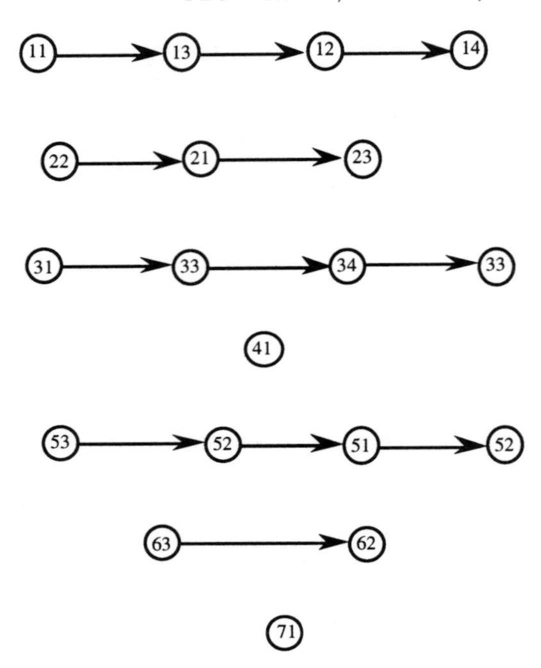

Figure 5.14. *Job shop structure*

5.2.1 Intra-Job Precedence

Typical of this precedence structure for instances of $J||C_{\max}$ is the directed graph $(T, \prec)$ in Figure 5.14. Observe that a technically important (but ultimately uninteresting) distinction between the job-shop and the flow shop problems is that disjunctive resolutions can now lead to schedule inadmissibilities. For example, if job 2 is processed prior to job 1 on processor 1 in the sample structure of Figure 5.14, then job 2 must precede job 1 on processor 2. Otherwise (*e.g.*, job 1 prior to 2 on processor 2) an ambiguity is formed by the implied cycle in the processing of operations $\{(11)(13)(12)(22)(21)\}$. As we shall see later, however, guarding against these occurrences is not difficult.

The next result establishes that job shop problems become difficult very quickly.

Theorem 5.7 *The (recognition version) problem* $J2||C_{\max}$ *is* $\mathcal{NP}$-*Complete, even when each job has at most three operations.*

Proof. To be clear, note that for a 2-processor problem we are allowing at most one "revisit" to a processor. Now, to prove the result we establish a

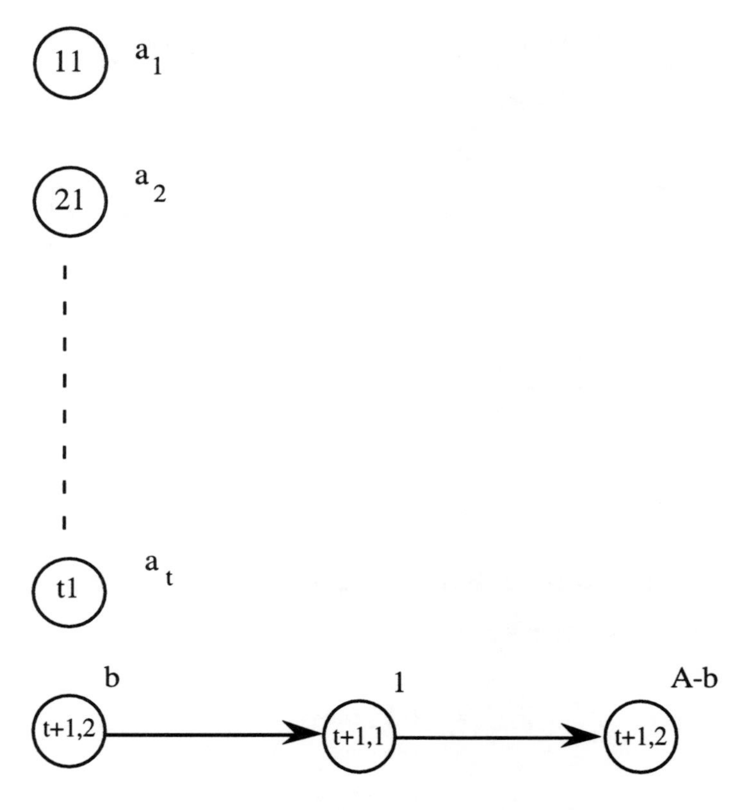

Figure 5.15. *$J2||C_{\max}$ structure for proof of Theorem 5.7*

reduction from the KNAPSACK problem described earlier (*i.e.*, Chapter 2). Following the construction in Rinnooy Kan (1976), our scheduling instance is formed by $t+1$ jobs and $t+3$ total operations. The precedence structure is shown in Figure 5.15. Let $A = \sum_{i=1}^{t} a_i$ and set $k = A+1$. Then it is easy to see that a schedule exists with $C_{\max}$ no greater than $A+1$ if and only if KNAPSACK has a solution. As before, details are left to the reader but the timing diagram in Figure 5.16 provides the idea. It is also easy to see that the value $A+1$ fixed in the mapping is a lower bound on the makespan of any admissible schedule. □

To show how sharp the result of Theorem 5.7 is consider that if we allow at most two operations per job, our problem is easily solved. We have from Jackson(1956):

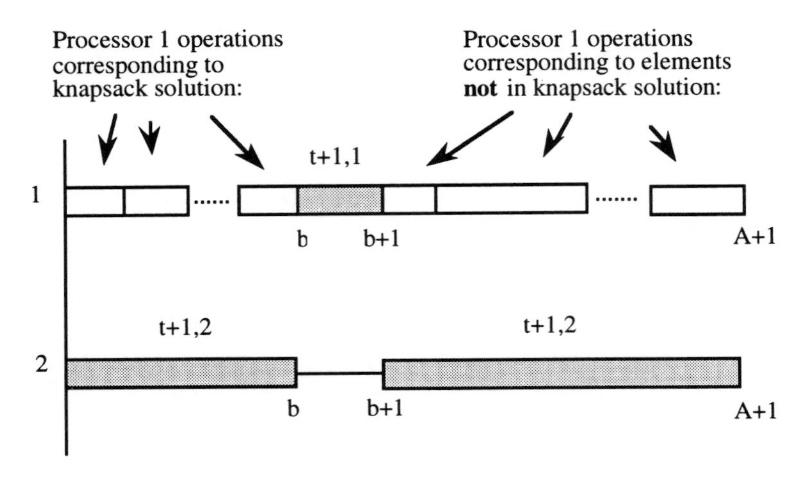

Figure 5.16. *Relationship between a suitable schedule and a knapsack solution*

A_{Ja}: Two-Processor Job Shop Algorithm

Step 1: Construct from $(T, \prec)$ the sets:

$$\begin{aligned} J_{12} &= \{j|(j1) \prec (j2)\} \\ J_{21} &= \{j|(j2) \prec (j1)\} \\ J_1 &= \{j|(j1) \text{ is independent}\} \\ J_2 &= \{j|(j2) \text{ is independent}\} \end{aligned}$$

Step 2: Solve the subproblems $F2||C_{\max}$ defined by sets J_{12} and J_{21}. Let the optimal permutations be given as $\hat{J}_{12}$ and $\hat{J}_{21}$.

Step 3: Create a schedule on processor 1 as

$$\{\hat{J}_{12}, J_1, \hat{J}_{21}\}$$

and on processor 2 as

$$\{\hat{J}_{21}, J_2, \hat{J}_{12}\}$$

□

Example 5.9 Let a two-processor job shop instance be given by the graph shown in Figure 5.17. Duration times are indicated by each operation. The sets constructed per the algorithm arise as follows: $J_{12} = \{1\}, J_{21} = \{2, 4, 6\}, J_1 = \{3\}$, and $J_2 = \{5\}$. Then, $\hat{J}_{12} = \{1\}$ and $\hat{J}_{21} = \{6, 4, 2\}$ and we produce the schedule shown. □

The validity of A_{Ja} is established by the following:

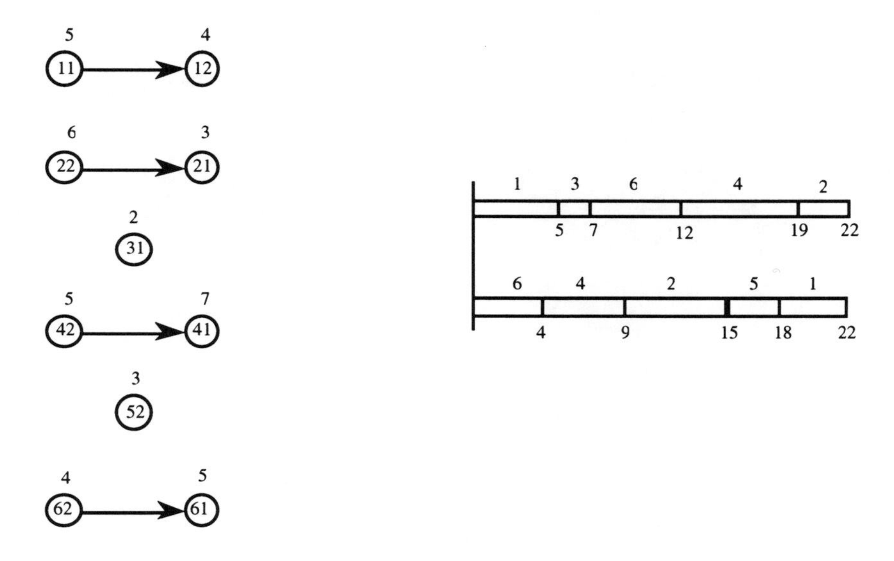

Figure 5.17. *Example 5.9*

Theorem 5.8 *Algorithm* $\boldsymbol{A_{Ja}}$ *will, in polynomial time, solve* $J2||C_{\max}$ *where each job has at most two operations per job.*

Proof. That the algorithm constructs a schedule in polynomial time is obvious. Now, let us assume that in the constructed schedule there is some idle time between operations on either of the two processors; otherwise, the optimality of the schedule follows trivially. But the only idle time possible is that generated by an operation having its start time on the (required) second processor blocked exclusively by the completion time of the predecessor operation on the respective first processor. Hence, if a better schedule exists, this idle time must be reduced but this denies the correct application of the flow shop algorithm in Step 2 relative to sets J_{12} or J_{21}. This completes the proof. □

5.2.2 Inter-Job Precedence

In this section, we shall describe an enumerative procedure for the more general job-shop problem where precedence is permitted between operations from different jobs. Certainly, the results of the ensuing treatment are applicable to the more structured problem of the previous section. We begin with a useful dominance property. As before our aim is to minimize makespan.

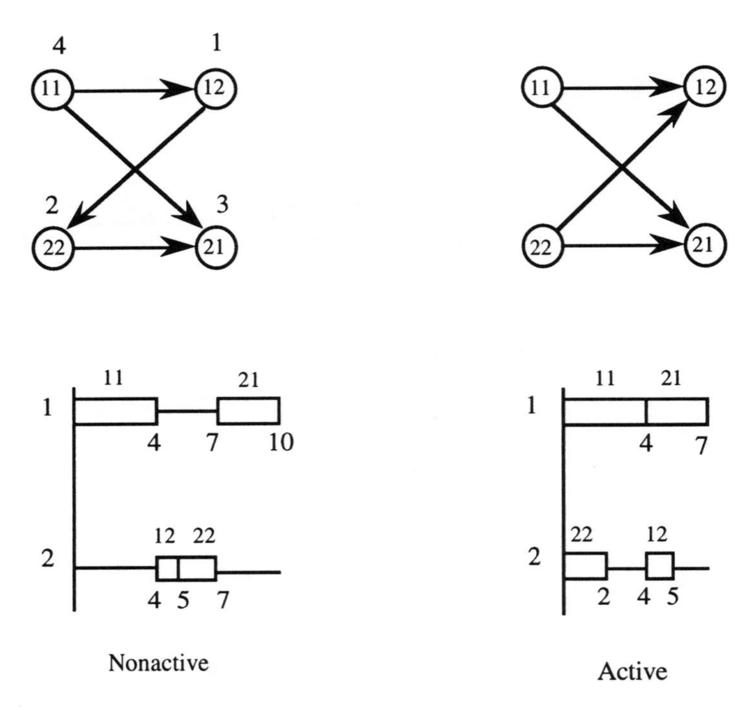

Figure 5.18. *Nonactive and active schedules*

We say that a schedule is *active* if no operation can be started earlier without delaying at least one other operation. Otherwise, the schedule is *nonactive*. Alternately, a schedule is active if there exists no continuous span of idle time on any processor great enough to process a delayed operation. The point is demonstrated in Figure 5.18, where a nonactive schedule on the left, *i.e.*, operation (22), can be scheduled earlier in the interval [0, 4] on processor 2. The result is the active schedule on the right.

It is immediate that we need only consider the subset of active schedules in our search for an optimum one. That is, an optimal active schedule will suffice in solving our problem. To begin, let us specify $H_k = \{T_j |$ job T_j is processed on processor $k\}$. Observe that we have abandoned the previous job/operation notation and will presently denote an instance by a set T of tasks (or jobs), each given by T_j. While creating a slight abuse of our earlier notation, this change is adopted here in order to make easier the current presentation as well as that in Chapter 7 (the clarity of the context should prevent any confusion with this task specification of T_j and the latter's

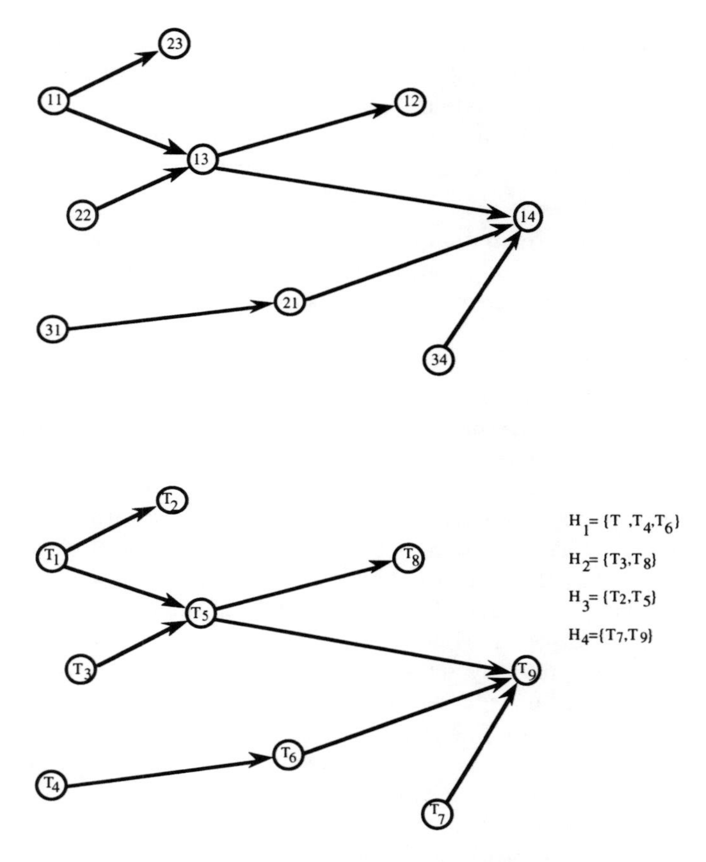

Figure 5.19. *General job shop structure*

previous use to symbolize tardiness). To illustrate, the first precedence structure in Figure 5.19 would now be given by the directed graph below it with sets H_k specified as shown. Note also that we will take $H_k \cap H_{k'} = \phi$ for all pairs of processors k and k'.

Now let $\hat{C} = (\hat{c}_j)$ be a $|T| = n$-tuple with $\hat{c}_j$ the early completion time of job T_j. Then the following procedure for generating active schedules can be stated.

A_{AS} : Active Schedule Generation

Step 1: Let β be the set of unscheduled jobs (equivalently, β implies some partial schedule). Let $\triangle = \min_{T_i \in \beta}(\hat{c}_i)$ and denote by i^* a job completing at $\triangle$. If $\beta = \phi$, go to step 3.

Step 2: Assume $i^* \in \hat{H}$. For all $T_x \in \hat{H} \cap \beta$, $T_x \neq i^*$, evaluate $\Delta + \tau_x \leq \hat{c}_x$. If this inequality holds for all jobs in $\hat{H} \cap \beta$ other than i^* (or if $\{\hat{H} \cap \beta\} \backslash i^* = \phi$), schedule i^*, set $\beta \leftarrow \beta - i^*$ and return to Step 1. Otherwise, from those jobs where the inequality does not hold, select one job to schedule, update β and the respective completion time, and go to Step 1.

Step 3: To generate another active schedule, select an unextended partial schedule, update β and completion time and go to Step 1. If no unextended partial schedule exists, stop. □

The algorithm is easy to demonstrate.

Example 5.10 We can apply $\boldsymbol{A_{AS}}$ to the six-job, two-processor instance shown in Figure 5.20. Accordingly, the initial early completion times are measured as $\hat{C} = (3, 5, 6, 7, 7, 10)$ with $\beta = T$. We have $\Delta = \hat{c}_1 = 3$ and as we would expect (*i.e.*, T_1 precedes T_5 and T_6), no blocking exists and so job T_1 is scheduled and removed from β. Operating with the same $\hat{C}$, we have next that $\Delta = \hat{c}_2 = 5$. Then, testing the inequalities defined by $\{H_2 \cap \beta\} \backslash T_2 = \{T_3, T_4\}$, we see that $\Delta + \tau_3 = 8 \not\leq \hat{c}_3 = 6$ and $\Delta + \tau_4 = 9 \not\leq \hat{c}_4 = 7$; thus T_2, T_3 and T_4 are in "conflict," meaning that if T_2 is scheduled, the other two jobs will necessarily be delayed. So, selecting *any* of the three to be presently scheduled would result in an active schedule extension. Let us arbitrarily select job T_2 and continue. We first update $\hat{C}$ as (3,5,8,9,9,12) as well as β by removing T_2. The next value of Δ is 8 corresponding to $\hat{c}_3$ and a conflict between T_3 and T_4 is found. Let us select arbitrarily 3, update $\hat{C}$ again, and continue whereupon no additional conflicts are detected and a complete, active schedule is ultimately formed.

In order to form another active schedule, we simply return to one of the conflict sets discovered earlier and make an alternative selection. For example, returning to the initial conflict detection involving jobs T_2, T_3, and T_4, we could now select job T_4 say. We would create the respective $\hat{C}$ as (3,9,10,7,11,13), set $\Delta = \hat{c}_2 = 9$, and continue just as before. In fact, the entire process, repeated until every conflict possibility has been exhausted, can be handily depicted by the tree indicated in Figure 5.21. Observe that our first construction corresponds to the left-most branch and also, that while every node in the tree corresponds to a partial, active schedule, we have only labeled each by the job in whose favor the respective conflict is to be resolved. Readers interested in further practice with algorithm $\boldsymbol{A_{AS}}$ are urged to generate other branches in the tree on their own. □

Our claim is that algorithm $\boldsymbol{A_{AS}}$ yields a valid generation of active schedules. That it does so is easy to see. We leave the proof of the next result as an exercise.

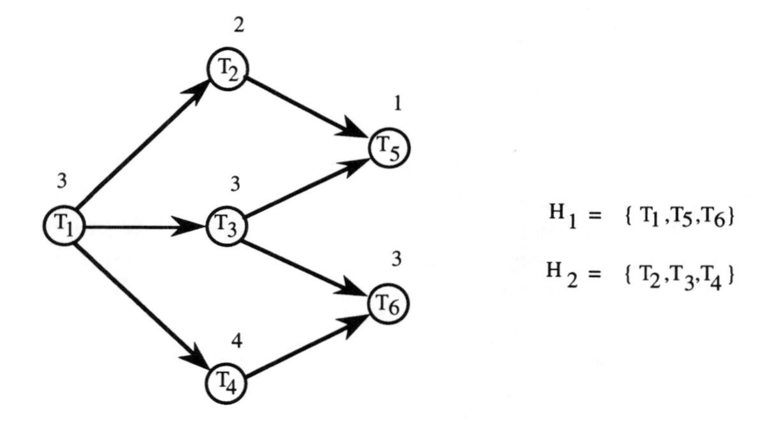

Figure 5.20. *Example 5.10*

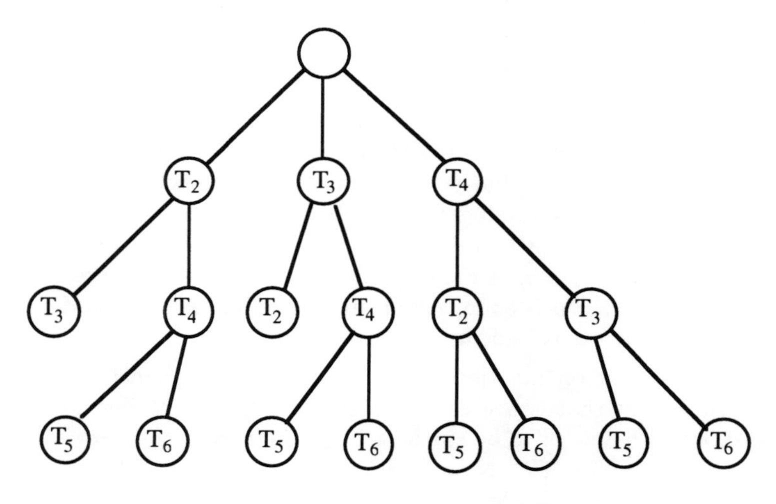

Figure 5.21. *Tree of active schedules for Example 5.10*

Theorem 5.9 *Algorithm* $\boldsymbol{A_{AS}}$ *will generate without omission or repetition the set of active schedules.* □

Required, as before, is a strategy for searching the tree of active schedules generated by algorithm $\boldsymbol{A_{AS}}$. For S_p a partial schedule, we will again denote by $B(S_p)$ a lower bound on an optimal completion of S_p. For this much-studied problem, there are a variety of options. Two are obvious (recall the proof of Theorem 5.7):

$$B_1(S_p) = \max_{T_j \in \beta}(\hat{c}_j)$$
$$B_2(S_p) = \max_k \left(\hat{s}_k + \sum_{T_j \in H_k \cap \beta} \tau_j\right)$$

Here, $B_1(S_p)$ is simply a relaxation bound derived by considering no processor conflict and assuming that jobs wait only as dictated by $\prec$. On the other hand, $B_2(S_p)$ assumes the opposite; precedence is relaxed and we enforce the requirement that processing cannot overlap on any processor. In the B_2 computation, $\hat{s}_k$ denotes the earliest time at which remaining (*i.e.*, unscheduled) work can begin on processor k. It should be clear that B_1 and B_2 are valid bounds. The next example demonstrates their application.

Example 5.11 Let us operate on the instance used in Example 5.10. Accordingly, at the initial conflict level, three partial schedules are indicated by the nodes at the first level of the tree as $\{T_1, T_2\}$, $\{T_1, T_3\}$, and $\{T_1, T_4\}$. Let us consider the second of these and compute B_1 and B_2 respectively. For B_1 we have 13 which is easy and for B_2, we determine a value of 12 arising as $\max(\hat{s}_1 + \tau_5 + \tau_6, \hat{s}_2 + \tau_2 + \tau_4) = \max(8 + 4, 6 + 6)$. Note that we have determined $\hat{s}$ values in a simple way (possibly not the best) as the minimum value over early start times of relevant, unscheduled jobs. Regardless, for $\{T_1, T_2\}$ we obtain B_1 as 12 and B_2 as 12 while for $\{T_1, T_4\}$ we have B_1 at 13 while B_2 is 14. Moreover, if B_1 and B_2 are valid then $\max\{B_1, B_2\}$ is also valid. Indeed, our strategy then is to compute for each partial schedule, corresponding B_1 and B_2 bounds; take the greater of the two and use this composite value to prune the active schedule tree. The result is shown in the tree of Figure 5.22 with the stated composite bound values given next to each node. An optimal schedule and the node/partial schedule producing it is indicated. □

An improvement on functions B_1 and B_2 is easy to construct. Needed is a measurement that captures the essence of the two earlier bounds. To this end, let us define ℓ_j to be the length of a longest path in the subgraph induced by successors of T_j. Now, for each H_k such that $H_k \cap \beta \neq \phi$, arrange jobs in nondecreasing ℓ-order and let these ordered sets be given as $\bar{H}_k$.

For each $\bar{H}_k$ constructed as stated and for each $j = 1, 2, \ldots, |\bar{H}_k|$, compute:

$$\lambda_j = \max(\lambda_{j-1}, \ell_{[j]}) + \tau_{[j]},$$

where $\lambda_0 = 0$ and $[j]$ denotes the jth job in $\bar{H}_k$. Set $W_k \leftarrow \lambda_{|\bar{H}_k|}$. We have

$$B_3(S_p) = \max_k(\hat{s}_k + W_k).$$

Following, we can demonstrate this bound on the current instance.

Example 5.12 Let us compute the ℓ-values initially. Accordingly, we have $\ell_1 = 7, \ell_2 = 1, \ell_3 = \ell_4 = 3$, and finally, ℓ_5 and ℓ_6 are 0. Now, again

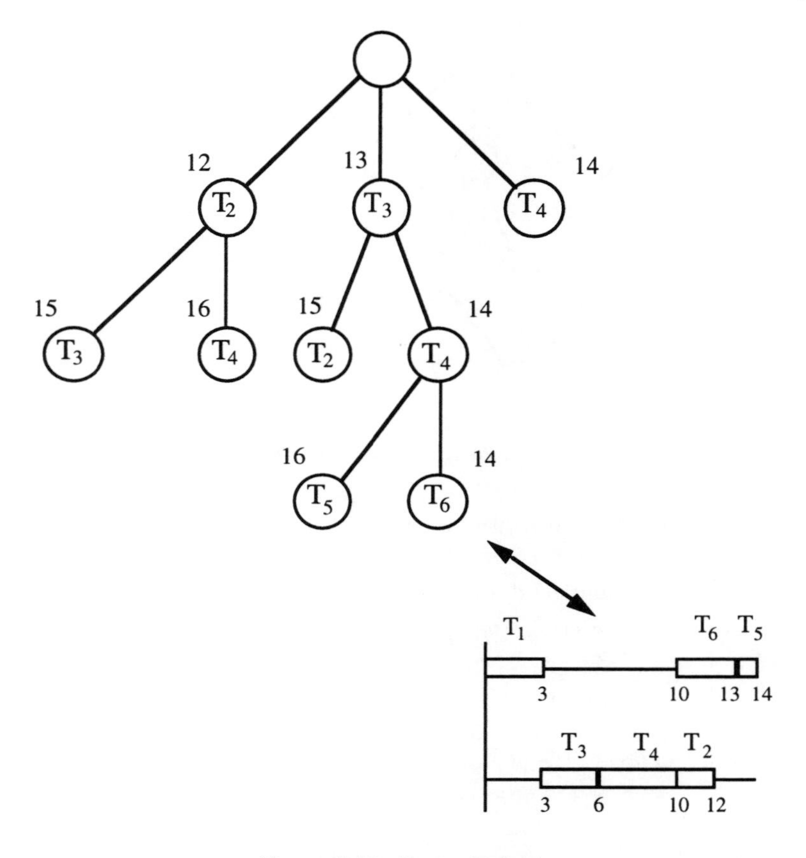

Figure 5.22. *Example 5.11*

we will perform the computation at the first conflict level and specifically consider $\{T_1, T_2\}$. First, relative to processor 2, we have $\bar{H}_1 = \{T_5, T_6\}$ and so $\lambda_1 = 1, \lambda_2 = 4$ yielding W_1 as 4. Similarly, we form $\bar{H}_2$ as $\{T_3, T_4\}$ and compute $\lambda_1 = 6, \lambda_2 = 10$ and hence $W_2 = 10$. Then, we have $B_3 = \max(\hat{s}_1 + W_1, \hat{s}_2 + W_2) = 15$. (Observe the improvement over the value from the composite of B_1 and B_2). For the partial schedule defined by $\{T_1, T_3\}$ we have $\bar{H}_2 = \{T_2, T_4\}$ and W_2 arises as 7; W_1 is unchanged since $\bar{H}_1$ is the same. Thus B_3 in this case is computed to be 13. Finally, for $\{T_1, T_4\}$ we find B_3 to be 14 and our branching choice now is the second option consisting of partial schedule $\{T_1, T_3\}$. Continuing produces the tree shown in Figure 5.23. Note that with bound function B_3 we were able to avoid the "false start" in the direction of $\{T_1, T_2\}$ which occurred with the

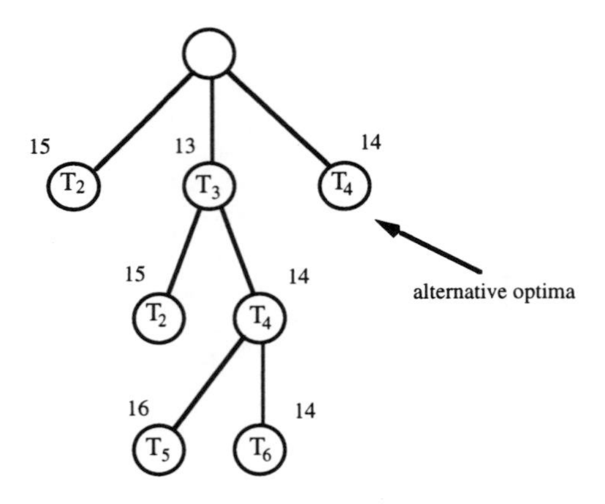

Figure 5.23. *Reference Example 5.12*

other bound alternative. Observe also that in this particular instance, the schedule found does not constitute a unique optima. □

Before establishing the dominance of B_3, let us be clear that it is valid.

Theorem 5.10 *Function $B_3(S_p)$ is a valid lower bound on the optimal completion of S_p.*

Proof. The only thing to prove is that the computation of the stated W_k values are valid. To this end, let us consider, for each k, a single processor problem of the form $1|\prec|f_{\max}$ where $f_{\max} = \max_i(c_i + \ell_i)$. Then, we have that $(i,j) \in \prec \Rightarrow \ell_i \geq \ell_j$ and certainly $c_i + \ell_i$ is nondecreasing in completion time. But then we can apply the result of Theorem 3.15 (regarding the resolution of problem $1||f_{\max}$) which shows that this modified problem solves by arranging jobs in nonincreasing ℓ-order. Recall that the sequential positions are filled out in last-to-first manner which is accomplished by the stated nondecreasing ℓ-ordering of members in the respective $\bar{H}_k$. The value of W_k is simply the outcome of the recursive evaluation of $f_{\max}$ for the relevant ordering. □

Consider an easy illustration of the sense of the previous proof. Suppose three jobs having duration times τ_1, τ_2, and τ_3 are given along with values ℓ_1, ℓ_2, and ℓ_3 respectively. Let us assume for ease that $\ell_3 \geq \ell_2 \geq \ell_1$. Then applying $\boldsymbol{A}_{\boldsymbol{LF_m}}$, job 1 would be processed third (with completion time $\tau_1 + \tau_2 + \tau_3$), job 2 second (with completion time $\tau_2 + \tau_3$), and job 3 would be first in the ordering (completing at time τ_3). The $f_{\max}$ value of the resultant schedule is, by definition, $\max(\tau_1+\tau_2+\tau_3+\ell_1, \tau_2+\tau_3+\ell_2, \tau_3+\ell_3)$ which, upon rewriting, is just $\max(\tau_1+\tau_2+\ell_1, \tau_2+\ell_2, \ell_3)+\tau_3$. Let us denote this by λ_3. But then we can simplify again by letting $\lambda_2 = \max(\tau_1 +$

$\ell_1, \ell_2) + \tau_2$ and finally, even further by setting $\lambda_1 = \max(0, \ell) + \tau_1$. That is, $\lambda_3 = \max(\ell_3, \lambda_2) + \tau_3$; $\lambda_2 = \max(\ell_2, \lambda_1) + \tau_2$; $\lambda_1 = \max(0, \ell) + \tau_1$ all of which is precisely what is produced by the recursion specified as part of B_3.

Finally, the superiority of bound B_3 relative to B_1 and/or B_2 is also easy to see. The following result captures this dominance.

Theorem 5.11 *Let S_p be any partial schedule, then*

$$B_3(S_p) \geq \max(B_1(S_p), B_2(S_p)).$$

□

We will leave the proof of this theorem as an exercise for the reader.

5.3 Open Shops

Let us examine again the instance illustrated in Figure 5.19. Implicit there is that operations involving the same job can be scheduled in any order (unless prohibited by precedence) *and* simultaneously when different processors are involved. That is, it is permissible for an operation (im) to be processed at the same time as an operation (im'). For example, as shown in the stated figure, it is assumed allowable for operations (31) and (34) to be processed at the same time on processors 1 and 4 respectively. Moreover, any sense of a relationship at all between operations of a given job is obscured, if not lost altogether, when the instance is encoded in the task/processor-grouping context employed earlier. Of course, this may not be a relevant concern; however, it could also be the case that a "job" is truly represented by some self-contained piece upon which work is performed and accordingly, at most one of its constituent operations can be performed at a time. This is, of course, not an issue with the classic flow and job shop models since the well-defined precedence structure among operations of a given job in both settings precludes, naturally, any possibility of the aforementioned simultaneous processing.

But suppose that we do enforce the requirement that operations of a given job cannot be scheduled simultaneously while still leaving flexible the order in which the operations for the job are processed. Such a configuration is commonly referred to as an *open shop* and clearly possesses attributes of the classic models examined previously. Figure 5.24 illustrates.

Expectedly, just as the open shop problem is closely related in structure to the flow shop and job shop problems, so too are its complexity attributes. The following result is not surprising nor is it difficult to establish.

Theorem 5.12 *The (recognition version) problem $O||C_{\max}$ is $\mathcal{NP}$-Complete.*

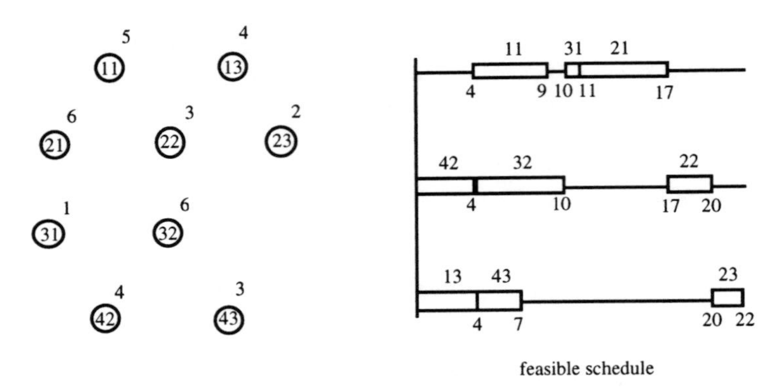

Figure 5.24. *Open shop illustration*

Proof. The reduction is from PARTITION which was defined at the beginning of Chapter 4. Accordingly, from an instance of PARTITION given by elements a_i with weights denoted by w_i, $1 \leq i \leq t$, we can construct an open shop instance on $3t+1$ jobs and three processors with duration times as follows:

$$\begin{aligned}
\tau_{i1} &= w_i, 1 \leq i \leq t \\
\tau_{i-t,2} &= w_{i-t}, t+1 \leq i \leq 2t \\
\tau_{i-2t,3} &= w_{i-2t}, 2t+1 \leq i \leq 3t \\
\tau_{3t+1,1} &= \tau_{3t+1,2} = \tau_{3t+1,3} = W/2,
\end{aligned}$$

where $W = \Sigma_{i=1}^{t} w_i$. Then it is easy to see that PARTITION is solved if and only if there exists a suitable schedule of the stated $3(t+1)$ operations which completes no later than $3W/2$. □

The idea of the proof of Theorem 5.12 is fairly straightforward (see Gonzalez and Sahni, 1976) and not unlike that employed in earlier contexts, *e.g.*, previous reductions employing KNAPSACK. In the present case, the notion is that on some processor, a "blocking" operation has to be scheduled (in order to achieve the stated threshold value on $C_{\max}$) in such a way that a pair of disjoint, closed intervals of time are created which, in turn, would have to be filled *exactly* by certain (nonpreemptible) operations and moreover, that this is possible if and only if these operations correspond to a suitable partition. This exact, packing-like phenomenon is depicted in Figure 5.25. Observe that the blocking phenomenon is created by one of the operations (shaded) of job $3t+1$. More importantly, however, it is also easy to see that the mapping specified in the proof of Theorem 5.12 reveals that an even stronger statement regarding difficulty can be made. We have:

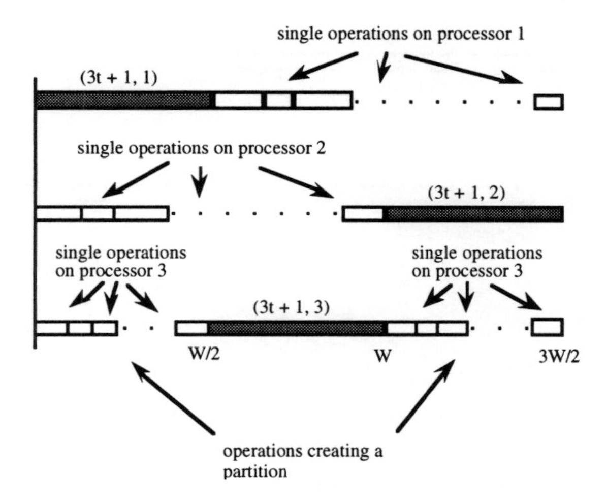

Figure 5.25. *Concept of proof of Theorem 5.12*

Corollary 5.13 *The (recognition version) problem $O3||C_{\max}$ remains $\mathcal{NP}$-Complete even when no more than one job requires in excess of one operation.* □

Also consistent with similar results for the basic flow shop and job shop problems, minimizing makespan in the open shop with two processors turns out to be easy. That is, there is a fast algorithm for its solution (as usual, in employing this language, we do not intend to imply that its proof of correctness is necessarily easy). Consider the following clever procedure by Gonzalez and Sahni (1976).

A_{GS}: Two-Processor Open Shop Algorithm

Step 0: Initialize by setting $\Delta_1 = \Delta_2 = \ell = r = \tau_{01} = \tau_{02} = 0$; $\pi = \emptyset$; $i = 1$.

Step 1: Compute $\Delta_1 \leftarrow \Delta_1 + \tau_{i1}$; $\Delta_2 \leftarrow \Delta_2 + \tau_{i2}$. If $\tau_{i1} \geq \tau_{i2}$, go to 2; otherwise go to 3.

Step 2: If $\tau_{i1} \geq \tau_{r2}$, then extend π by concantenating as πr and set $r \leftarrow i$; otherwise, concantenate as πi. If $i = n$, go to 4; otherwise set $i \leftarrow i + 1$ and return to 1.

Step 3: If $\tau_{i2} \geq \tau_{\ell 1}$ then extend π by concantenating as $\ell\pi$ and set $\ell \leftarrow i$; otherwise, concantenate as $i\pi$. If $i = n$, go to 4; otherwise set $i \leftarrow i + 1$ and return to 1.

Step 4 If $\Delta_1 - \tau_{\ell 1} < \Delta_2 - \tau_{r2}$ set $\pi_1 \leftarrow \pi r \ell$ and $\pi_2 \leftarrow \ell \pi r$; otherwise set $\pi_1 \leftarrow \ell \pi_r$ and $\pi_2 \leftarrow r \ell \pi$. Remove all 0s from the permutations π_1

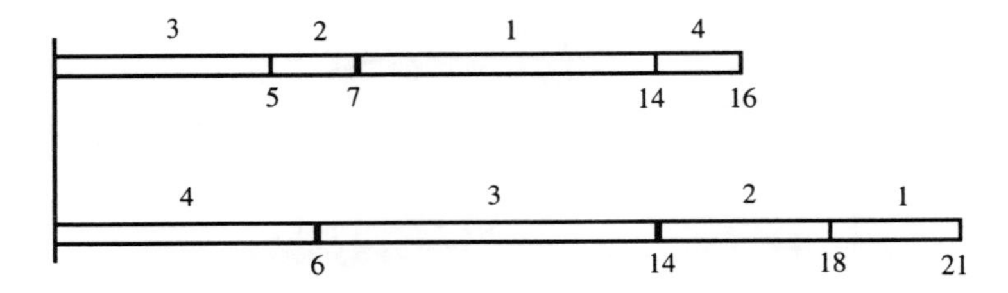

Figure 5.26. *Final schedule for Example 5.13*

and π_2 and schedule in these orders on respective processors. □

We will not take space to establish the correctness of $\boldsymbol{A_{GS}}$; however, its application is illustrated with the following example.

Example 5.13 Consider the instance below involving four jobs and operation duration times as indicated.

$$\begin{matrix} 1 \\ 2 \\ 3 \\ 4 \end{matrix} \begin{bmatrix} 7 & 3 \\ 2 & 4 \\ 5 & 8 \\ 2 & 6 \end{bmatrix}$$

The computation can be easily summarized in the following table:

After iteration	π	r	ℓ
1	0	1	0
2	00	1	2
3	200	1	3
4	3200	1	4

Applying Step 4 of $\boldsymbol{A_{GS}}$ we form $\pi_1 \leftarrow \pi r \ell = \{320014\}$ and $\pi_2 \leftarrow \ell \pi r = \{432001\}$ or upon eliminating dummy 0s we obtain $\pi_1 = \{3214\}$ and $\pi_2 = \{4321\}$. The corresponding schedule is shown in Figure 5.26. □

We point out that by the nature of the algorithm, π_1 and π_2 are constructed in such a way that we can simply schedule the respective jobs consecutively without conflict and hence without fear of generating any internal idle time. That this is the case is formalized in the stated Gonzalez and Sahni reference.

Interestingly, if we are prepared to allow operations to be interrupted or preempted, there is a complete solution to the basic open shop problem employing a bipartite matching formulation. In fact, it is not too difficult to see how this can be done. Following, we will describe the key elements of the construction. Details are left in Gonzalez and Sahni (1976).

We may assume that an instance is defined by n jobs and m processors; however, each job need not involve m operations. Trivially, we can see that an optimal makespan value is bounded from below by the larger of the maximum, total processing requirement taken over all jobs and the maximum, total processor running time taken over all processors. Let this value be α. Then, we can construct a bipartite graph with bipartition $\{X, Y\}$ where X consists of vertices denoting jobs, n of which are "real" and m are artificial. Similarly, Y contains vertices pertaining to m real processors and n which are artificial, *i.e.*, $|X| = |Y| = n + m$. Edges are created between the real job-vertices and real processor-vertices if and only if there is an actual operation defined by the pair. These edges are weighted by the respective operation processing time. Additional edges incident to the job-vertices representing real jobs are created as follows: For a given job i whose total processing time is less than α, construct an edge (x_i, y_{m+i}) and fix its weight at this difference. Similarly, add additional edges incidence to real processor-vertices, y_j, if the respective total processor running time is less than α and weight these edges, (x_{n+j}, y_j), by the difference. After these two constructions, the total weight of edges incident to every vertex x_i, $1 \leq i \leq n$, and incident to every vertex y_j, $1 \leq j \leq m$, is exactly α. Finally, we add edges (x_{n+i}, y_{m+j}), $1 \leq i \leq m$ and $1 \leq j \leq n$, as needed in order to force every one of the artificial vertices to also satisfy the weighted incidence requirement of α. It's not difficult to argue that this is always possible.

Then upon completion of this construction, we have a weighted $(n + m) \times (n + m)$ bipartite structure with the additional attribute that the total weight of edges incident to each vertex is α. At this point, we simply produce a family of (possibly, nondisjoint) perfect matchings. Each such matching defines an assignment of $n + m$ operations, some of which will be fictitious (*i.e.*, involving artificial jobs and/or processors). For a given such matching, suppose a least weight edge value is $\bar{t}$. Then we simply schedule accordingly, the operations denoted by the matching, each for a period of $\bar{t}$ time units. The weight of each edge in this matching is reduced by $\bar{t}$ which has the additional effect of eliminating at least one edge from the original graph, another perfect matching is found, and the process is repeated. This scheme is repeated until no edges are left at which point a complete schedule results. All artificial jobs and processors are elminiated, leaving a feasible schedule of the original operations which, importantly, completes at time α and hence, represents an optimal schedule. Let us demonstrate with an illustration.

Example 5.14 Suppose three jobs are to be scheduled on three processors requiring a total of seven preemptible operations with duration times as shown in Figure 5.27. Observe that $\alpha = \max\{\max_i \Sigma_j \tau_{ij}, \max_j \Sigma_i \tau_{ij}\} = 11$. The initial, bipartite graph including edge weights appears on the right. Now, an initial matching is selected that we note by M_1 and that is shown in Figure 5.28. The value of $\bar{t}$ is 1 and so the six jobs (three real and three artificial) are scheduled on the corresponding processors *vis-a-vis* the edges in M_1, each for a duration of 1 time unit. The weight of each edge in M_1 is reduced by 1 which in effect eliminates edge (x_2, y_2) from the graph. Returning to the original graph with the updated weights, we find another matching. One possibility is M_2, also shown in Figure 5.28 and which produces a $\bar{t}$ value of 2 and corresponding to edge (x_2, y_1). The six operations are scheduled in the period 1 to 3, weights of M_2-edges are reduced, and we look for a third matching. Repeating this process produces matchings $M_3, \ldots, M_7$, each of which is shown in the same figure. The final timing diagram is presented in Figure 5.29. Note on the latter, there are (expectedly) seven intervals within which the operations have been scheduled relative to the seven matchings. The shaded segments correspond to artificial operations. We remove dummy processors 4, 5, and 6 as well as operations corresponding to dummy jobs 4, 5, and 6 and we are left with the timing diagram of a feasible schedule as shown at the bottom of Figure 5.29. Note that on the first processor, jobs 1 and 3 have been preempted while on the third processor, job 2 is split; both jobs requiring the second processor are run without interruption. As claimed, the schedule completes at the least possible time of 11. □

Arguments that establish the validity of the procedure described and demonstrated above are constructive. These amount to a verification that the stated strategy will always produce a feasible open shop schedule and, importantly, one that always completes at the lower bound value α and is thus optimal. It turns out that both of these attributes are fairly easy to substantiate by simply arguing the respective points from the classic graph-theortic perspective of what is being captured by the structure of the perfect matchings that are selected. We invite the reader to pursue this as an exercise and in particular, by recalling the notion of 1-*factors* in r-regular, bipartite graphs. In any event, the strategy described somewhat informally above leads directly to the following result:

Theorem 5.14 *The open shop problem $O|pmtn|C_{\max}$ is solvable in polynomial time.* □

We finish this section by noting that the open shop (as with its flow and job shop counterparts) is also quite well studied with a host of results in-

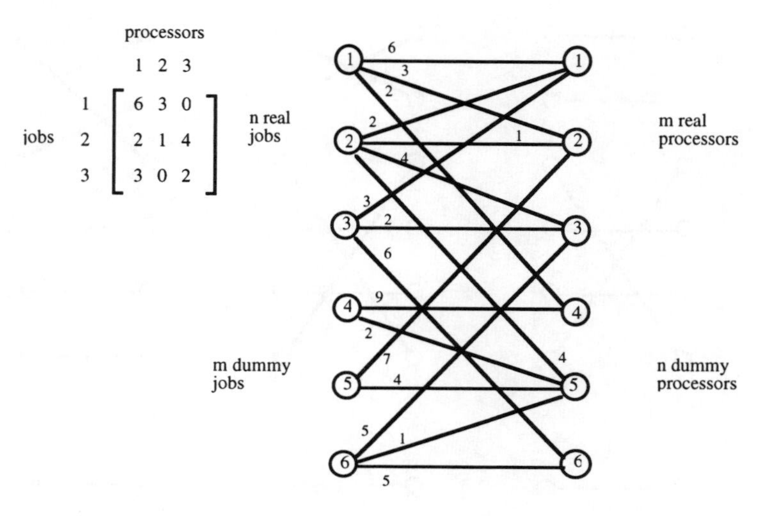

Figure 5.27. *Example 5.14*

volving the standard extensions to alternative performance measures, nonzero release times, precedence structures, and bounds on processing times including *unit execution time* problems. Among these are works such as Achugbue and Chin (1982a), Lawler *et al.* (1981), Adiri and Amit (1983), Cho and Sahni (1981), Fiala (1983), Gonzalez (1979), Gonzalez (1982), and Liu and Bulfin (1985, 1988).

5.4 Exercises

5-1 Give a nonpermutation solution that is optimal for an instance of $F3||C_{\max}$.

5-2 Either provide (fast) algorithms or complexity ($\mathcal{NP}$-Completeness) proofs for problems $F2||T_{\max}$, $F2||\Sigma T_j$, and $F2||\Sigma c_j$.

5-3 Complete the details of the proof of Theorem 5.1.

5-4 Apply $\boldsymbol{A_{J_2}}$ to the flow shop instance in Figure 5.30.

5-5 Given an instance of $F2||C_{\max}$ where an optimal solution exists and that does not satisfy Johnson's property.

5-6 Complete the proof of Theorem 5.3 [*i.e.*, cases (ii) and (iii)].

5-7 Apply $\boldsymbol{P_4}$ to the instance of Exercise 5-4.

5-8 Apply the result of Theorem 5.4 to the following instance of $F3||C_{\max}$:

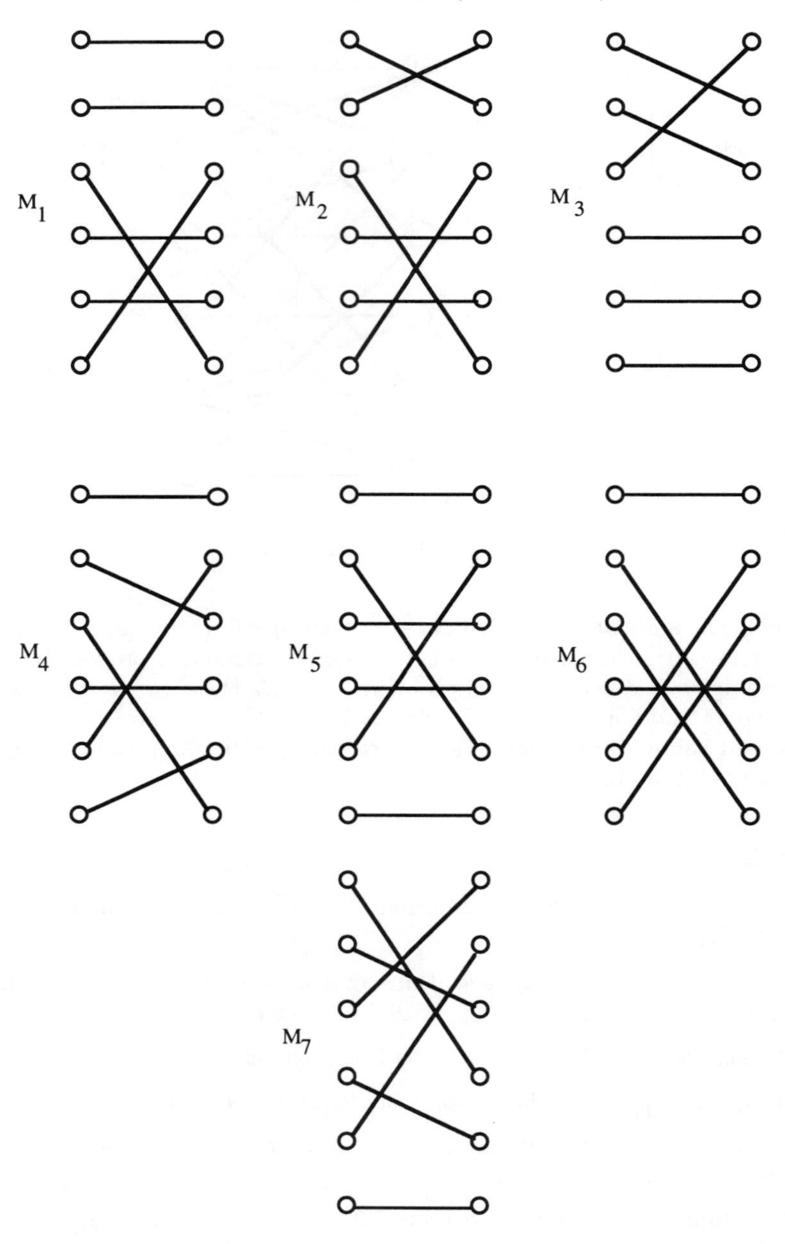

Figure 5.28. *Matchings of Example 5.14*

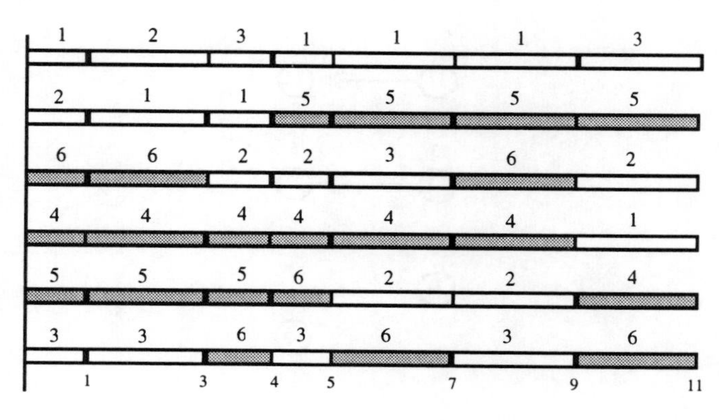

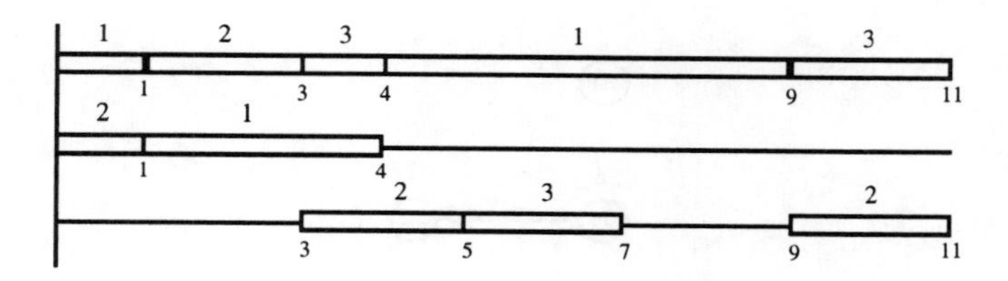

Figure 5.29. *Schedule construction of Example 5.14*

$$\begin{array}{c} 1 \\ 2 \\ 3 \\ 4 \\ 5 \end{array} \begin{bmatrix} 6 & 4 & 7 \\ 4 & 2 & 3 \\ 8 & 1 & 5 \\ 5 & 2 & 2 \\ 7 & 3 & 9 \end{bmatrix}$$

5-9 Apply the result of Theorem 5.5 to the following instance of $F3||C_{\max}$:

$$\begin{array}{c} 1 \\ 2 \\ 3 \\ 4 \\ 5 \end{array} \begin{bmatrix} 6 & 3 & 2 \\ 7 & 6 & 1 \\ 5 & 4 & 1 \\ 3 & 5 & 2 \\ 4 & 8 & 3 \end{bmatrix}$$

5-10 Construct another instance for $F3||C_{\max}$ that demonstrates the notion captured in Example 5.6.

5-11 Solve the following instance of $F3||C_{\max}$ using the bound function of Section 5.1.3.

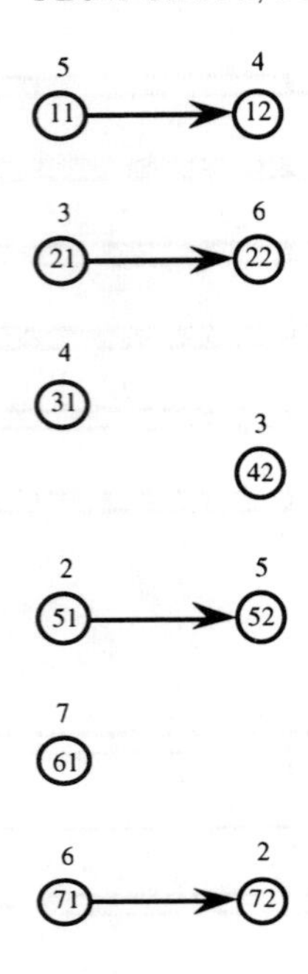

Figure 5.30. *Exercise 5-4*

$$\begin{array}{c} 1 \\ 2 \\ 3 \\ 4 \end{array} \begin{bmatrix} 3 & 4 & 9 \\ 7 & 1 & 8 \\ 5 & 8 & 2 \\ 2 & 6 & 8 \end{bmatrix}$$

5-12 Repeat Exercise 5-11 on the reversal of the instance shown.

5-13 Provide the details of the proof of Theorem 5.7.

5-14 Are there any interesting ways to extend $\boldsymbol{A_{Ja}}$ to the problem $J3||C_{\max}$? Discuss.

5-15 Apply $\boldsymbol{A_{Ja}}$ to the instance in Figure 5.31.

5-16 Apply $\boldsymbol{A_{AS}}$ and generate all active schedules for the instance in Figure 5.32.

5-17 From the generation in Exercise 5-16, construct a nonactive schedule. Show why it is nonactive.

5-18 Apply branch-and-bound to the instance of Exercise 5-16; use bounds $B_1(S_p)$ and $B_2(S_p)$ in *composite* fashion (*i.e.*, $\max(B_1(S_p), B_2(S_p))$).

5-19 Repeat Exercise 5-18 using $B_3(S_p)$.

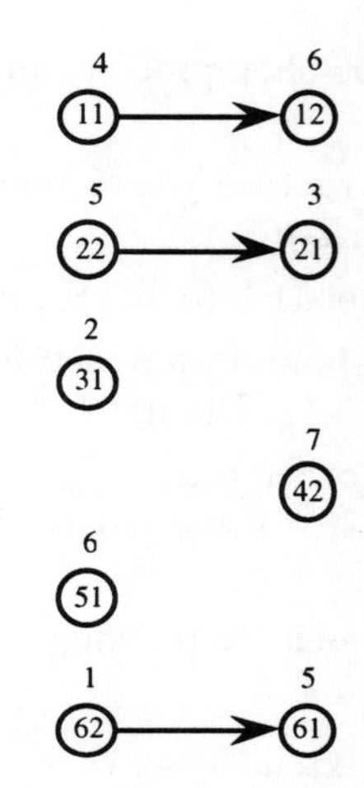

Figure 5.31. *Exercise 5-15*

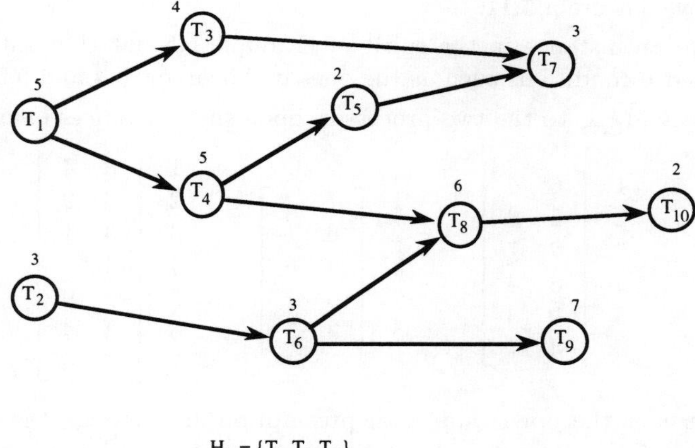

$H_1 = \{T_1, T_2, T_9\}$

$H_2 = \{T_3, T_4, T_7\}$

$H_3 = \{T_5, T_6, T_8, T_{10}\}$

Figure 5.32. *Exercise 5-16*

5-20 Give an instance of a job-shop problem and a nonactive schedule that is optimal.

5-21 Describe a job-shop structure where bound function $B_1(S_p)$ would tend to dominate function $B_2(S_p)$.

5-22 Repeat Exercise 5-21 relative to $B_2(S_p)$ dominating $B_1(S_p)$.

5-23 Apply the branch-and-bound approach for the job-shop to the flow shop instance of Exercise 5-11. Discuss.

5-24 Give other measures (other than $C_{\max}$) for which the set of active schedules are sufficient. Give some where they are not. In each case, provide the argument.

5-25 For the instance in Exercise 5-16, compute $B_3(\phi)$.

5-26 Prove Theorem 5.9.

5-27 Can Theorem 5.6 be extended so that the conditions defining sets Π_{12} and Π_{23} are relaxed to include optimal solutions (not necessarily satisfying Johnson's rule)? Discuss.

5-28 Prove Theorem 5.11.

5-29 Give an instance in the spirit of Example 5.5 but that satisfies no dominance conditions such as the ones in Theorems 5.4 and 5.5.

5-30 Apply $\boldsymbol{A_{GS}}$ to the two-processor, open-shop instances below:

$$\begin{matrix}1\\2\\3\\4\\5\\6\end{matrix}\begin{bmatrix}6 & 4\\5 & 3\\2 & 7\\8 & 4\\9 & 3\\4 & 7\end{bmatrix}\qquad\begin{matrix}1\\2\\3\\4\\5\end{matrix}\begin{bmatrix}7 & 2\\6 & 1\\4 & 8\\5 & 7\\3 & 3\end{bmatrix}\qquad\begin{matrix}1\\2\\3\\4\\5\\6\\7\end{matrix}\begin{bmatrix}8 & 4\\6 & 9\\3 & 1\\4 & 1\\2 & 9\\3 & 4\\7 & 6\end{bmatrix}$$

5-31 Consider the open shop with preemption and assume the data for Exercises 5-8, 5-9, and 5-11 represent instances accordingly. Solve each using the matching procedure demonstrated in Example 5.14.

5-32 Repeat Exercise 5-31 relative to the instances below:

$$\begin{matrix}1\\2\\3\\4\end{matrix}\begin{bmatrix}6 & 5 & 7\\2 & 1 & 0\\0 & 4 & 5\\2 & 0 & 0\end{bmatrix}\qquad\begin{matrix}1\\2\\3\\4\\5\end{matrix}\begin{bmatrix}7 & 4 & 6 & 3\\2 & 0 & 0 & 1\\5 & 4 & 0 & 3\\0 & 2 & 9 & 4\\5 & 4 & 0 & 1\end{bmatrix}$$

CHAPTER 6

NONSTANDARD SCHEDULING PROBLEMS

From our adopted convention in Chapter 1 regarding what constitutes a "scheduling problem" and, indeed, from the coverage in the preceding five chapters, the prevailing assumption has been that both the level of work (*i.e.*, tasks, jobs, etc.) *and* that of resources (*i.e.*, processors) is fixed (relative to a given instance). In the present chapter, we will examine some legitimate as well as prominent scheduling problems that arise under a slight liberalization of this convention. In particular, we will now consider settings where the "resource" level is not fixed per se, but rather is the subject of an optimization/feasibility evaluation in and of itself. We will look at three such problems.

6.1 The Classroom Assignment Problem

Our first problem was previously described in Chapter 1. Known as the *classroom assignment problem* the aim is to seek, as the name suggests, an assignment or allocation of rooms to a set of classes. These classes meet at different time intervals and any such room allocation must be cognizant of certain basic constraints. For example, two classes may not meet simultaneously in the same room. Similarly, a given class cannot meet in two different rooms, etc. One objective in the problem is to find a schedule of all classes that requires the fewest number of rooms overall. On the other hand, it may be that we are limited in the number of rooms available and the problem becomes an admissibility test. That is, we would simply seek any room assignment satisfying the stated availability. The Earth Day example of Chapter 1 provided an illustration of the latter.

As we would expect, the generic classroom assignment problem arises in other settings as well. Among these are problems involving hotel room assignments, shipyard scheduling, and the assignment of airport gates to flights. For the sake of convenience, however, we will continue to employ the language of classroom assignment throughout our coverage.

6.1.1 Vertex Coloring and the Fundamental Problem

The Earth Day illustration in Chapter 1 required a schedule of fourteen classes, labeled as $A, B, \ldots, N$, over an eight-period day (each period is one hour) subject to an availability of only five classrooms. Also prespecified were intervals in which each class was to be scheduled. Accordingly, we saw that a feasible five-room schedule was in fact possible. For ease, the schedule is exhibited again below:

Period	1	2	3	4	5	6	7	8
Room 1	D	D		C	C	F		B
Room 2	I	I	E	E	E		G	G
Room 3		H	H		J	K	K	
Room 4		N	N	N				M
Room 5		A	L	L				

But while five rooms suffice in this case, a smaller number is not possible since there are five courses that require, simultaneously, period 2 for at least some portion of their respective intervals.Thus, if we wanted more than simply a test of room feasibility and, instead, had sought the *minimum* number of rooms needed to accommodate all of the sessions, the five-room schedule shown above would still provide the answer.

But of course in this particular example, we are fortunate in that our claim of having found an optimum number of rooms is based on the obvious outcome; a feasible solution is known that has value (yielding an upper bound) equivalent to a trivial lower bound. On the other hand, these two values can be expected to be generally different and even then, deciding the so-called feasibility issue might not be an easy task itself.

In order to formalize matters, let us assume that there are a total of n classes and for each of these, which we denote as c_i, let us specify the set of periods in which the class is to be taught as P_i. Now, we can create a graph $G = (V, E)$ with vertices corresponding to classes and edges formed such that $(i, j) \in E$ if class c_i and class c_j have a required period in common, *i.e.*, if $P_i \cap P_j \neq \phi$. Relative to the Earth Day illustration, the graph shown in Figure 6.1 would be constructed. Here, vertices are labeled by the letters corresponding to the fourteen classes and note, for example, that an edge exists between vertices representing classes F and K since $P_F \cap P_K = \{6, 7\} \cap \{6\} \neq \phi$.

The implication of an edge in our graph G should be evident. In particular, if $(i, j) \in E$, then, by construction, classes c_i and c_j require at least one common period, and as such must be in separate rooms during that period (recall that, by assumption, if a class runs for more than one period, it must stay in its assigned room throughout all periods). Conversely, two vertices

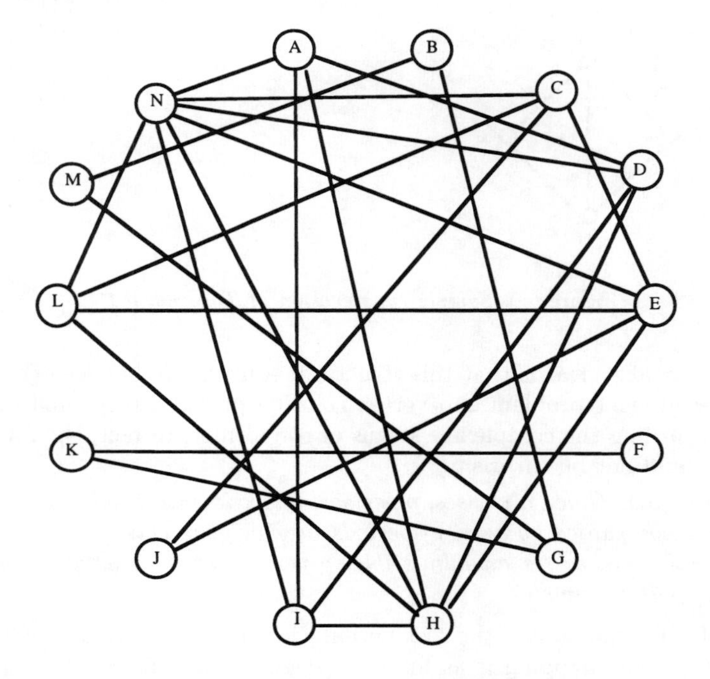

Figure 6.1. *Graph of Earth Day illustration*

that are independent or nonadjacent in G represent a pair of classes that can be scheduled in the same room without fear of conflict. It follows then that any subset of vertices C in G that are mutually independent or equivalently, form a *stable set* in G, represent a set of classes that can share a single, common room throughout any schedule. Moreover, if V can be partitioned as $\{C_1, C_2, \ldots, C_k\}$, where each C_j is a stable set, then we could schedule all courses using at most k rooms. That is, classes represented by vertices in C_1 are conducted in room 1, those represented by vertices in C_2, room 2, and so forth. Such a partition exhibiting the smallest k would correspond to a schedule using the fewest number of rooms.

But seeking the smallest k-partition of the vertex set of a graph into stable sets is easily recognized as equivalent to determining the *chromatic number* of the graph and accordingly, the resulting coloration of the graph would yield a feasible, minimum-room assigment. Thus, deciding if a five-room schedule exists for the fourteen Earth Day seminars is the same as asking if the graph given above is 5-colorable. The schedule shown earlier affirms the existence of such a coloring.

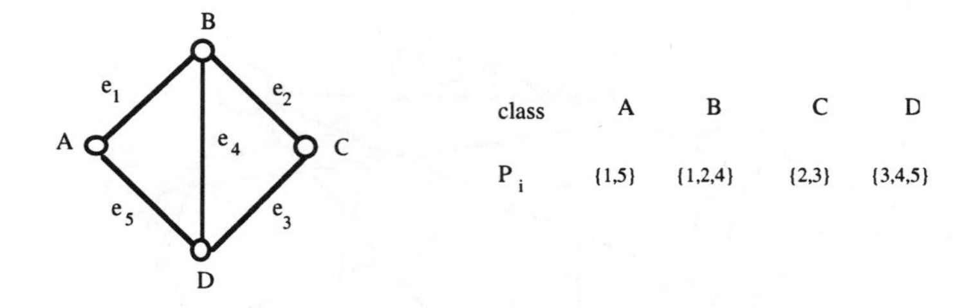

Figure 6.2. *Sample construction of Theorem 6.1*

However, also clear is that this structural equivalence between the classroom assignment problem and vertex coloring provides incriminating evidence regarding the complexity status of the former. In fact, the following result should not be surprising.

Theorem 6.1 *Given n classes, p periods, and k identical rooms where each class is accompanied by a set of periods in which the class is to be taught, deciding if there exists an admissible, p-period schedule using at most k rooms is $\mathcal{NP}$-Complete.*

Proof. Understandably, the reduction is from GRAPH k-COLOR ABILITY. The mapping is as follows: Given a graph $G = (V, E)$ and an integer k for the coloring instance, form an $|E|$-period scheduling instance by creating $|V|$ classes with respective sets P_i constituted such that if edge e_ℓ in E is defined by vertex set $\{i, j\}$, then classes c_i and c_j must meet during period ℓ. It is an easy exercise to see that G is k-colorable if and only if an admissible k-room schedule exists. □

Example 6.1 For the graph shown in Figure 6.2 and from the given vertex and edge labels, the corresponding classroom assignment instance would consist of four classes with period-sets as indicated. But the graph is obviously 3-colorable which means that a three-room assignment is possible. These outcomes are shown in Figure 6.3. □

There is also an easy corollary of the theorem which suggests that the classroom assignment admissibility test remains hard even for a small number of rooms.

Corollary 6.2 *The classroom assignment problem stated in Theorem 6.1 remains $\mathcal{NP}$-Complete even for the case of three identical rooms.*

Proof. The proof is immediate following the $\mathcal{NP}$-Completeness of deciding if a graph is 3-colorable (*cf.* Garey and Johnson, 1979). □

So, given an arbitrary set of classes and a set of prespecified periods within which the classes are to be scheduled, simply deciding if it can be

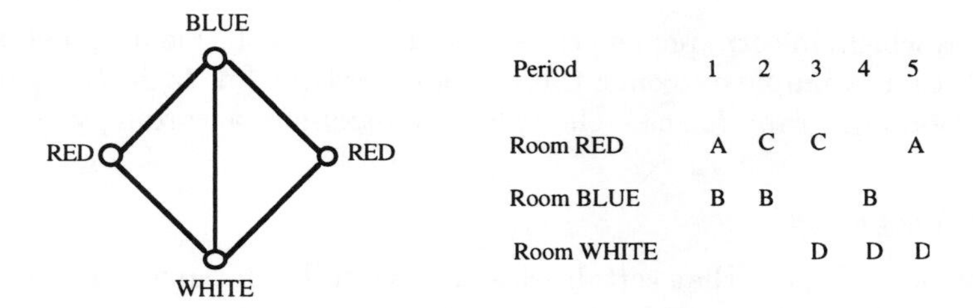

Figure 6.3. *Schedule generated from the coloring of Example 6.1*

done at all subject to some fixed room availability is (in the complexity sense) a difficult exercise. But if we examine the sense of this claim, as well as the problem statement of Theorem 6.1, and do so particularly from the perspective of the Earth Day planners, a (possibly important) subtlety might pique our interest.

In the $\mathcal{NP}$-Completeness result of Theorem 6.1, there are no stipulations regarding the complexion of the prespecified scheduling periods, P_i. However, in reality, a legitimate restriction would seem to be that for those classes requiring multiple periods, these periods at least be contiguous. Convening a two-period class for one period followed by an interruption prior to convening the final period would not only seem unnatural but could also be ineffective in a pedagogical sense. As we observed in the Earth Day illustration, all of the sets P_i had this contiguous-period or *interval* scheduling attribute (*e.g.*, Carter and Tovey, 1992).

But does a requirement that an instance of the classroom assignment problem have so-called interval sets P_i make a difference? Happily, it does.

Theorem 6.3 *The admissibility version of the classroom assignment problem with sets P_i satisfying a contiguous-period or interval property is solvable in polynomial time.*

Proof. The graph constructed from an instance of the classroom assignment problem is an interval graph. But interval graphs are known to be easily colorable and the proof is complete. □

Interval graphs are also known to be *perfect*, which means that they have chromatic number equal to the size of their largest clique. In addition, recall that in graphs formed from a classroom assignment instance, cliques correspond to those subsets of classes that have a scheduling period in common, thus providing a trivial lower bound on the minimum number of rooms required in any assignment. But this is the same as employing the size of any clique in a graph as a lower bound on its chromatic number. So, the graph for the Earth Day instance is an interval graph and thus is perfect, which, in turn, renders the stated five-room solution an expectation rather than a simple outcome; the subset of vertices $\{A, D, H, I, N\}$ in the corresponding graph forms a clique and is maximum accordingly.

6.1.2 Modifications

Let us now suppose that certain classes have to be in particular rooms as a matter of either preference or necessity. Recall that this was the notion described in the Earth Day illustration when we fixed the room assignments of classes A, D, I, and E to rooms 2, 3, 1, and 4 respectively. Further, we then claimed that no five-room, eight-period schedule was possible at all. In the following, we show how this "inadmissibility" claim can now be easily substantiated in our adopted graph-theoretic context.

The equivalent question is the following: Given a graph $G = (V, E)$, an integer k, and a (possibly empty) subset $V' \subseteq V$ of vertices that are colored in a feasible way using no more than k colors, can the vertices in $V \backslash V'$ be colored yielding a feasible total coloring and so that, in all, no more than k colors are used? That is, with rooms identifiable by colors, we would assume certain class-vertices to be specified accordingly and then seek an appropriate "color completion" of the rest of the graph. Let us examine the Earth Day graph again, and further, suppose we encode rooms to colors as follows:

Room 1	=	RED
Room 2	=	BLUE
Room 3	=	GREEN
Room 4	=	WHITE
Room 5	=	ORANGE

Thus, we would fix the color of vertex A as BLUE, of D as GREEN, of I as RED, and vertex E would be WHITE. Now, in examining the partially colored graph, we observe that vertex H is adjacent to vertices A, D, I, and E so H cannot be one of the aforementioned colors. Let it be colored ORANGE. But vertex N, which is adjacent to H, and hence cannot be colored ORANGE, is also adjacent to the four other precolored vertices, meaning that it will require a sixth color. Thus, no five-color completion of the given partially colored graph is possible, or, stated equivalently, no five-room assignment of the Earth Day classes is possible given the prespecified room requirements.

Certainly, this preroom assignment, or, equivalently, k-color completion version of the classroom assignment problem, is hard in general (sets P_i do not necessarily form intervals). This follows trivially since the conventional version, dealt with in the previous section, is a special case with $V' = \phi$. But suppose now, rather than fixing a subset of specific classes to specific rooms, we allow some flexibility by associating a subset of classes with room "preferences." That is, for each class i, and a set of rooms S_i that are satisfactory relative to the conduct of the class, is there an admissible assignment that places each class in an acceptable room? This has been called the classroom assignment *satisfice* problem and is clearly a generalization of the so-called preroom assignment version just addressed.

Actually, this satisfice notion of having room preferences but some flexibility accordingly reflects a more realistic view of actual classroom assignment tasks. Unfortunately, the underlying combinatorial problem is difficult even under fairly restrictive conditions including adherence to the interval property.

Theorem 6.4 *The satisfice version of the classroom assignment problem is $\mathcal{NP}$-Complete even when the interval property holds.*

Proof. The reduction is from a special version of the satisfiability problem (SAT) which itself follows from a transformation from 3-satisfiability (3-SAT). In particular, Carter and Tovey show that an arbitrary instance of 3-SAT can be reduced to an instance of SAT where:

1. Each clause has either two or three literals.
2. No pair of 3-clauses (clauses with three literals, resp. 2-clauses) have a common literal.
3. No 2-clauses have a common literal.

To see this, consider an instance of 3-SAT and let x be a variable having a duplicated literal in the expression. Assume x appears in u clauses as x and is complemented in v clauses. Let $k = \max\{u, v\}$ and create new variables as $y_1, y_2, \ldots, y_k$. Now, replace the ith appearance of x with y_i for $1 \leq i \leq u$. Similarly, substitute $\bar{y}_j$ for the ith occurrence of $\bar{x}$ for $1 \leq j \leq v$. Add to the original expression clauses $\{y_i \vee \bar{y}_{i+1}\}$ for $1 \leq i \leq k-1$ and the clause $\{y_k \vee \bar{y}_1\}$. In this updated expression, it is easy to see that conditions 1–3 are met and that the overall transformation is polynomial. It is also easy to see that clause $\{y_i \vee \bar{y}_{i+1}\}$ implies that if y_i is false, $\bar{y}_{i+1}$ must also be false. Further, the cyclic nature of the clauses forces variables y_i to be either all true or all false.

Now, from an instance of this special version of SAT, let us construct an instance of this satisfice classroom assignment problem. In this regard, assume there are n variables x_i each of which represents two classrooms which we label as T_i and F_i. This yields $2n$ total rooms. Now, for each 2-clause in our special SAT expression, let us define a class c_j^2 that must be scheduled in period 1. Each of these has a two-room preference corresponding to the two literals in the stated clause. For each 3-clause, define a class c_j^3 requiring period 2 and having a three-room preference. Note that none of the period-1 classes have a preferred room in common, and similarly period-2 classes. This follows directly from conditions 2 and 3 above.

Define now, n classes, L_i that meet for both of the two periods and prefer rooms T_i and F_i. The role of these L_i classes is to force each variable in the SAT expression to be always true (if L_i uses F_i) or always false (if L_i uses T_i). So, if class L_i gets room T_i then every other class c_j^2 and c_j^3 must only use room F_i.

This completes the construction. It is easy to see that a feasible solution exists for the satisfice classroom assignment instance if and only if the corresponding SAT instance is satisfiable. □

From the proof of Theorem 6.4, we also have an immediate corollary:

Corollary 6.5 *The satisfice version of the classroom assignment problem subject to the interval property remains hard even when instances are restricted to two periods.* □

With this outcome, we conclude our coverage of the classroom assignment problem. It is worth pointing out, however, that some interesting special cases of the problem (submitting to polynomial-time algorithms) are also described in the previously cited work of Carter and Tovey along with a nice summary of what is known generally about various aspects of the basic problem.

6.2 Staffing Problems

In this section we consider the so-called *staffing problem*. Here, the objective is to size and schedule a minimum cost workforce in such a way that sufficient workers are employed during each time period of some fixed, planning cycle. Again, side constraints may be present. For example, we might be interested in producing a least cost assignment of workers to shifts of a weekly or seven-day schedule so that enough are present each day in order to meet predetermined daily requirements and, in addition, so that each worker is given a fixed number of days off. The resultant "schedule" of workers then repeats in this way from week to week.

More formally, suppose each cycle consists of m periods and with b_i workers required in period i. Each worker is assigned a "shift" consisting of a fixed number of, say $k < m$ work days with the remaining $m - k$ to be non-work or idle days. Then, letting x_j be the number of workers assigned to shift j at a cost c_j per worker assigned, the problem can be expressed by P_{SC} as shown below:

$$\begin{aligned} P_{SC}: \quad \min \quad & \boldsymbol{cx} \\ s.t. \quad & \boldsymbol{Ax} \geq \boldsymbol{b} \\ & \boldsymbol{x} \geq 0 \text{ and integer} \end{aligned}$$

Here, $\boldsymbol{A}$ is a 0–1 matrix with each column representing a shift; a 1 indicates a work day and a 0 denotes a non-work or idle day.

Sometimes the problem is stated in a more combinatorial context: Given a collection C of m-tuples, each having k 1s and $m-k$ 0s, we want to find a least cost selection of these, say $f^* : C \rightarrow \mathbb{Z}^{\geq 0}$ such that $\sum_{\bar{c} \in C} f^*(\bar{c}) \cdot \bar{c} \geq \boldsymbol{b}$. In addition, it should come as no surprise that the problem is $\mathcal{NP}$ -Hard (*cf.* Garey and Johnson, 1979). An easy illustration follows.

Example 6.2 Suppose four periods constitute a work cycle with shifts and daily requirements defined as follows:

shift	work periods	requirements
1	1, 3	5
2	1, 2	3
3	2, 3	3
4	1, 4	2

Let the shift costs be given as $\mathbf{c} = (2, 2, 2, 6)$. Then P_{SC} results as

$$\begin{array}{ll} \min & 2x_1 + 2x_2 + 2x_3 + 6x_4 \\ s.t. & \\ & \begin{bmatrix} 1 & 1 & 0 & 1 \\ 0 & 1 & 1 & 0 \\ 1 & 0 & 1 & 0 \\ 0 & 0 & 0 & 1 \end{bmatrix} \boldsymbol{x} \geq \begin{bmatrix} 5 \\ 3 \\ 3 \\ 2 \end{bmatrix} \\ & \boldsymbol{x} \geq 0 \text{ and integer} \end{array}$$

having optimal solution $\boldsymbol{x}^* = (2, 2, 1, 2)$. That is, there will be two workers assigned to shifts 1, 2, and 4 while only one worker is relegated to the third shift. The cost of this assignment is 22. The schedule is presented in Figure 6.4. Note that there is at least one other optimal staffing: (1,2,2,2). □

The solution to the instance of Example 6.2 was found in an elementary way, namely by a standard branch-and-bound approach with linear relaxations acting as subproblems and branching defined by fractional-valued variables accordingly (we certainly cannot expect solutions of the LP relaxation of P_{SC} to be integral). But while there may be more clever ways to implement this sort of (exponential) procedure, its general application is not called into question owing to the problem's $\mathcal{NP}$-Hard status. Still, this does not mean that there are no interesting solvable cases. The next example is designed to motivate this possibility.

Example 6.3 Let us assume a seven-day cyclic schedule is to be created where each shift worker is to be given two days off, and further, let these two days be consecutive. Suppose we want to use the minimum number of workers while meeting the daily requirements given by $\boldsymbol{b} = (7, 7, 7, 7, 7, 7, 7)$.

4			
4			
2			
2	3	3	
1	2	1	4
1	2	1	4
day 1	day 2	day 3	day 4

Figure 6.4. *Staffing of Example 6.2*

In the notation above, $m = 7$, $k = 5$, and $\boldsymbol{c}$ is a vector of 1s. The matrix $\boldsymbol{A}$ is shown below.

$$\boldsymbol{A} = \begin{bmatrix} 0 & 1 & 1 & 1 & 1 & 1 & 0 \\ 0 & 0 & 1 & 1 & 1 & 1 & 1 \\ 1 & 0 & 0 & 1 & 1 & 1 & 1 \\ 1 & 1 & 0 & 0 & 1 & 1 & 1 \\ 1 & 1 & 1 & 0 & 0 & 1 & 1 \\ 1 & 1 & 1 & 1 & 0 & 0 & 1 \\ 1 & 1 & 1 & 1 & 1 & 0 & 0 \end{bmatrix}$$

Now, if we solve the corresponding linear relaxation of P_{SC} relative to this instance, we obtain the solution $\bar{\boldsymbol{x}} = (1.4, 1.4, 1.4, 1.4, 1.4, 1.4, 1.4)$ which, naturally enough, makes no sense in the context of a staff schedule. But let us perform the following "round-off" computation using $\bar{\boldsymbol{x}}$: Set $x_1^* = \lceil \bar{x}_1 \rceil$ and $x_j^* = \lceil \bar{x}_1 + \bar{x}_2 + \ldots + \bar{x}_j \rceil - \lceil \bar{x}_1 + \bar{x}_2 + \ldots + \bar{x}_{j-1} \rceil$ for $j = 2, 3, \ldots, n$. In fact, in the current instance, this process produces an optimal solution, $\boldsymbol{x}^* = (2, 1, 2, 1, 1, 2, 1)$. The 10 workers are scheduled throughout the cycle as shown in Figure 6.5. In fact, if this simple LP-rounding idea could be shown to be correct and, in particular, if it is correct for any interesting class of staffing problems, we would have a nice outcome. Fortunately, the procedure *is* valid as we will show, following work described in an interesting paper of Bartholdi *et al.* (1980). □

Suppose $\boldsymbol{a}$ is a 0-1 vector. Then we say that $\boldsymbol{a}$ is *circular* if its 1s appear consecutively. Note that we consider the first and last positions in the vector to be consecutive, *e.g.*, the vectors (0,1,1,1,1,1,0) and (1,1,0,0,0,1,1) are circular while (0,1,1,1,0,0,1) is not. A matrix is *column circular* (resp., *row circular*) if its columns (resp., rows) are circular. The matrix $\boldsymbol{A}$ in Example 6.3 is both row and column circular.

				7		
6	7	7	7	6	7	5
6	6	6	6	6	4	4
5	6	6	6	3	3	3
4	5	5	5	3	3	3
3	4	4	2	2	2	2
3	3	1	1	1	1	1
2	3	1	1	1	1	1
day 1	day 2	day 3	day 4	day 5	day 6	day 7

Figure 6.5. *Staffing of Example 6.3*

Now, let us assume P_{SC} to be the same as before but with $\boldsymbol{A}$ in row-circular form (whether or not arising from a so-called m, k-staff scheduling problem) having at least one 1 in each row. We also assume $\boldsymbol{c}$ and $\boldsymbol{b}$ to be integral vectors. Now, let $\boldsymbol{T}$ be a transformation matrix that is nonsingular and unimodular (every basis submatrix has determinant +1, -1, or 0), and define a change of variables as $\boldsymbol{x} = \boldsymbol{T}\boldsymbol{y}$. Then P_{SC} can be restated as P_{SCT} below:

$$P_{SCT}: \quad \min \quad (\boldsymbol{CT})\boldsymbol{y}$$
$$s.t.$$
$$\begin{bmatrix} \boldsymbol{AT} \\ \boldsymbol{T} \end{bmatrix} \boldsymbol{y} \geq \begin{bmatrix} \boldsymbol{b} \\ \boldsymbol{0} \end{bmatrix}$$
$$\boldsymbol{y} \text{ unrestricted, integer}$$

Following from the properties of $\boldsymbol{T}$, it is easy to see that $\boldsymbol{x} = \boldsymbol{T}\boldsymbol{y}$ solves P_{SC} if and only if $\boldsymbol{y} = \boldsymbol{T}^{-1}\boldsymbol{x}$ solves P_{SCT}.

Let $\boldsymbol{T}$ be a matrix of the following form:

$$\begin{bmatrix} 1 & 0 & \cdots & \cdots & 0 \\ -1 & 1 & & & \vdots \\ 0 & -1 & & & \vdots \\ \vdots & \vdots & & & 0 \\ 0 & 0 & \cdots & \cdots & 1 \end{bmatrix}$$

Note that such a $\boldsymbol{T}$ is nonsingular and unimodular and results in the change

$$x_1 = y_1; x_j = y_j - y_{j-1} \quad \text{for } j = 2, 3, \ldots, n.$$

Conversely, from the inverse $\boldsymbol{T}^{-1}$ we have

$$y_j = x_1 + x_2 + \ldots + x_j \quad \text{for } j = 1, 2, \ldots, n,$$

with $y_n = x_1 + \ldots + x_n$ representing the total workforce size.

Now, letting the column in the constraint matrix of $\begin{bmatrix} \boldsymbol{AT} \\ \boldsymbol{T} \end{bmatrix}$ corresponding to y_n be $\bar{\boldsymbol{a}}_n$, we can rewrite the matrix as $\begin{bmatrix} \boldsymbol{AT} \\ \boldsymbol{T} \end{bmatrix} = [\bar{\boldsymbol{A}}, \bar{\boldsymbol{a}}_n]$. Similarly, we can rewrite $\boldsymbol{CT}$ as $[\bar{\boldsymbol{c}}, \bar{\boldsymbol{c}}_n]$ and $\boldsymbol{y} = (\bar{\boldsymbol{y}}, y_n)$ where $\bar{\boldsymbol{y}} = (y_1, y_2, \ldots, y_{n-1})$. The full problem, under this restructuring, becomes:

$$\begin{array}{lll} P_{\bar{S}\bar{C}\bar{T}} & \min & \bar{c}\bar{y} + \bar{c}_n y_n \\ & s.t. & \bar{A}_{\bar{y}} + \bar{a}_n y_n \geq \begin{bmatrix} \boldsymbol{b} \\ \boldsymbol{0} \end{bmatrix} \\ & & \bar{\boldsymbol{y}}, y_n \text{ unrestricted and integer.} \end{array}$$

But if we observe the matrix $\boldsymbol{A}$ above, we see that it has at most one 1 and at most one -1 in each row with all other elements equal to 0. This is particularly fortunate for we know that such an $\boldsymbol{A}$ is the vertex-arc incidence matrix of a digraph and is thus totally unimodular. This means that if we fix y_n in $P_{\bar{S}\bar{C}\bar{T}}$, then solving the corresponding LP-relaxation will result in an integer solution. From this we can, in turn, produce a suitable $\boldsymbol{x}$ by the stated transformation.

Example 6.4 Consider the instance of Example 6.3 where $\begin{bmatrix} \boldsymbol{AT} \\ \boldsymbol{T} \end{bmatrix}$ results as shown in Figure 6.6. The last column in the array is $\bar{a}_n$ and $\begin{bmatrix} \boldsymbol{b} \\ \boldsymbol{0} \end{bmatrix}$ is shown on the right. Let us set $y_n = 10$. Then, solving the LP given by $P_{\bar{S}\bar{C}\bar{T}}$ produces values of 1,2,4,5,7, and 8 for variables y_1 through y_6 respectively. Finally, solving for the shift variables as $\boldsymbol{x} = \boldsymbol{Ty}$ produces the values (1,1,2,1,2,1,2). Note that comparison of this vector with the outcome in the earlier illustration reveals an alternative optimal solution. □

Of course, the demonstration in the example shown in Figure 6.6 contains a bit of "cheating" in that the value of y_n was fixed ahead of time. In fact, we would have to search for the smallest value of y_n which when fixed in $P_{\bar{S}\bar{C}\bar{T}}$ yields a feasible solution. This can be accomplished by performing a binary search over the closed interval $[0, \mathbf{1b}]$. There are some (fairly standard) details involved in verifying the correctness of this process but we will not consider them here.

In any event, it turns out that we can solve the present problem without this sort of parametric search. The strategy is predicated upon the following result:

Theorem 6.6 *If $(\bar{y}'^*, y_n'^*)$ is an optimal solution for the linear relaxation of $P_{\bar{S}\bar{C}\bar{T}}$, then there exists an integer-optimal solution $(\bar{y}^*, y_n^*)$ with $|y_n'^* - y_n^*| < 1$.* □

$$\begin{bmatrix} AT \\ T \end{bmatrix} = \begin{bmatrix} -1 & 0 & 0 & 0 & 0 & 1 & 0 \\ 0 & -1 & 0 & 0 & 0 & 0 & 1 \\ 1 & 0 & -1 & 0 & 0 & 0 & 1 \\ 0 & 1 & 0 & -1 & 0 & 0 & 1 \\ 0 & 0 & 1 & 0 & -1 & 0 & 1 \\ 0 & 0 & 0 & 1 & 0 & -1 & 1 \\ 0 & 0 & 0 & 0 & 1 & 0 & 0 \\ 1 & 0 & 0 & 0 & 0 & 0 & 0 \\ -1 & 1 & 0 & 0 & 0 & 0 & 0 \\ 0 & -1 & 1 & 0 & 0 & 0 & 0 \\ 0 & 0 & -1 & 1 & 0 & 0 & 0 \\ 0 & 0 & 0 & -1 & 1 & 0 & 0 \\ 0 & 0 & 0 & 0 & -1 & 1 & 0 \\ 0 & 0 & 0 & 0 & 0 & -1 & 1 \end{bmatrix} \quad \begin{bmatrix} b \\ 0 \end{bmatrix} = \begin{bmatrix} 7 \\ 7 \\ 7 \\ 7 \\ 7 \\ 7 \\ 7 \\ 0 \\ 0 \\ 0 \\ 0 \\ 0 \\ 0 \\ 0 \end{bmatrix}$$

Figure 6.6. *Constraint matrices of Example 6.4*

We leave the proof of this theorem in Bartholdi *et al.* (1980) but it should not be difficult to see that it leads to a nice round-off result.

A_{SC}: Round-Off Algorithm for the Cyclic Staffing Problem

Step 1: Solve the linear relaxation of P_{SC} and let this solution be x'. If x' is integer we are done; otherwise, proceed to Step 2.

Step 2: Create two linear programs from the relaxation of Step 1. In the first of these, add the constraint $x_1+x_2+\cdots+x_n = \lfloor x'_1+x'_2+\cdots+x'_n \rfloor$ and in the second, add $x_1+x_2+\cdots+x_n = \lceil x'_1+x'_2+\ldots+x'_n \rceil$. The solutions to these linear programs can be taken to be integral with the better of the two yielding a solution to P_{SC}. □

The reader may want to apply A_{SC} to the instance of Example 6.2. The two constraints to be added to the linear relaxation there are $x_1+x_2+x_3+x_4 = 6$ and $x_1+x_2+x_3+x_4 = 7$.

Interestingly, when c is a 1-vector (the problem is to find a minimum size work force), our strategy is simpler still. This follows from an easy result:

Theorem 6.7 *Let y' be an optimal solution to the linear relaxation of P_{SCT} where $\mathbf{c}$ is a vector of unit costs. Then an optimal solution to the original integer problem is given by*

$$y^* = (\lceil y'_1 \rceil, \lceil y'_2 \rceil, \ldots, \lceil y'_n \rceil).$$

Proof. The objective function is $y_n = x_1+x_2+\ldots+x_n$. Hence, the minimum objective function value for P_{SCT} is at least $\lceil y^*_n \rceil$ and so if we can show that $\lceil y_n^* \rceil$ is feasible we are done. Recall that by the construction of $\boldsymbol{T}$ we have at most three forms for the constraints in P_{SCT}. These are:

- $y_j \geq b_i$
- $y_i - y_k \geq b_i$
- $y_j - y_k + y_n \geq b_i$.

But since the y-values are real numbers, we have

- $\lceil y_j^* \rceil \geq y_j^* \geq b_i$
- $\lceil y_j^* \rceil - \lceil y_k^* \rceil \geq \lfloor y_i^* - y_k^* \rfloor \geq b_i$
- $\lceil y_j^* \rceil - \lceil y_k^* \rceil + \lceil y_n^* \rceil \geq \lceil y_j^* + y_n^* \rceil - \lceil y_k^* \rceil \geq \lfloor y_j^* - y_k^* + y_n^* \rfloor \geq b_i$,

and the desired feasibility is verified. This completes the proof. □

So, in the case of unit costs, we can modify algorithm $\boldsymbol{A_{SC}}$ by replacing the previous Step 2 with the following version:

> **Step 2:** Determine an optimum integer solution as $x_1^* = \lceil x_1' \rceil$ and $x_j^* = \lceil x_1' + x_2' + \ldots + x_j' \rceil - \lceil x_1' + x_2' + \ldots + x_{j-1}' \rceil$ for $j = 2, 3, \ldots, n$. □

The reader can verify that this application was enforced regarding the outcome in Example 6.3.

In concluding this section, it is worth noting that the nice solution capabilities described above are directly dependent on the row-circularity property exhibited by our cyclic-scheduling constraint matrix $\boldsymbol{A}$. But this property follows for the so-called m, k-shift scheduling problem directly from its definition. (This is analogous to claiming a bipartite property for the underlying graph of an assignment problem.) Indeed, the theorems and algorithms of this section are certainly valid for any such $\boldsymbol{A}$. The only catch is that we must be able to recognize (quickly) if an arbitrary 0-1 matrix is row-circular (under some permutation of columns). For example, the matrix on the left in Figure 6.7, while not in row-circular form, can be made so by a rearrangement of columns resulting in the array shown on the right. Fortunately, the issue of deciding if the columns of an $m \times n$, 0-1 matrix can be permuted to produce the row-circular property can be resolved efficiently following an $O(m^2 n)$ algorithm of Tucker (1971).

6.3 Timetabling Problems

In this final section, we consider scheduling problems arising in a setting frequently referred to as *timetabling.* As before, it is easiest to describe the problem by employing an illustration. Suppose we are given a set of tutors and another set consisting of distinct "lessons," where the intent is

$$\begin{bmatrix} 0 & 1 & 1 & 1 & 0 & 0 & 1 & 1 \\ 1 & 0 & 1 & 1 & 0 & 1 & 0 & 1 \\ 1 & 0 & 0 & 0 & 1 & 1 & 0 & 1 \\ 0 & 1 & 1 & 0 & 1 & 1 & 1 & 0 \end{bmatrix} \begin{bmatrix} 1 & 1 & 0 & 0 & 0 & 1 & 1 & 1 \\ 1 & 1 & 1 & 1 & 0 & 0 & 0 & 1 \\ 0 & 1 & 1 & 1 & 1 & 0 & 0 & 0 \\ 0 & 0 & 0 & 1 & 1 & 1 & 1 & 1 \end{bmatrix}$$

Figure 6.7. *Row circularity determination*

to produce a schedule or "timetable" matching the tutors to lessons and hence to their pupils for one-on-one instruction in periods of fixed length (*e.g.*, 1-hour lessons, etc.). Inherent in this description is the notion that a given lesson, *e.g.*, beginning violin instruction, advanced cello, etc.) may be required by several students; however, each student meets individually with the particular tutor. Thus, each lesson is conducted by exactly one tutor and also, in any period, we assume that at most one session of a given lesson type can be scheduled.

More formally, let the tutors be denoted by $x_1, x_2, \ldots, x_m$, and the lessons by $y_1, y_2, \ldots, y_n$. In addition, let the number of students who are to be tutored by teacher x_i in lesson y_j be given by r_{ij} where all such requirements are captured by the matrix $\boldsymbol{R} = [r_{ij}]$ for $1 \leq i \leq m$ and $1 \leq j \leq n$. The objective is to create a timetable for the tutelage of all students using the fewest number of 1-hour periods. A numerical example may serve to clarify.

Example 6.5 Assume four tutors must offer instruction in five lessons with specific requirements shown below:

$$\boldsymbol{R} = \begin{array}{c} \\ x_1 \\ x_2 \\ x_3 \\ x_4 \end{array} \begin{array}{c} \begin{array}{ccccc} y_1 & y_2 & y_3 & y_4 & y_5 \end{array} \\ \begin{pmatrix} 1 & 0 & 1 & 0 & 2 \\ 1 & 0 & 1 & 0 & 0 \\ 0 & 1 & 0 & 1 & 0 \\ 1 & 1 & 0 & 1 & 1 \end{pmatrix} \end{array}$$

For example, tutor x_1 must instruct one student in lesson y_1, another one in lesson y_3, and two students in lesson y_5, while not giving instruction in subjects y_2 or y_4. Now, if we examine the row requirements in $\boldsymbol{R}$ we see that at least four periods will be required in any timetable, *i.e.*, tutors x_1 and x_4 must give a total of four periods of instruction. Thus any timetable using only four periods must be optimal since this value (the maximum of row and column sums in $\boldsymbol{R}$) provides a lower bound. Fortunately, we can form such a timetable as exhibited in Figure 6.8. In this specific case, note also that within the four-period schedule there is some flexibility in how instructors x_2 and x_3 may conduct their tutorials. □

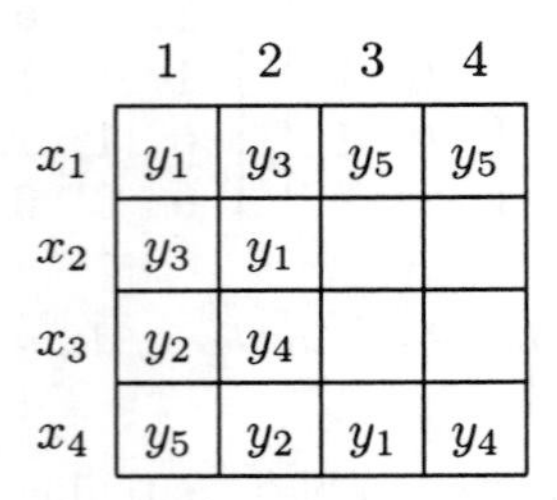

	1	2	3	4
x_1	y_1	y_3	y_5	y_5
x_2	y_3	y_1		
x_3	y_2	y_4		
x_4	y_5	y_2	y_1	y_4

Figure 6.8. *Timetable for Example 6.5*

6.3.1 Edge-Coloring and Class-1 Graphs

As it turns out, there is a complete solution to the timetabling problem– at least for the version described above. Helpful in recognizing as well as appreciating this result is (i), a (somewhat obvious) reformulation of the basic problem and (ii), a little knowledge from the theory of edge-colorings. We examine first the reformulation.

Let us construct a bipartite graph G having bipartition $\{X, Y\}$ where $X = \{x_1, x_2, \ldots, x_m\}$ and $Y = \{y_1, y_2, \ldots, y_n\}$. Edges in G are formed from $\boldsymbol{R}$ such that between vertex x_i and vertex y_j, r_{ij} edges are created. Accordingly, with the instance of the previous illustration, the graph in Figure 6.9 would result.

Now, under our assumption that each tutor can teach at most a single lesson in any given period and each lesson is taught no more than once in any period by at most one tutor, it should be clear that every admissible schedule corresponds to a matching in the corresponding bipartite graph and conversely. It follows then that our problem is identical to finding a partition of the edge set of our bipartite graph into as few matchings as possible. This equivalence is demonstrated for the current example by listing four matchings in Figure 6.10 (trivially, this is minimum since the degree of vertices x_1 and x_4 is four). The reader should have no difficulty in seeing how the schedule shown previously was constructed from these matchings.

But thus far our achievement has been essentially cosmetic. We've changed only the language of the problem where rather than looking for a schedule of tutors to lessons *per se*, we now seek a collection of edge-disjoint matchings in a particular bipartite graph. However, if we recall the problem of edge-colorings in graphs, and specifically, edge-coloring of bipartite graphs, we should begin to see how our timetabling problem might be resolved. Specifically, any admissible edge-coloring of a graph clearly induces

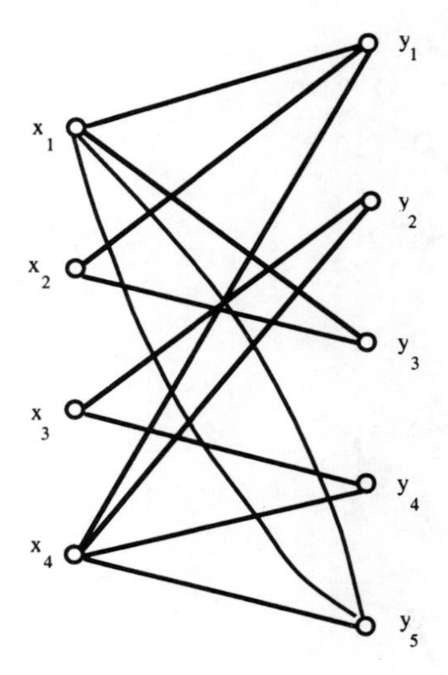

Figure 6.9. *Sample graph formed from R*

a collection of matchings. Thus, the least number of colors required for an admissible edge-coloring corresponds to a smallest collection of matchings accordingly. Hence, the fewest number of periods required to schedule the students with their tutors is exactly the chromatic index of the relevant bipartite graph. Clearly, in the example above, any assignment of colors to the matchings shown (one color per matching) yields a legitimate edge-coloring of the original bipartite graph indicating its chromatic index to be four.

But Vizing showed that the chromatic index of any graph is always restricted to one of two values; either it takes the value of the graph's maximum vertex degree or one more than this. We thus have a graph-theoretic interpretation of the earlier lower bound claim regarding the minimum number of timetable periods required, *i.e.,* the maximum degree in the stated bipartite graph constructed from $\boldsymbol{R}$ is

$$p = \max\left(\max_i \sum_j r_{ij}, \max_j \sum_i r_{ij}\right).$$

Now, among edge-colorists, a graph that has chromatic index exactly equal to the maximum degree is called a *class-1* graph. Graphs which require the higher value are, naturally enough, referred to as *class-2.* The important observation for us, however, is that all bipartite graphs are class-

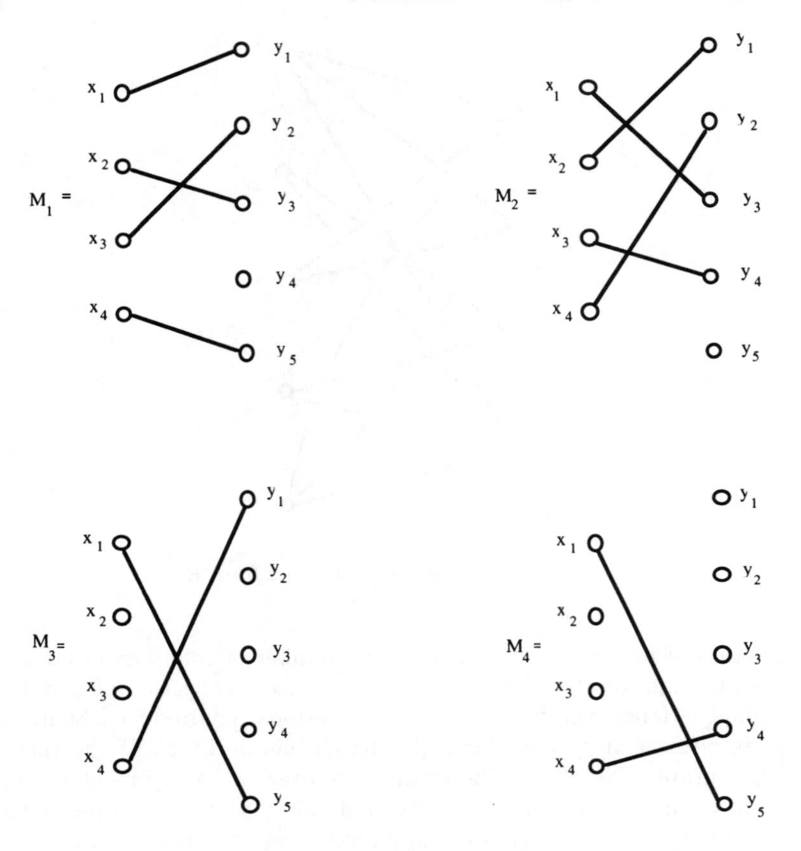

Figure 6.10. *Disjoint matchings of graph in Figure 6.9*

1. (See Exercise 6-26). So, from the timetabling perspective, if no tutor is committed to more than p periods and if no lesson must be taught more than p times, then a p-period timetable is *always* possible. Also important is that finding such an edge coloring in a bipartite graph and thus producing the resultant timetable is easy. Again, readers are asked to establish this for themselves (see Exercise 6-27).

6.3.2 Modifications

Let us now complicate matters by placing a limit on the number of tutoring sessions that can be conducted simultaneously. This sort of consideration might arise if the number of rooms available for tutoring is restricted (we assume that each room is reserved for one tutor-student lesson per period). For example, in our illustration, suppose only three rooms are available rather than four. In this case, the schedule during the first two periods as indicated is not admissible.

It turns out that our good fortune still holds, however. To see this, consider the following schedule. We can still complete all sessions in a total of four periods but now only three rooms will be required.

x_1	y_1	y_5	y_3	y_5
x_2	y_3		y_1	
x_3		y_2		y_4
x_4	y_5	y_1	y_4	y_2

But the schedule just constructed was not unanticipated. In fact its existence was guaranteed by the following result.

Theorem 6.8 *Let G be a bipartite graph with edge set E and with maximum degree $\triangle$. Then if $p \geq \triangle$, there exists p edge-disjoint matchings of G say $M_1, M_2, \ldots, M_p$ such that*

$$E = M_1 \cup M_2 \cup \ldots \cup M_p$$

and, for $1 \leq i \leq p$

$$\left\lfloor \frac{|E|}{p} \right\rfloor \leq M_i \leq \left\lceil \frac{|E|}{p} \right\rceil .$$

Now, we know that $|E|$ tutor–pupil sessions are required overall and so for a p-period timetable, we must have at least $\left\lceil \frac{|E|}{p} \right\rceil$ rooms available in some period. But it follows from the theorem that if this number of rooms is available, it is always possible to find a feasible p-period schedule in which at most $\left\lceil \frac{|E|}{p} \right\rceil$ rooms are simultaneously utilized.

We will give the proof of this theorem following that in Bondy and Murty (1976). Accordingly, we first need a technical lemma:

Lemma 6.9 *Suppose M and $\bar{M}$ are disjoint matchings in G and where $|M| > |\bar{M}|$. Then there are disjoint matchings M' and $\bar{M}'$ of G such that $|M'| = |M| - 1$, $|\bar{M}'| = |\bar{M}| + 1$ and $M' \cup \bar{M}' = M \cup \bar{M}$.*

Proof. Let M and $\bar{M}$ be two matchings with $|M| > |\bar{M}|$. Now, if we form the graph $M \cup \bar{M}$ we see that each connected component of the resultant structure is either an even cycle or a path with edges alternately in M and $\bar{M}$. But by hypothesis $|M| > |\bar{M}|$ so some component that is a path must be an $\bar{M}$-augmenting path (begins and ends with edges in M). If we alter

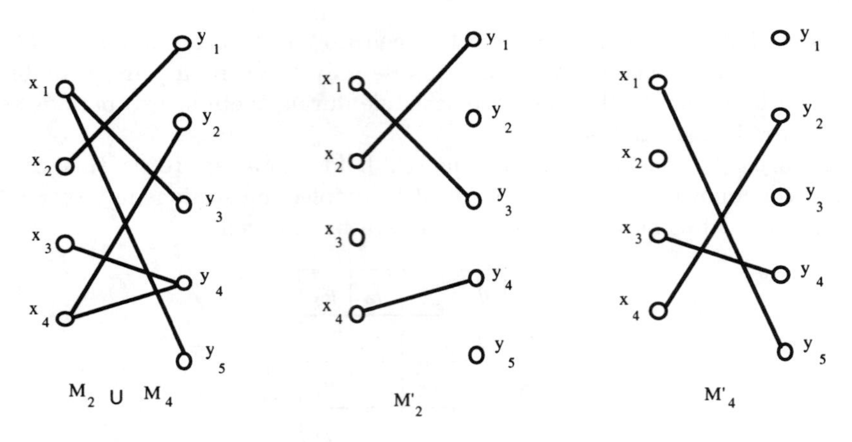

Figure 6.11. *Construction of Lemma 6.9*

matching edges in this path accordingly, we will form two new matchings M' and $\bar{M}'$ with $|M'| = |M| - 1$ and $|\bar{M}'| = |\bar{M}| + 1$ as claimed. □

We can illustrate the result of this lemma by returning to the four matchings given earlier. If we form $M_2 \cup M_4$ as shown in Figure 6.11, we quickly see that an augmenting path of the sort described in the lemma is given by the vertex sequence x_3, y_4, x_4, y_2 [the edge (x_2, y_1) forms another]. Upon altering the composition of this path as indicated, we are able to produce two new matchings both having three edges as anticipated by the lemma. These are also shown in Figure 6.11. We then can repeat the construction using M_1 and M_3, producing again, a pair of three-edge matchings which along with the other pair give us the four matchings that created the correct timetable shown above. We now return to the main result.

Proof of Theorem: Since G is bipartite we know that it is Class-1 and its edge set can be partitioned into $\triangle$ edge-disjoint matchings. Thus, for any $p \geq \triangle$ there is a partition of the edge with p such matchings. That is, we can form

$$E = M_1' \cup M_2' \cup \ldots \cup M_p',$$

where some M_i may be empty (*i.e.*, $i > \triangle$). But the bounds given above on the size of any M_i indicate that any two such matchings differ in cardinality by at most 1. Thus, we can repeatedly apply Lemma 6.9 to any pair of matchings M_i' and M_j' that differ in size by more than 1 and we will ultimately produce the desired matchings $M_1, M_2, \ldots, M_p$. □

Let us demonstrate the sense of the theorem with the current instance.

Example 6.6 Since $|E| = 12$ in our bipartite graph and for $p = 4$, the result of Theorem 6.8 shows that the four-period schedule with three lessons in each period was to be expected. Now, suppose we constrain the same instance even further by allowing only two rooms for any timetable period. Clearly, any admissible table will need at least six periods. But by Theorem 6.8, however, we see that it is possible to create such a timetable since

$$\left\lfloor \frac{12}{6} \right\rfloor \leq M_i \leq \left\lceil \frac{12}{6} \right\rceil .$$

Following, we exhibit a suitable schedule directly. The interested reader may wish to form the actual matchings first.

x_1	y_1		y_3		y_5	y_5
x_2		y_3		y_1		
x_3		y_2	y_4			
x_4	y_5			y_2	y_1	y_4

□

The reader will note that Theorem 6.8 establishes an *existence* property. As intimated, it remains that the stated matchings and thus the relevant timetable has to ultimately be exhibited if we are to actually resolve the scheduling problem. But this can easily be done following the inherently constructive nature of the proof of the theorem as well as the preceding lemma. Again, we leave this as an exercise for the reader.

We conclude the section by making note of a particularly interesting "real-world" application of the timetabling model(s) just presented. Specifically, in Bartholdi and McCroan (1990), a problem of scheduling interviews for a job fair is described. The setting is public interest law (*e.g.*, legal services for the poor, public defense, etc.) which tends to involve firms generally unable, financially, to support substantial independent recruiting, and in turn that attract different sorts of students than are likely to frequent so-called corporate job fairs. The subject of the Bartholdi–McCroan study is the Southeastern Public Interest Job Fair which is held annually in Atlanta and that is sponsored jointy by the Atlanta Legal Aid Society and the Atlanta Bar Association. The fair typically draws 25–50 law firms from across the country and 100–200 law students from all over the Southeast.

A salient attribute of this work is its evident "systems" view of the particular, job-fair timetabling interpretation. While predicated upon a structural perspective not unlike that described in this section (*i.e.*, edge-coloring) a distinguishing characteristic here is the development of a user-friendly computer program that produces, as the authors' suggest, "convenient" schedules for law firm-student interviews. The efficacy of the model is corroborated by its continued use at the aforementioned Atlanta fair since

1985 with the exception of 1993, when the market for lawyers was so poor that the fair was cancelled; some things are apparently beyond the help of even a clever algorithm.

6.4 Exercises

6-1 A set of classes and their respective period requirements is shown below. Find a classroom assignment using the fewest number of rooms. Give a convincing (and correct) argument that the minimum has been achieved.

Class	A	B	C	D	E	F	G
Periods	$(1, 2)$	$(1, 4, 8)$	$(3, 7, 8, 10)$	$(4, 5, 6, 7)$	$(5, 9)$	$(2, 3)$	$(6, 9, 10)$

6-2 Repeat Exercise 6-1 on the instance below:

Class	A	B	C	D	E	F	G	H
Periods	1	$(6, 7, 8)$	$(3, 4, 5, 6, 7)$	$(8, 9, 10)$	$(7, 8, 9, 10)$	6	$(5, 6)$	3

6-3 Prove that at least three rooms are needed to schedule the five classes in the periods indicated as follows:

Class	A	B	C	D	E
Periods	$(1, 4)$	$(1, 3)$	$(2, 4)$	$(3, 5)$	$(2, 5)$

6-4 Relative to the instance of Exercise 6-3, suppose we add a sixth class which requires periods 3 and 6 and alter the period requirement of Class D to (6,5). Show that with this change, a six-period schedule is possible using only two rooms.

6-5 Give a three-room schedule for the instance below:

Class	A	B	C	D	E	F	G
Periods	$(1, 6, 7)$	$(2, 4, 5)$	$(3, 5, 8, 9)$	$(1, 2)$	$(3, 6)$	$(7, 8)$	$(4, 9)$

6-6 Assume that classes D, E, and F of the instance in Exercise 6-5 must be in different rooms. Is a three-room schedule possible now?

6-7 Consider the following instance:

Class	A	B	C	D	E	F	G
Periods	1	4	$(2, 6, 7)$	$(1, 2, 8)$	$(6, 8, 9)$	5	$(4, 5, 7)$

Now, suppose that classes A, C, F, and B must be in the same room. Does a three-room schedule exist? If so, give it; if not, indicate why.

6-8 In Section 6.1.2, a so-called k-color completion problem was described and also stated to be hard. In this regard, show that the problem remains hard even when k is fixed at 4 and when instances are restricted to planar graphs.

6-9 Give a perfect graph and indicate why it is perfect. Give a graph that is not perfect, showing why.

6-10 From the perfect graph formed in Exercise 6-9, form a contiguous period, classroom assignment instance.

6-11 Given a list of classes and their required periods, we defined a graph construction that rendered the question of deciding if a k-room schedule is possible equivalent to that of asking if the relevant graph is k-colorable. Give some graph classes for which the coloring question is well-solved. Then, from a member of each, create a corresponding class/period instance and solve it by determining the chromatic number of the graph; give a suitable schedule.

6-12 Illustrate the construction of Theorem 6.4 on the 3-SAT instance below:

$$(x_1 \vee \bar{x}_2 \vee x_3) \wedge (\bar{x}_1 \vee x_2) \wedge (\bar{x}_2 \vee x_3)$$

6-13 Repeat Exercise 6-12 relative to the following 3-SAT instance:

$$(\bar{x}_1 \vee x_3) \wedge (x_2 \vee \bar{x}_3) \wedge (x_1 \vee x_2 \vee \bar{x}_3) \wedge (\bar{x}_1 \vee \bar{x}_2 \vee x_3) \wedge (x_1 \vee \bar{x}_2)$$

6-14 Suppose in Exercise 6-5, four rooms, numbered as 1, 2, 3, and 4, are available. Classes A, B, C, and G can be in any of the four; however, classes D, E, and F have limited flexibility as follows:

Class D: rooms 1 or 4

Class E: rooms 1, 2, or 3

Class F: rooms 2 or 4.

Given these requirements, find an optimal (*i.e.*, least room) assignment.

6-15 Solve the five-day staffing instance given below with $\mathbf{c} = (3, 2, 5, 4, 3)$:

shift	work periods	daily requirements
1	1,2,4	6
2	1,3,4	5
3	2,3,5	7
4	2,3,4	6
5	1,3,5	4

6-16 Find the minimum number of workers to satisfy the seven-day staffing requirements shown below:

shift	work periods	daily requirements
1	1,2,3,5,6	8
2	1,4,5,6,7	9
3	3,4,5,6,7	5
4	1,4,5,6,7	7
5	1,2,3,4,7	6
6	2,3,4,5,6	7
7	2,3,5,6,7	8

6-17 Consider the cyclic instance in Example 6.3. Keep parameters m and k fixed at 7 and 5 but change $\boldsymbol{b}$ to (6,5,7,4,7,5,4). Solve using the round-off computation indicated in the example.

6-18 Repeat Exercise 6-17 with $\boldsymbol{b} = (7,7,7,7,7,7,7)$ but with $m = 7$ and $k = 4$.

6-19 Use the outcome of Exercise 6-17 by fixing y_n and resolving as in Example 6.4.

6-20 Repeat the previous exercise but now by using the result from Exercise 6-18.

6-21 Apply $\boldsymbol{A_{SC}}$ to the instance below:

$$\boldsymbol{A} = \begin{bmatrix} 1 & 1 & 1 & 1 & 1 & 0 & 0 \\ 0 & 1 & 1 & 1 & 1 & 1 & 0 \\ 0 & 0 & 1 & 1 & 1 & 1 & 1 \\ 1 & 0 & 0 & 1 & 1 & 1 & 1 \\ 1 & 1 & 0 & 0 & 1 & 1 & 1 \\ 1 & 1 & 1 & 0 & 0 & 1 & 1 \\ 1 & 1 & 1 & 1 & 0 & 0 & 1 \end{bmatrix}$$

$\boldsymbol{c} =$ (3,4,1,2,5,3,2)
$\boldsymbol{b} =$ (8,6,9,5,4,6,10)

6-22 Apply $\boldsymbol{A_{SC}}$ to the following instance:

$$\boldsymbol{A} = \begin{bmatrix} 1 & 1 & 1 & 0 & 0 & 1 \\ 1 & 1 & 0 & 0 & 1 & 1 \\ 1 & 0 & 0 & 1 & 1 & 1 \\ 0 & 0 & 1 & 1 & 1 & 1 \\ 0 & 1 & 1 & 1 & 1 & 0 \\ 1 & 1 & 1 & 1 & 0 & 0 \end{bmatrix}$$

$\boldsymbol{c} =$ (1,1,1,1,1,1)
$\boldsymbol{b} =$ (7,4,5,9,7,5)

6-23 Given an $m \times n$ matrix $\boldsymbol{A}$ of 1s and 0s, we know that deciding if $\boldsymbol{A}$ has (or could be made to have by permuting columns) the row-circularity property is easy. Suppose we are to decide the same issue for an $m \times k$ submatrix of $\boldsymbol{A}$. What is the status of the problem now?

6-24 Five faculty have been lined up to conduct review sessions during 1-hour periods in various topics prior to departmental comprehensive examinations. The relevant subjects and the required number of periods by each faculty member are shown in the table below. Subject to the same assumptions as those considered in Example 6.5, find a schedule of these sessions using the fewest number of periods.

		linear programming	nonlinear programming	combinatorial optimization	queueing theory
	1	2	0	1	0
	2	0	1	0	2
Faculty	3	1	1	0	1
	4	0	1	0	1
	5	1	0	1	0

6-25 Assume that only three rooms are available for the review sessions referred to in Exercise 6-24. Give a suitable timetable using the fewest number of periods. Repeat the exercise considering only a two-room availability.

6-26 Prove that bipartite graphs are Class-1.

6-27 Give a fast algorithm for producing a feasible edge-coloring in a bipartite graph.

6-28 The *marriage theorem* is often stated as follows:

If $G = (V, E)$ is a k-regular bipartite graph with $k > 0$, then G possesses a perfect matching.

Prove this.

6-29 Describe how the result captured by the marriage theorem can be used to resolve Eercise 6-26.

6-30 Provide some realistic illustrations of the timetabling problem.

CHAPTER 7

PROJECT SCHEDULING

In 1966, the Milwaukee Braves were no more, having changed homes, becoming the Atlanta Braves. Purportedly, a key component of the enticement for the Milwaukee club to move at all was the promise of a new stadium facility in the city of Atlanta (a reduction in the frequency of early spring snowstorms could have been a factor as well). But building a stadium is neither a minor nor a short-term undertaking and even after the inevitable political wrangling concludes, votes on bond issues tallied, and contract negotiations settled, the physical construction of such facilities can be a mammoth task. As a consequence, the management of these sorts of projects becomes crucial in order that deadlines be met (or at least the effects of missed deadlines be minimized) and costs be controlled. Indeed, a one million dollar bonus was reportedly offered if the Atlanta planners could complete their project in one year.

Addressing the types of issues that are inherent in the planning and completion of activities comprising large-scale projects such as the construction of Atlanta–Fulton County stadium lie at the heart of *project scheduling.* Among these issues are some of the same ones that, although now couched in practical language, have already been raised in more abstract contexts throughout this book: dealing with task dependencies, fixing task start times, and resolving task conflicts in the face of limited resource availability. But there are also other aspects of the so-called project scheduling "approach" that have typically been associated with it alone and which are responsible for preserving its somewhat separate identity within the realm of scheduling as interpreted in a broad, general sense.

It is not our aim in this chapter to present anything approaching a self-contained treatise of the topic of project scheduling. As a consequence, our brief and introductory coverage here will do little justice to those works often credited with the origination and development of the field (*cf.* Elmaghraby, 1977). Still, three meaningful topics will be addressed: the construction of "project networks," basic scheduling computations, and time–cost optimization models.

7.1 Project Network Construction

As suggested, the kinds of settings generally identified with project scheduling often involve large-scale undertakings such as the construction of a swimming and diving facility for the 1996 Olympic games, the transport of troops and equipment for military operations such as Operation Desert Shield/Storm, or planning alternative (*i.e.*, off-campus) summer school arrangements for Georgia Tech in 1996 in order that the campus be used for the Olympic Village. Whatever the project, let us assume it to be defined in terms of constituent activities or tasks that comprise it. The length of this list is not relevant *per se* nor is it unique in any sense, depending rather upon the level of detail desired. In addition, for each constituent activity let us also list those other activities, if any, that are required to have been completed before the current one can begin, *i.e.*, precedence relationships or, in project scheduling terminology, these are commonly referred to as activity dependencies.

A typical illustration of this was provided by the bicycle assembly example given in Chapter 1. Note, however, that in the Chapter 1 illustration, the bicycle assembly precedence structure was captured at the outset in the form of a digraph. Indeed, throughout the text we have taken these digraphs to be part of our instance construction. But if all we have are the two activity lists described above (the list of activities comprising the project and the list of activity dependencies), we would have to construct the corresponding digraph ourselves. In the project scheduling lexicon, these are sometimes called *project networks* or *activity networks*.

Now, from a given list of precedence relationships, there are two types of project networks that can be formed. One of these is trivial to construct while the other is not. The first (and the easy one) is coincident with the adopted convention that we have employed throughout; a digraph or directed network captures the precedence requirements by identifying activities with vertices or nodes while the arcs, by definition, depict technological precedence (note that in this chapter, we will blur the boundary between graphs and networks since the prevailing literature in project scheduling traditionally adopts the language of the latter). In the parlance of project scheduling, these networks are referred to as "activity-on-node" structures and are symbolized by AON.

The second alternative is more interesting. Referred to (expectedly) as AOA structures, the notion here is to depict activities by arcs with nodes playing only the natural role of defining initial and terminal points for the arcs. Consider the bicycle assembly illustration again.

Example 7.1 Below is a statement of the precedence requirements for the ten bike assembly tasks given now in the list form just described. The first

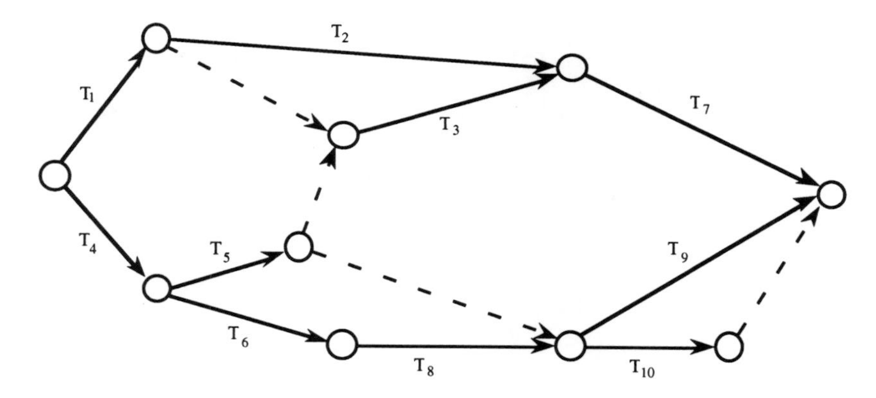

Figure 7.1. *AOA depiction of the bicycle assembly requirements*

row lists the tasks and the second lists the dependencies.

T_1	T_2	T_3	T_4	T_5	T_6	T_7	T_8	T_9	T_{10}
-	T_1	$T_1T_5T_4$	-	T_4	T_4	$T_1T_2T_3T_5T_4$	T_4T_6	$T_5T_8T_6T_4$	$T_5T_8T_6T_4$

Observe that in our list, we have included predecessor requirements that may otherwise be deducible by transitivity, *i.e.*, we show $T_4 \prec T_8$ and $T_6 \prec T_8$ where only the latter is needed since $T_4 \prec T_8$ is implied by $T_4 \prec T_6$ which is shown earlier in the list.

The AON network format was depicted in Chapter 1 and is not repeated here. The reader should agree that its construction is easy to the point of being uninteresting. On the other hand, the AOA version is trickier. To see why, consider a (correct) AOA representation shown in Figure 7.1. Note that in this network we have introduced some arcs (the broken ones) that do not represent real activities but that are needed in order to create a valid depiction of the precedence requirements. These are referred to as *dummy activities*; they depict precedence but consume no time. Obviously, AON networks don't require dummy activities at all while they will often be needed (their use is not unique accordingly) in the AOA case. It should also be clear that from a given AOA structure, the corresponding AON representation is formed from the line digraph of the AOA network. □

Now as implied above, there may be several ways to construct a correct AOA network from a fixed list of precedence requirements. This follows from the possibility of alternative dummy activity use. We can illustrate this with the simple example in Figure 7.2; both networks capture exactly the same precedence structure but employ different numbers of dummy activities. In this regard, a realistic assumption is that we would prefer

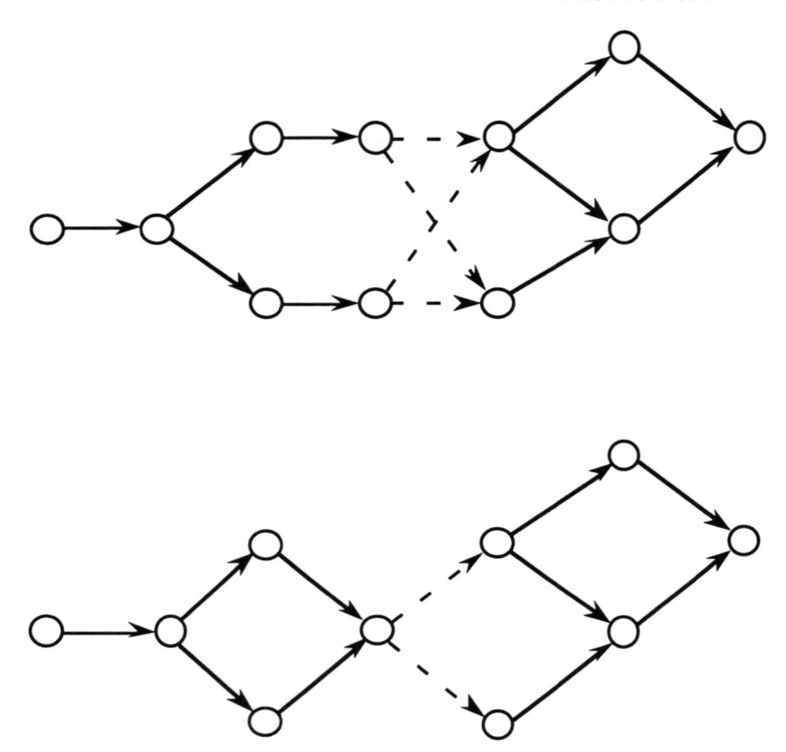

Figure 7.2. *Alterative dummy activity use*

minimum size AOA network representations, meaning ones that exhibit the fewest number of dummy activities.

Unfortunately, the task of finding an efficient, general way to form so-called minimum dummy-activity AOA networks appears to be a difficult one:

Theorem 7.1 *The minimum dummy-activity problem is $\mathcal{NP}$-Hard.*

Proof. The proof is from Krishnamoorthy and Deo (1979) and follows by a reduction from the vertex cover problem. Recall that the latter (stated in recognition format) seeks, for a graph $G = (V, E)$, a decision regarding the existence of a subset $V' \subset V$ with $V' \cap \{i, j\} \neq \phi$ for all edges $\{i, j\}$ in E, and which has size bounded from above by some input threshold value k. The mapping is as follows: letting vertices in G be given by $v_1, v_2, \ldots, v_n$ and edges in E by $e_1, e_2, \ldots, e_m$ create a set of activities as $\{v_1, v_2, \ldots, v_n, e_1, e_2, \ldots, e_m, x\}$. Precedence relationships are given by pairs $\{(e_i, v_j) | e_i \text{ is incident to } v_j \text{ in } G\} \cup \{(e_i, x) | 1 \leq i \leq m\}$. It should

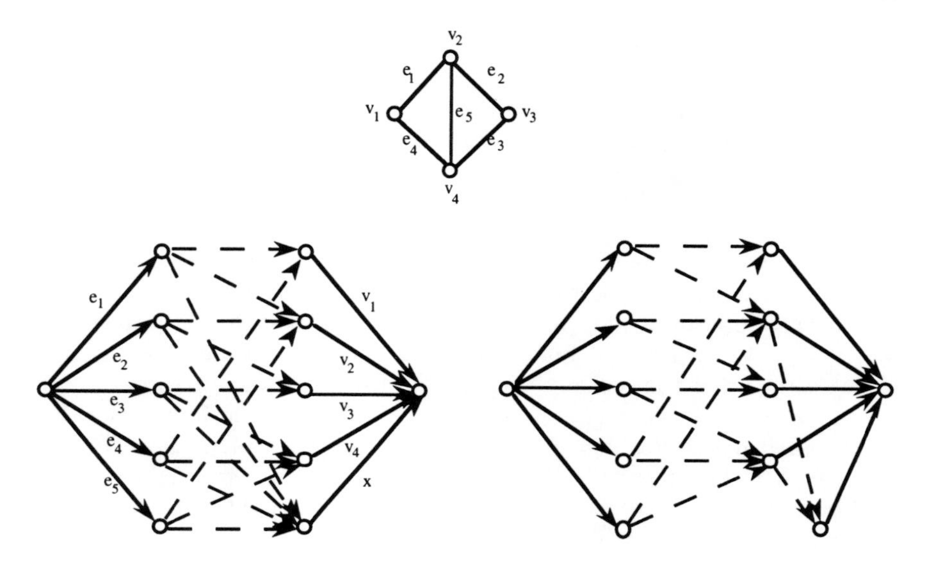

Figure 7.3. *Illustration of the construction of Theorem 7.1*

be easy to see that G possesses a suitable vertex cover exactly when the respective AOA network requires no more than $2|E| + k$ dummy activities. We will leave details of this outcome as an exercise. □

Example 7.2 We can illustrate the mapping of Theorem 7.1 with a small example. Consider the graph at the top of Figure 7.3. Clearly the vertex subset $\{v_2, v_4\}$ constitutes a cover and is minimum in this regard. Now, as shown on the left, a corresponding AOA network can be drawn using $3|E|$ dummies ($2|E|$ of these are necessary in any construction). On the other hand, the structure on the right uses only $2|E| + 2$ and, more importantly, any smaller number is invalid, resulting in a false precedence structure. The identification of the dummies in the second AOA network with the optimum vertex cover in G should be evident. □

The complexity status of the minimum dummy-activity problem calls for the usual interpretation that comes with the worst case perspective. That is, while there are some very complicated instances of the problem (which may or may not be contrived), many others exist where the number of dummies is easily minimized or at least reduced to some number that is verifiably close to the optimum. In fact, it is fairly easy to formulate heuristic rules for constructing AOA networks and that yield relatively efficient structures (*e.g.*, Wiest and Levy, 1977).

7.2 Basic Scheduling Calculations

7.2.1 Critical Paths

Throughout the remainder of this chapter, and as a matter of convenience, we will consider all project networks to exist in AOA format. Accordingly, we shall refer to activities in their network context where (i, j) signifies the activity represented by arc (i, j) and we will denote its duration time by τ_{ij}. Following, we consider what must be one of the most primitive questions in project scheduling: What is the least project duration?

But this question is also particularly easy to answer. For a project network $G = (V, A)$, let us define for each vertex $i \in V$, a variable, s_i, that denotes the start time of any activity (arc) directed out of vertex i. Let us also assume, without loss of generality, that vertices have been labeled in so-called *acyclic* fashion, *i.e.*, for every arc $(i, j) \in A$, $i < j$. We also assume that the network has exactly one unpreceded vertex labeled by 1 and only one unsucceeded vertex, labeled by $n = |V|$. Then the following linear program (P_{CP}) will work:

$$
\begin{array}{lll}
P_{CP}: & \min & s_n - s_1 \\
 & \text{s.t.} & s_j - s_i \geq \tau_{ij}, \quad (i,j) \in A
\end{array}
$$

As an exercise, the reader may want to form the dual of P_{CP} and consider its interpretation (see Exercise 7-11) in a network flow context. In any event, note that our recommendation here is not to solve P_{CP} explicitly in order to determine least project duration time. We only provide the formulation now so that subsequent models and computations might be facilitated.

Clearly, the minimum amount of time required to complete all activities in a project is determined by the length of the longest path(s) of activities in the respective project network. By the length of a path we mean the total duration time of all activities comprising it. These longest paths are referred to as *critical paths* which follows from the "criticality" associated with how activities on such paths are scheduled. Indeed, if any are delayed from their earliest possible start times, the duration of the entire project is delayed as well.

Determining the length of a longest path in an acyclic network is particularly easy by obvious labeling methods. Following, we state a handy way of performing this bookkeeping using fundamental matrix manipulations (*cf.* Lawler, 1976b). From the project network $G = (V, A)$ let us construct a matrix $Q = (q_{ij})$, $1 \leq i, j \leq n$ where

$$
q_{ij} = \begin{cases} \tau_{ij}, & (i,j) \in A \\ 0, & i = j \\ -\infty, & \text{otherwise.} \end{cases}
$$

Recall that dummy activities have duration times of zero. Also, let us construct a vector $S^0 = (s_i^0)$, $1 \leq i \leq n$ where

$$s_i^0 = \begin{cases} 0 & \text{if vertex } i \text{ is unpreceded} \\ -\infty, & \text{otherwise.} \end{cases}$$

Now, beginning with S^0 we form recursively vectors S^k as

$$S^k = S^{k-1} \cdot Q, \; k = 1, 2, \ldots \bar{k} \leq n$$

where operations of addition ($*$) and multiplication ($\#$) are defined as "max" and "+" respectively, *i.e.*, a $* \; b = \max(a, b)$, a $\# \; b = a + b$. Thus, the computation of s_j^k, under $*$ and $\#$, is determined as follows:

$$s_j^k = \max\left((s_1^{k-1} + q_{1j}), (s_2^{k-1} + q_{2j}), \ldots, (s_n^{k-1} + q_{nj})\right).$$

The recursion stops when either two successive vectors are identical or when $k = n$, whichever occurs first. Since we have assumed the input network to be acyclic, this process is finite and stopping must occur ultimately. The values in the final S-vector yield start times identified with each vertex (*i.e.*, a solution to P_{CP}) and hence start times of the activities directed from each. Generally, the value of s_i^k corresponds to the length of a longest path to vertex i using at most k arcs. Thus, the value of $s_n^{\bar{k}}$ yields the critical path length, and hence the minimum project duration.

We will not take space here to justify the validity of the stated procedure, again leaving this as an easy exercise for the interested reader. Following, we give an illustration.

Example 7.3 The matrix Q is formed for the network shown in Figure 7.4 and the S-vector computation is summarized as indicated. The minimum project duration time is nine time units.

$$Q = \begin{array}{c} \\ 1 \\ 2 \\ 3 \\ 4 \\ 5 \\ 6 \end{array} \begin{array}{c} \begin{array}{cccccc} 1 & 2 & 3 & 4 & 5 & 6 \end{array} \\ \begin{pmatrix} 0 & 3 & 2 & -\infty & -\infty & -\infty \\ -\infty & 0 & -\infty & 1 & 4 & -\infty \\ -\infty & -\infty & 0 & 3 & -\infty & 2 \\ -\infty & -\infty & -\infty & 0 & -\infty & 2 \\ -\infty & -\infty & -\infty & -\infty & 0 & 2 \\ -\infty & -\infty & -\infty & -\infty & -\infty & 0 \end{pmatrix} \end{array}$$

$$\begin{array}{rcccccccc} & & & 1 & 2 & 3 & 4 & 5 & 6 \\ S^0 & = & (& 0 & -\infty & -\infty & -\infty & -\infty & -\infty &) \\ S^1 & = & (& 0 & 3 & 2 & -\infty & -\infty & -\infty &) \\ S^2 & = & (& 0 & 3 & 2 & 5 & 7 & 4 &) \\ S^3 & = & (& 0 & 3 & 2 & 5 & 7 & 9 &) \\ S^4 & = & (& 0 & 3 & 2 & 5 & 7 & 9 &) \end{array}$$

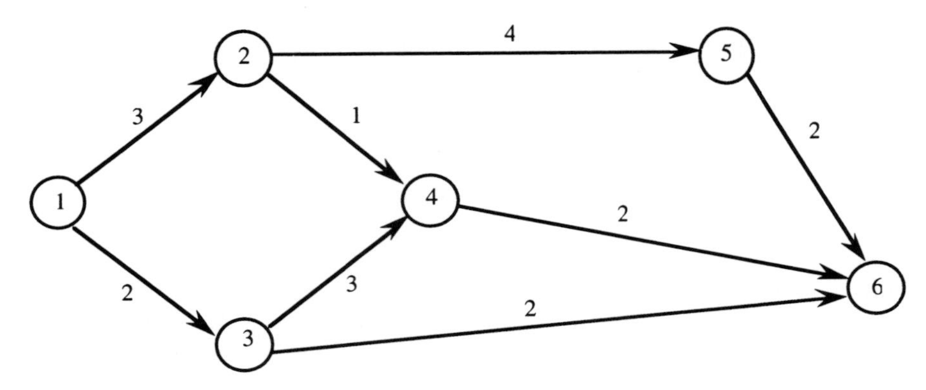

Figure 7.4. *AOA network of example*

Note that the explicit identification of the critical path, and thus of the critical activities, is easily determined in the usual manner by backtracking through the s-label computation. We simply begin with the label of the terminal vertex n (recall that for convenience we have assumed that the network possesses one terminal vertex and one initial vertex) and proceed recursively; given a vertex j that is known to be on a longest path, we need only determine those vertices i for which the inequality $s_j^{\bar{k}} - q_{ij} \geq s_i^{\bar{k}}$ is binding. For those (at least one such inequality must hold without slack) that are, the respective arc/activity (i, j) must be on a longest path to vertex j. The process repeats in this way until the initial vertex is reached. In the present illustration, this computation produces a single critical path given by the vertex sequence $\{1, 2, 5, 6\}$. □

It should be clear that the matrix procedure just described can be easily modified to handle computations on AON structures. In Exercise 7-6 we suggest readers convince themselves of this.

7.2.2 Late Start and Slack Times

The computations described above yield activity early start times by beginning at unpreceded vertices in the project network and growing (longest) paths from these until terminal vertices are reached. The resulting process is sometimes referred to as the *forward pass computation.* If we presently let the early start time of an activity directed from vertex j be given by ES_{jk}, then $ES_{jk} = \max_i(EF_{ij})$ where EF_{ij} denotes the *early finish* times of activities directed into vertex j. Clearly, $EF_{ij} = ES_{ij} + \tau_{ij}$.

Table 7.1. *Example 7.4*

Activity	ES_{ij}	LS_{ij}	EF_{ij}	LF_{ij}	S_{ij}	FS_{ij} (if $\neq S_{ij}$)
(1, 2)	0	0	3	3	0	–
(1, 3)	0	2	2	4	2	–
(2, 4)	3	6	4	7	3	1
(2, 5)	3	3	7	7	0	–
(3, 4)	2	4	5	7	2	–
(3, 6)	2	7	4	9	5	–
(4, 6)	5	7	7	9	2	–
(5, 6)	7	7	9	9	0	–

Now, consider the analogous, *backward pass computation.* If the allowable or *late finish* time of an activity directed into a vertex j is known, then subtraction of the respective activity's duration time would yield its *late start* time. Letting these late start and late finish times be denoted by LS_{ij} and LF_{ij} respectively, we have $LS_{jk} = LF_{jk} - \tau_{jk}$ where $LF_{ij} = \min_k(LS_{jk})$.

If we know the late and early start times (equivalently, late and early finish times) of an activity, we have a measurement of the activity's so-called *slack* time. Denoted by S_{ij}, activity slack time is given as $LS_{ij} - ES_{ij} = LF_{ij} - EF_{ij}$ and derives its name from its measurement as the length of time that the start of an activity can be delayed in order that the minimum project duration not be exceeded. A more "local" form of slack that is sometimes useful is *free slack*, FS_{ij}, which measures delay that an activity can sustain before any other activity start time is increased. That is, $FS_{ij} = ES_{jk} - EF_{ij}$. It is also clear that free slack can never exceed the conventional slack value for an activity and when free slack (different from zero) occurs, it does so at vertices in the project network having in-degree in excess of 1 *and* that are not on a critical path.

We are thus led to an alternative specification of the critical path in a project network as a path of least total slack. If backward pass computations are performed from an initial specification of the allowable project finish time having value identical to the least project finish time, then a critical path is one of zero total slack. The next example illustrates all of these computations.

Example 7.4 We consider the same project network as in the previous example with the relevant computations summarized in Table 7.1. □

At this point, an obvious consideration is how the matrix procedure specified earlier can be employed in the backward pass mode (that it can be used at all is not in question). But this is also easy to resolve: one quick way is to simply negate the values in the Q matrix and take its transpose. Letting this matrix be $\bar{Q}$, we then can create the analog of the S-vector by forming $\bar{S}^0 = (\bar{s}_i^0)$ $1 \leq j \leq n$ and where

$$\bar{s}_j^0 = \begin{cases} f^*, & \text{if } j = n \\ +\infty, & \text{otherwise.} \end{cases}$$

Here, f^* is a parameter specifying allowable project finish time. At this point, the same recursive procedure can be applied where successive $\bar{S}^k$ are formed as

$$\bar{S}^k = \bar{S}^{k-1} \cdot \bar{Q}, \quad k = 1, 2, \ldots, k' \leq n$$

with a $\# \ b = a + b$ and a $* \ b = \min(a, b)$. The values in the final $\bar{S}$-vector yield late finish times directly. Stopping is defined as before.

Example 7.5 Continuing with the same illustration, the matrix-based computation of late finish times is summarized as follows.

$$\bar{Q} = \begin{array}{c} \\ 1 \\ 2 \\ 3 \\ 4 \\ 5 \\ 6 \end{array} \begin{array}{c} \begin{array}{cccccc} 1 & 2 & 3 & 4 & 5 & 6 \end{array} \\ \begin{pmatrix} 0 & \infty & \infty & \infty & \infty & \infty \\ -3 & 0 & \infty & \infty & \infty & \infty \\ -2 & \infty & 0 & \infty & \infty & \infty \\ \infty & -1 & -3 & 0 & \infty & \infty \\ \infty & -4 & \infty & \infty & 0 & \infty \\ \infty & \infty & -2 & -2 & -2 & 0 \end{pmatrix} \end{array}$$

$$\begin{array}{lcl} & & \begin{array}{cccccc} 1 & 2 & 3 & 4 & 5 & 6 \end{array} \\ \bar{S}^0 & = & (\begin{array}{cccccc} \infty & \infty & \infty & \infty & \infty & 9 \end{array}) \\ \bar{S}^1 & = & (\begin{array}{cccccc} \infty & \infty & 7 & 7 & 7 & 9 \end{array}) \\ \bar{S}^2 & = & (\begin{array}{cccccc} 5 & 3 & 4 & 7 & 7 & 9 \end{array}) \\ \bar{S}^3 & = & (\begin{array}{cccccc} 0 & 3 & 4 & 7 & 7 & 9 \end{array}) \\ \bar{S}^4 & = & (\begin{array}{cccccc} 0 & 3 & 4 & 7 & 7 & 9 \end{array}) \end{array}$$

□

7.3 Time–Cost Optimization

7.3.1 Linear Time–Cost Data

One of the more interesting problems in project scheduling is also one of the most important ones. Often referred to as the *time–cost problem*, the notion now is that activity duration times are allowed to vary in the sense that they can be shortened as some nondecreasing function of cost. The problem is one of finding a least cost way of reducing activity duration times so as to meet a fixed project duration.

Example 7.6 Consider the project network in Figure 7.5, and for each activity, let us specify three parameters: ℓ_{ij}, u_{ij}, and s_{ij}. The value ℓ_{ij} is a lower bound on the duration of activity (i, j) or its *crash time*; u_{ij} is the upper bound on duration time known as *normal time*; and s_{ij} is a cost coefficient that reflects the cost per unit of time reduction on the activity (we are assuming a linear time–cost function for each activity). Now, suppose we would like to complete the four–activity project shown in no more than 10 time units. (Since the time–cost data is linear, there is no reason to complete earlier than this.) But the project requires 12 units if every activity is performed as cheaply as possible (normal time for each activity), so it is evident that some activity shortening must be done. Moreover, this shortening must be accomplished on the critial path(s) which in this case is given by the vertex sequence $\{1, 2, 3, 4\}$. In this regard, we need only examine the s-values for the activities comprising this path whereupon we see that activity (2, 3) exhibits the smallest such value and can be compressed accordingly to its crash time point of 3. The set of all activity duration times is shown to the right and we see that a project duration time of 10 results. The differential cost in compressing from 12 to 10 is 8 (a cost of 4 in going through the compression to 11 plus 4 more in moving to 10).

Suppose, however, that we want a further compression of project duration time to 9. Certainly, it is sufficient that we begin with the network duration times just obtained (with overall duration of 10) and seek a minimum cost compression of it. As before, we first locate the critical path or paths, but now we see that every path, and hence all activities are critical. Clearly the choice of a least cost combination of activity compressions is a little more interesting than it was previously. Indeed, there is now a real combinatorial exercise required in order that an optimum compression be identified. We leave it to readers to verify that the choice shown in Figure 7.6 is optimum with a differential cost of 8 relative to the duration-10 compression. We also leave it to readers to convince themselves that for networks even modestly larger than this, some ingenuity may be required if these optimum compression combinations are to be found efficiently. For completeness, we include the remaining compression all the way to project crash time of 7, and for each we give the differential cost incurred in moving from the previous shortening. □

Fortunately, there is a clever strategy for treating the time–cost problem. In order to motivate it, let us begin with a formulation for the case of linear time–cost functions. To this end, let us denote the duration time of an activity by x_{ij} and let the allowable project duration bound be part of the instance and specified by parameter θ. Then the following linear program

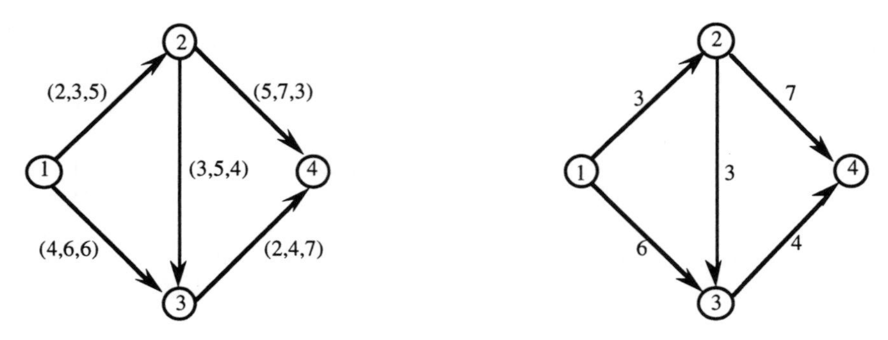

Figure 7.5. *Time–cost illustration*

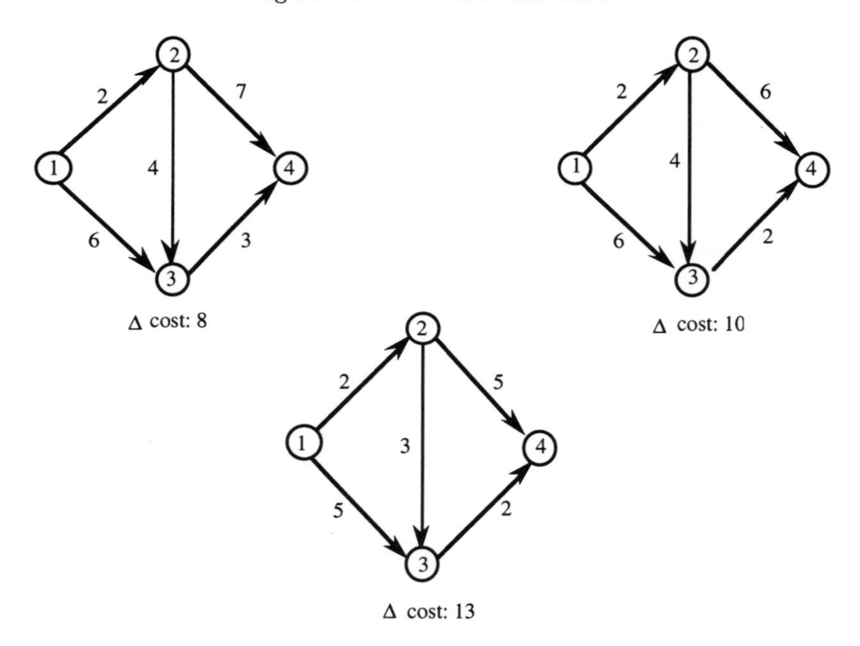

Figure 7.6. *Compressions for Example 7.6*

is sufficient for finding a minimum–cost set of duration times that produce an overall project duration bounded from above by θ.

$$
\begin{aligned}
P_{TC}: \qquad & \max \sum_{(i,j)\in A} s_{ij}x_{ij} & \\
s.t. \quad & s_i - s_j + x_{ij} \le 0 & \text{for all } (i,j) \in A \\
& x_{ij} \le u_{ij} & \text{for all } (i,j) \in A \\
& -x_{ij} \le -\ell_{ij} & \text{for all } (i,j) \in A \\
& s_n - s_1 \le \theta &
\end{aligned}
$$

The reader should have little trouble deriving the objective function of P_{TC}, and even less in interpreting it; *i.e.*, the aim is to perform activities at their "cheaper" times (in the direction of normal time) and so maximizing the stated linear combination clearly suffices. Note also that we have preserved the same notation as before in denoting activity start times by variables s_i. In addition, we have assumed every activity to have a time–cost relationship even though in reality, some activities may have a constant duration time; indeed some activities may be dummies. Dealing with these alterations is the subject of Exercise 7-17.

Now, for a given value of θ, we could solve P_{TC} and be done. On the other hand, we might want to find optimum durations for every possible value of θ over the interval defining project crash and normal times. Of course, there could be many such values. There is, however, a very clever and efficient way to handle this. Following, we give a procedure due to Fulkerson (1964) noting that efficient alternatives have since appeared. Our choice of the original work of Fulkerson is based upon its simplicity.

The algorithm employs network-flow language and concepts throughout but at this point in the presentation, we will be content to leave abstract the issue of what constitutes a "flow" in the time–cost sense. Rather, we will provide a more concrete interpretation of these notions after the procedure is stated and demonstrated.

A_{TC} : The Time–Cost Algorithm of Fulkerson

Step 0: Consider the input project network to be a flow network with source given by the unpreceded vertex in $G = (V, A)$ and sink, the unsucceeded one. Let these vertices, as before, be labeled as 1 and n respectively. Let the original flow values, f_{ij}, be zero on all arcs (i, j). (Naturally, we can start with nonzero flows if we are not interested in all project compression values. See Exercise 7-13.)

Step 1: Given a current set of flow values, f_{ij}, form a *modified network* following the constructions given in Table 7.2.

Step 2: On the modified network just constructed, determine the lengths of longest paths from vertex 1 to vertex j for $2 \le j \le n$ and let these values be denoted by s_j. The actual duration time for the activities in the project network are given by $x_{ij} = \min(u_{ij}, s_j - s_i)$.

Step 3: Construct the subgraph of the current modified network that corresponds to the longest paths from vertex 1 to vertex n just computed and define a maximum flow problem on this subgraph where arc capacities are given in Table 7.3. If a maximum flow (minimum cutset capacity) of value $+\infty$ results, stop. Otherwise, use this maximum flow solution to update total flows overall and return to Step 1. □

Example 7.7 Consider the instance used previously in Example 7.6. For ease of presentation, we will keep track of flow values on a separate network. Now, starting with zero flows everywhere, the first modified network results (see Table 7.2) as shown in Figure 7.7, part a. Here, the s_j values are given by each vertex and the subgraph of longest paths is denoted by bold arcs. The max-flow problem is defined on the latter [shown in (b)] where it is easy to see that a maximum flow value of 4 results. Current flows are augmented accordingly as shown in (c) and the procedure returns to Step 1 where a new modified network is constructed. Note that we have included actual duration times as specified within Step 2, and also that our max-flow solutions are determined by inspection, which in turn follows from the identification of a minimum capacity cut-set. The entire computation is summarized as shown. Observe the correspondence between the total "flow" values generated above and the successive differential cost values specified in Example 7.6 (*i.e.*, 4,8,10,13). □

Let us now consider what this algorithm is doing and why it is correct. Key in this regard will be an examination of the dual of the earlier time–cost formulation and, in particular, an appeal to basic primal–dual conditions in linear programming. In forming the dual, let us define (dual) variables f_{ij}, g_{ij}, h_{ij}, and v to coincide with the constraints as they appear in P_{TC}. Denoting the dual formulation as P_{DTC}, we have

$$
\begin{aligned}
P_{DTC}: \quad \min v\theta + \sum_{(i,j)\in A} g_{ij}u_{ij} \quad &- \quad \sum_{(i,j)\in A} h_{ij}\ell_{ij} \\
s.t. \quad f_{ij} + g_{ij} - h_{ij} \quad &= \quad s_{ij} \text{ for all } (i,j) \in A \\
\sum_k f_{jk} - \sum_i f_{ij} \quad &= \quad \begin{cases} v, & j = 1 \\ 0, & j \neq 1, n \\ -v, & j = n \end{cases} \\
f_{ij}, g_{ij}, k_{ij}, v \quad &\geq \quad 0
\end{aligned}
$$

Even the modestly observant reader will note a distinct flow-like structure in this formulation. On the other hand, that this is so should be expected

Table 7.2. *Modified Network Construction*

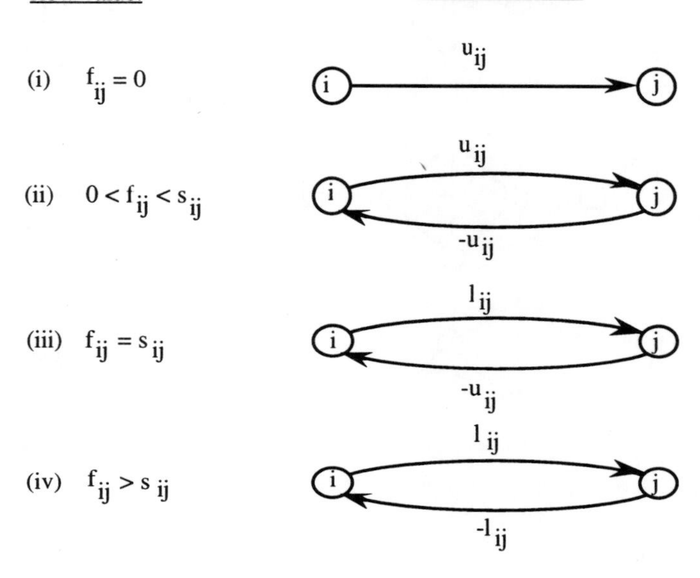

Table 7.3. *Capacity specification in the maximum flow network*

Forward arcs (arcs in the max-flow network of Step 3 which are directed consistent with the original instance, project network):

f_{ij} case	arc capacity
(i) (ii)	$s_{ij} - f_{ij}$
(iii)(iv)	$+\infty$

Reverse arcs (arcs directed opposite to the orientation in the original network):

f_{ij} case	arc capacity
(ii)(iii)	f_{ij}
(iv)	$f_{ij} - s_{ij}$

□

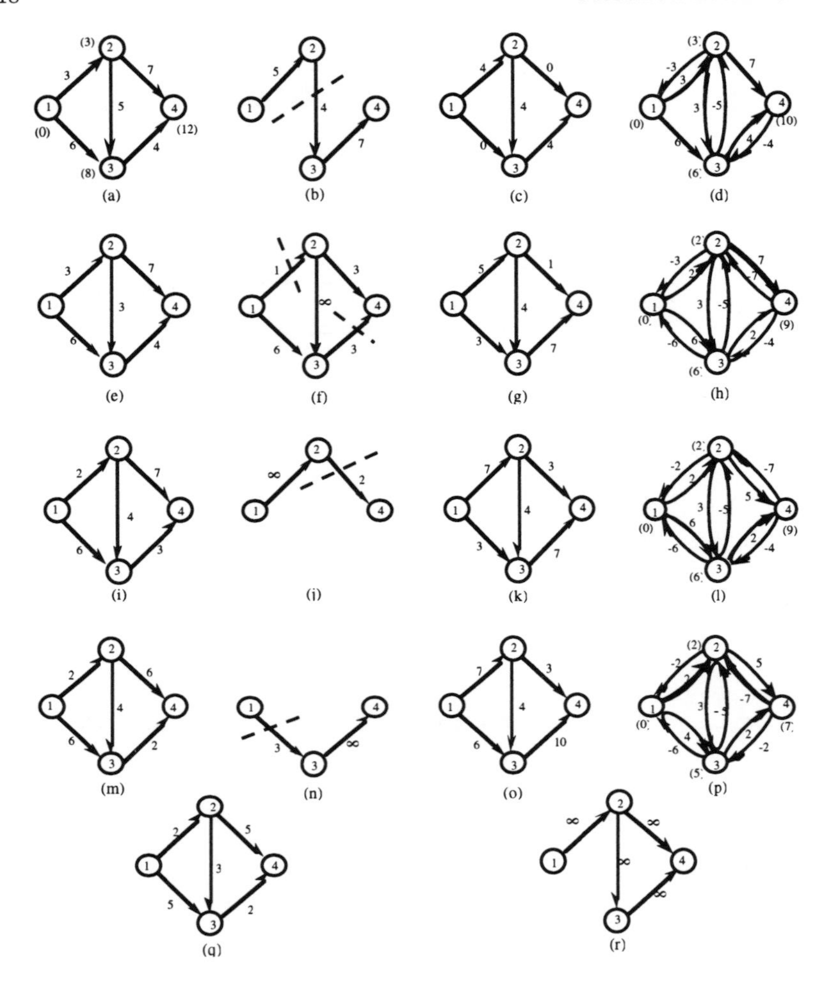

Figure 7.7. *A_{TC} computations for Example 7.7*

from observing and interpreting the dualization of the formulation P_{CP} which was suggestesd previously.

Now, recall that at any stage in the computation, a set of "flows" is given by the construction in Step 3. But these values can be assigned to the various f_{ij} in P_{DTC} with the total source-sink flow value assigned to the dual variable v. Clearly, these assignments satisfy the flow conservation-like constraints of P_{DTC}. Moreover, suitable (*i.e.*, feasible) values for g_{ij} and h_{ij} are easy to determine, yielding an admissible dual feasible solution

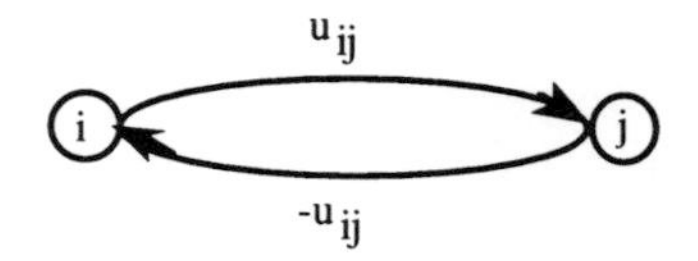

Figure 7.8. *Modified network structure for the case* $0 < f_{ij} < s_{ij}$

overall.

But the max-flow network is constructed from the so-called modified network formed in Step 2 and employs arc weights stipulated in Table 7.2. Note in the latter that four cases are specified in terms of the value of dual variable f_{ij}. Now, in any optimal solution to P_{DTC}, variables g_{ij} and h_{ij} cannot both be simultaneously different from zero; otherwise, the corresponding primal duration variable, x_{ij}, would have to be at once equal to both crash and normal time following complementary slackness conditions. Thus, for each f_{ij} category shown, we can specify the sense of values for g_{ij} and h_{ij}. Accordingly, for the cases of $f_{ij} = 0$ and $0 < f < s_{ij}$, we know that (in any pair of optimal solutions to P_{TC} and P_{DTC}) $g_{ij} > 0$ and $h_{ij} = 0$. When $f_{ij} = s_{ij}$ we must have $g_{ij} = h_{ij} = 0$ and for $f_{ij} > s_{ij}$ we would require $g_{ij} = 0$ and $h_{ij} > 0$. Similarly, in the first two cases, complementarity would requie that $x_{ij} = u_{ij}$, in the third that $\ell_{ij} \leq x \leq u_{ij}$, and in the last case, that $x_{ij} = \ell_{ij}$. By the fundamental theorem of linear programming, we know that any pair of admissible solutions to P_{TC} and P_{DTC} satisfying these slackness conditions must be optimal in their respective problems.

Returning to the table, let us now observe the structures that are called for in the stated modified network requirements. When $f_{ij} = 0$ we have only an arc directed from i to j and with weight u_{ij}. Any feasible set of start time values (called for in Step 2) must satisfy the inequality $s_j - s_i \geq u_{ij}$ and since we are asked to set $x_{ij} = \min(u_{ij}, s_j - s_i)$, we obtain $x_{ij} = u_{ij}$ as required by complementarity. In the case of nonzero $f_{ij} < s_{ij}$, the construction is shown in Figure 7.8. Again, suitable start time values (defined by a longest path computation on the modified network) must satisfy the pair of inequalities $s_j - s_i \geq u_{ij}$ and $s_i - s_j \geq -u_{ij}$ which is the same as $s_j - s_i \leq u_{ij}$ and upon combining, we have $s_j - s_i = u_{ij}$ where $x_{ij} = \min(u_{ij}, s_j - s_i) = u_{ij}$ as required. We will not repeat this verification for the other cases but the reader will find again that the corresponding network constructions are defined so as to guarantee that the relevant complementary slackness conditions hold.

In terms of the variables f_{ij}, a value strictly less than s_{ij} (which includes the case $f_{ij} = 0$) elicits the following interpretation: the "flow" is insufficient to reduce the duration of the respective activity, *i.e.*, we're not willing to pay the cost of s_{ij} to shorten it. On the other hand, when $f_{ij} = s_{ij}$, we have just enough flow to begin shortening the activity and may as well shorten it as far as possible since the price of reduction has now been paid. The last case ($f_{ij} > s_{ij}$) is similar in that we would like to keep shortening the respective duration but we can't reduce time below the crash point and so we will fix the duration at its lower bound in all subsequent compressions. Given particular flow values, the subgraph in the corresponding modified network (Step 3) upon which the next max-flow problem is defined simply reflects that portion of the project where time can still be "purchased." The capacity settings from Table 7.3 quantify the amount of such additional cost and a least capacity cutset (maximum flow value) measures the value of a least cost compression.

Observe that upon application of the algorithm, the duration of 11 was not produced explicitly. This is because the project time–cost function between duration 12 and 10 is linear. That is, the same compression in going from 12 to 11 was employed in going from 11 to 10.

7.3.2 Nonstandard Time–Cost Data

The procedure presented in the previous section is predicated, in no small measure, on the linearity assumption relative to each activity's time–cost function. In this section we consider relaxations of this assumption and suggest ways to alter the time–cost model accordingly.

Convex time–cost functions. The most harmless modification arises when time–cost functions are nonlinear but at least remain convex. In such a case, we can create the familiar approximation of the input time–cost function by a piecewise linear substitute. In essence, this substitution is reflected by replacing the original activity on the project network by k new ones (in series), each having a linear time–cost relationship. The magnitude of k depends on the degree of approximation proposed by the "k-piece" linearization. These new activities are referred to as *pseudo-activities.*

The primary issue in the use of pseudo-activities is in the specification of their respective time–cost data. That is, the new time–cost data must be fixed in such a way that it is consistent with the corresponding data for the underlying real activity. But this construction is straightforward. In Figure 7.9, we exhibit the conversion for an approximation involving four pseudo-activities.

Nonconvex time–cost data. Suppose an activity possesses a time–cost function that is nonconvex (but continuous). Again, we could linearize through the use of the aforementioned pseudo-activity replacement; the schematic of Figure 7.9 still applies. However, there is a new complication for if the linearized construction is formulated as P_{TC}, the order pertaining

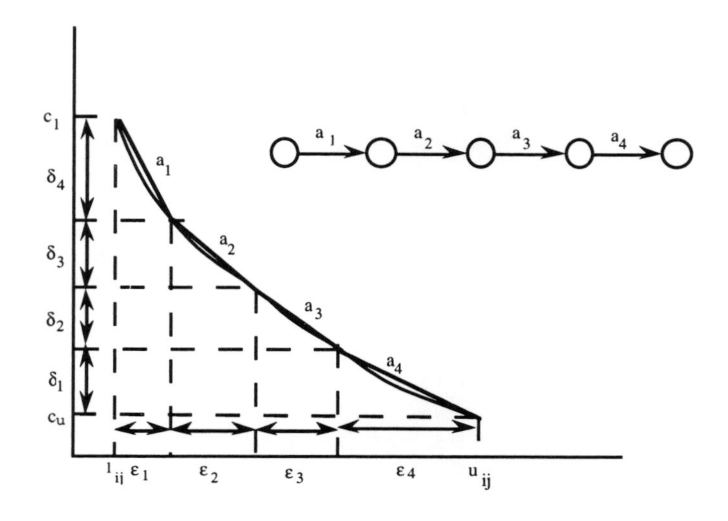

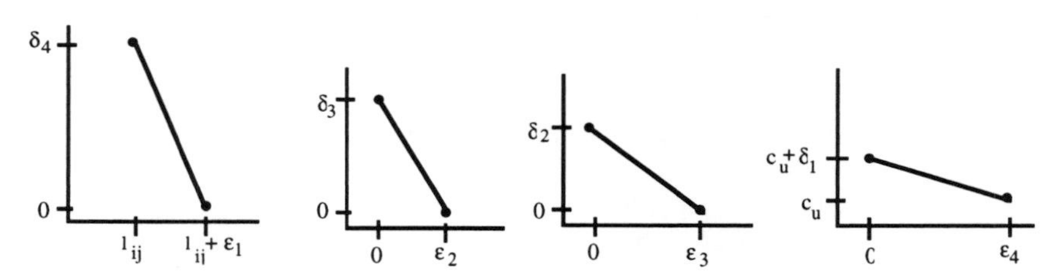

Figure 7.9. *Pseudo-activity construction*

to the shortening of pseudo-activities is reversed, a phenomenon that is nonsensical from the perspective of the real activity.

But there is a standard way to guarantee that pseudo-activities are crashed in the correct sequence. Consider the nonconvex function in Figure 7.10 and, in particular, a pseudo-activity breakdown as indicated. Further, let us suppose that the duration time variables for these pseudo-activities are given as $x_{a_1}, x_{a_2}, x_{a_3}$, and x_{a_4}. Now, it is easy to see that by imposing the following constraints, time compression on the pseudo-activities proceeds in an order that is consistent with time compression for the original activity. Note that each α_i is a binary variable which "couples" adjacent pieces in the pseudo-actrivity approximation.

$$\frac{x_{a_4}}{u_{ij} - \triangle_3} \leq \alpha_3 \leq \frac{x_{a_3}}{\triangle_3 - \triangle_2} \leq \alpha_2 \leq \frac{x_{a_2}}{\triangle_2 - \triangle_1} \leq \alpha_1 \leq \frac{x_{a_1}}{\triangle_1}$$

For example, no shortening of pseudo-activity a_3 can proceed until pseudo-

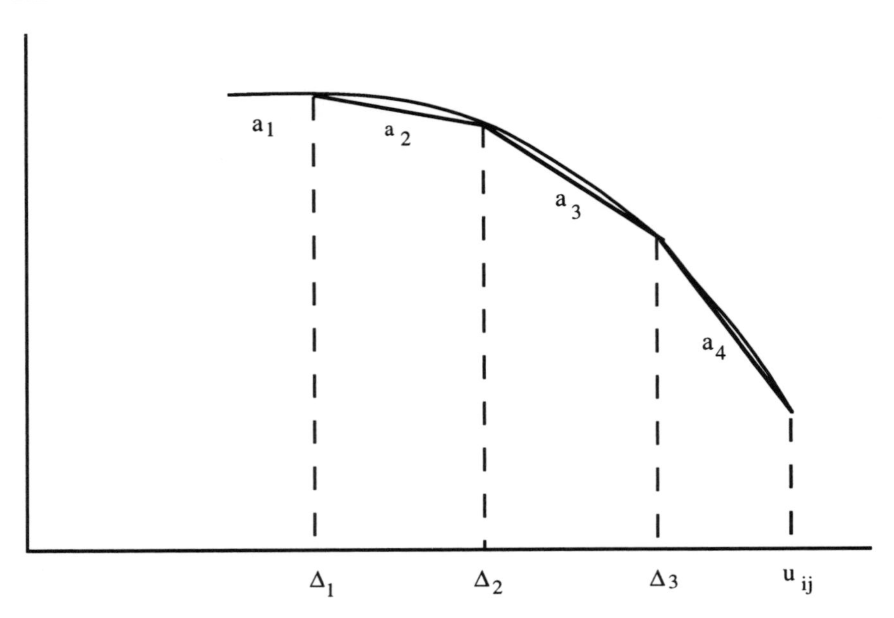

Figure 7.10. *Nonconvex time–cost functions*

activity a_4 is compressed to its crash time.

For our final case, suppose an activity a has discrete time–cost data such as that in Figure 7.11. Here, for a given duration time x_a^j, a cost of c_j is incurred. So, in order to guarantee that shortening moves from exactly one duration point to another, we again introduce a binary variable, one for each point. We then augment our model by the following constraints:

$$s_j - s_i \geq \sum_{j=1}^{r} \alpha_j x_a^j$$

$$\sum_{j=1}^{r} \alpha_j = 1$$

$$\alpha_j \in \{0,1\}, 1 \leq j \leq r$$

In addition, our objective function is modified to include a cost component of the form $\sum_{j=1}^{r} \alpha_j c_j$.

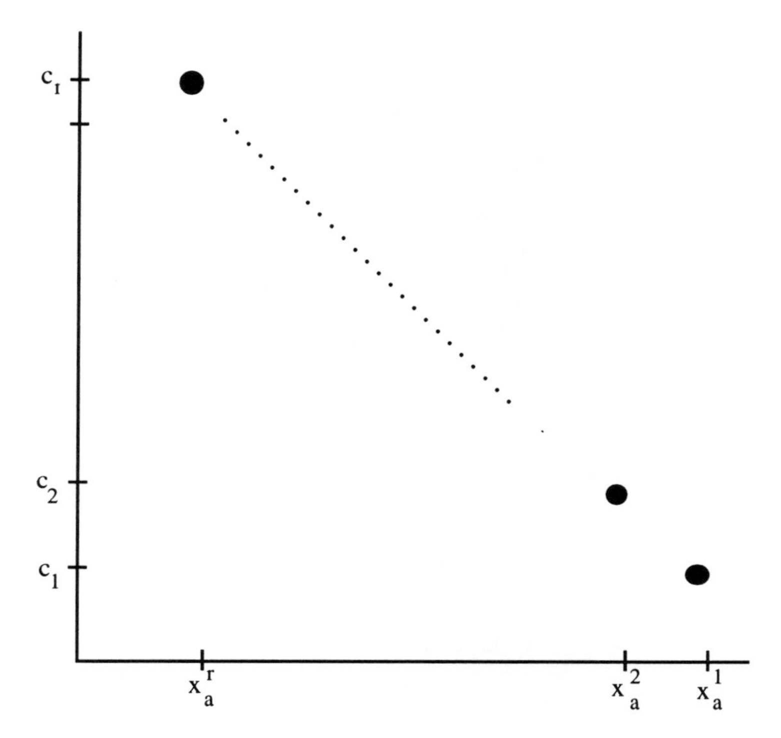

Figure 7.11. *Discrete time–cost data*

7.4 Exercises

7-1 Construct an AOA project network from the precedence list below:

Activity:	T_1	T_2	T_3	T_4	T_5	T_6	T_7	T_8	T_9
Preceded by:	–	–	T_1	T_2	T_2	T_1, T_4	T_5	T_3	T_3, T_6, T_7

7-2 Follow the construction of Theorem 7.1 and form the corresponding precedence structure from the graph G in Figure 7.12. From this precedence list, form an AOA project network using as small a number of dummies as possible.

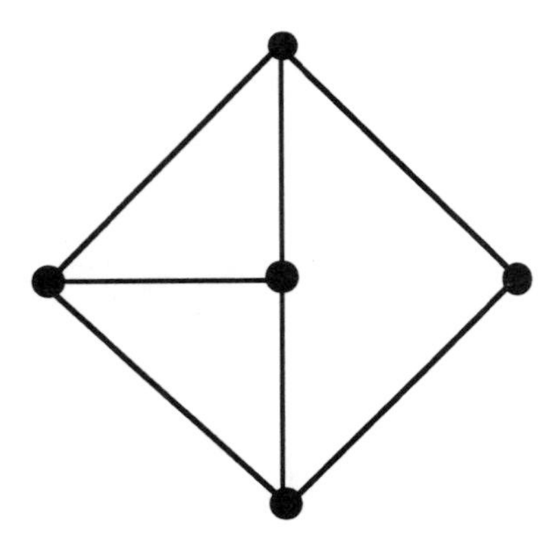

Figure 7.12. *Exercise 7-2*

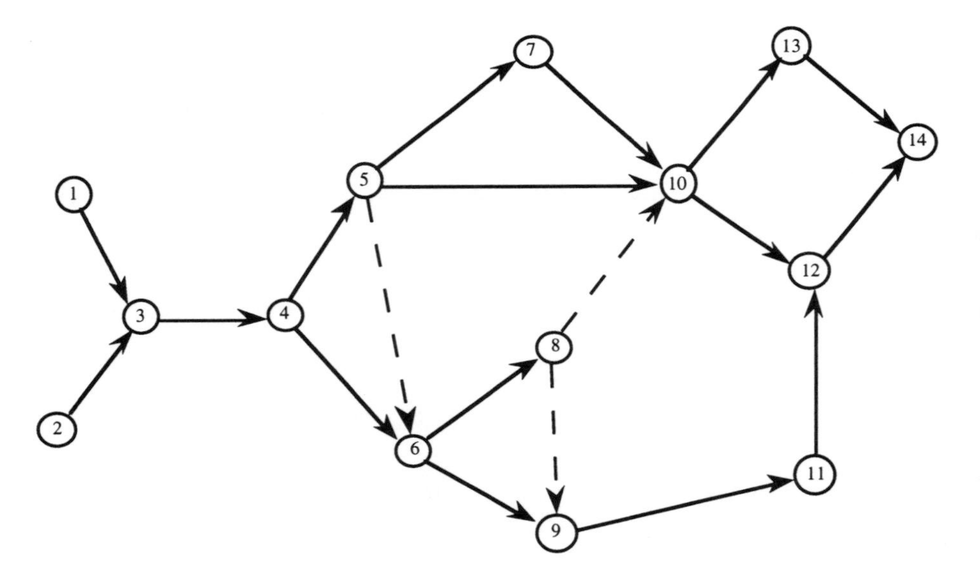

Figure 7.13. *Exercise 7-4*

7-3 Formulate a heuristic for the minimum dummy activity problem. State clearly the sense of its construction and argue its efficacy. Also, give an instance upon which it performs poorly.

7-4 Form an AON structure by constructing the line-digraph of the AOA network in Figure 7.13.

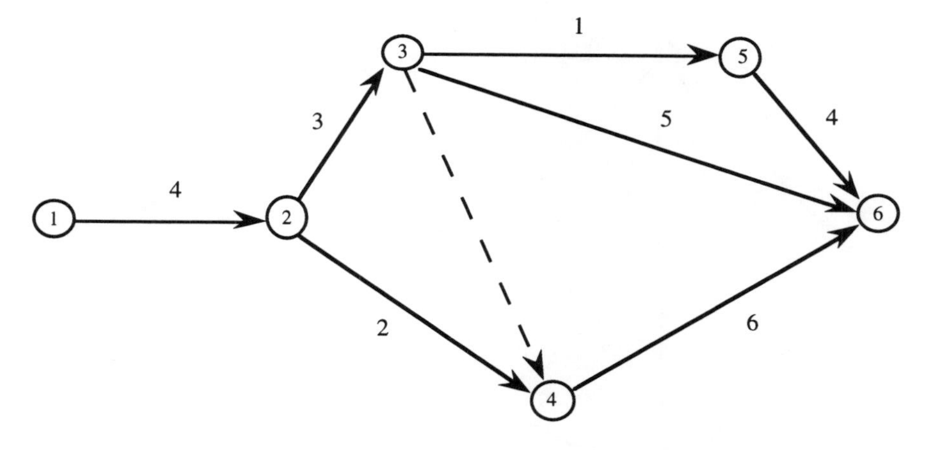

Figure 7.14. *Exercise 7-5*

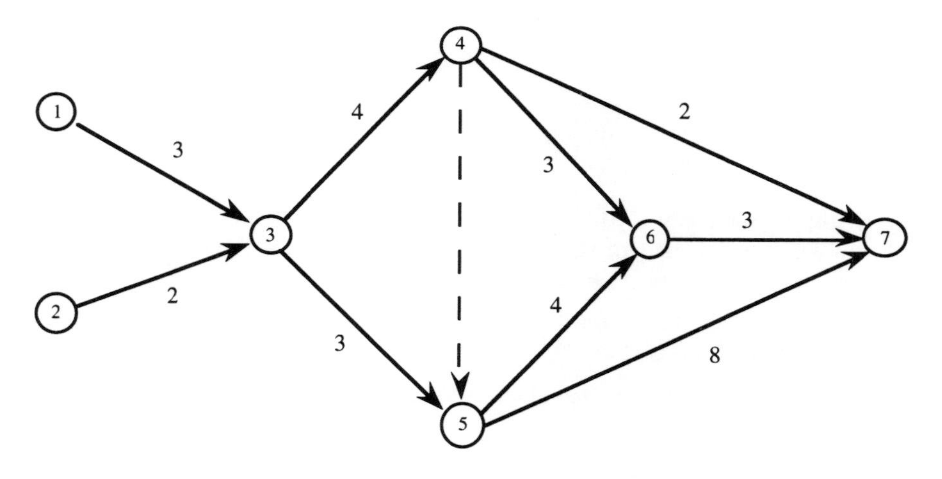

Figure 7.15. *Exercise 7-7*

7-5 Use the so-called matrix procedure to determine start times in the AOA network in Figure 7.14.

7-6 Form the line-digraph of the network in Exercise 7-5 and repeat the exercise (of 7-5) for the resulting AON network.

7-7 Use the matrix procedure to determine early start times and late finish times for the AOA network of Figure 7.15. Use the zero-slack convention to start the backward pass computations.

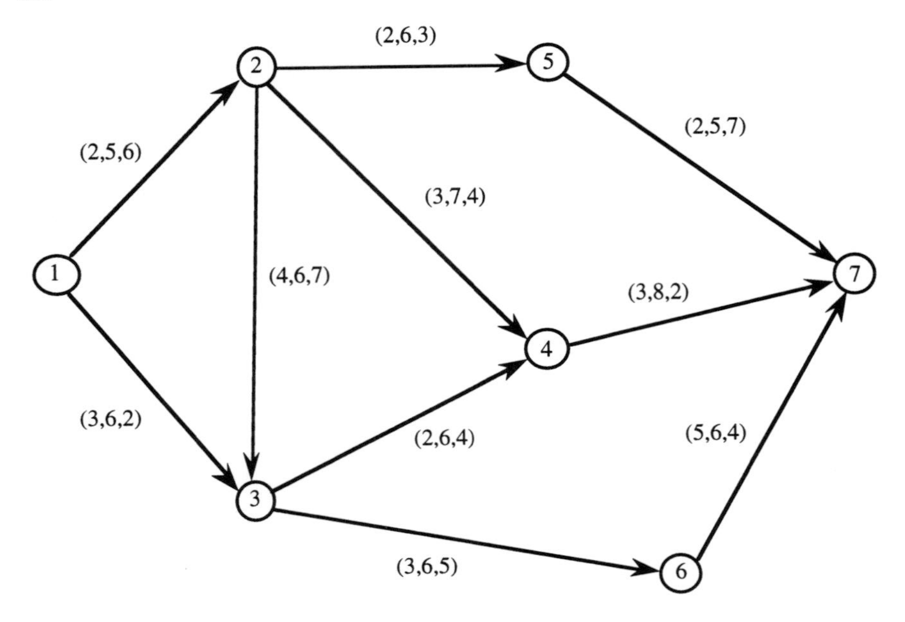

Figure 7.16. *Exercise 7-12*

7-8 Suppose for some AOA network, the matrix procedure iterates (and converges) k_1 times for the forward pass. Suppose for the backward pass, it iterates k_2 times. Is $k_1 = k_2$? Discuss.

7-9 Compute all slack values (including free slack) for the activities in the network of Exercise 7-7.

7-10 Show that activity free slack FS_{ij} could never exceed activity total slack S_{ij}.

7-11 Form and interpret the dual of P_{CP}.

7-12 Apply the time–cost algorithm $\boldsymbol{A_{TC}}$ to the instance shown in Figure 7.16.

7-13 Repeat Exercise 7-12 on the instance shown in Figure 7.17 and *starting* with the flows indicated.

7-14 Referring to the instance of Exercise 7-12, select an iteration other than that by the zero-flow solution, and show that primal and dual feasibility are satisfied in addition to complementary slackness.

7-15 Give a (nonzero) feasible flow solution that results in a suboptimal primal solution using the instance of Exercise 7-12.

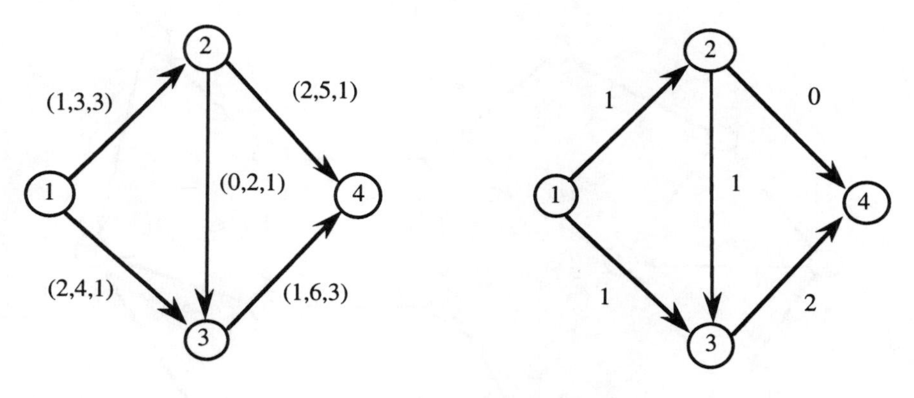

Figure 7.17. *Exercise 7-13*

7-16 Suppose that for every time unit change of project duration, an overhead cost of α dollars is incurred. Thus there is a *total cost* of project duration resulting as the sum of the usual compression costs and these indirect, overhead costs. Modify formulation P_{TC} to find a least total cost project duration.

7-17 Modify P_{TC} to deal with duration times that are constants.

7-18 Consider the instance in Figure 7.18. Apply algorithm $\boldsymbol{A_{TC}}$.

7-19 Suppose that for some project duration θ the activities given by $\mathcal{C}_\theta$ are critical. Now, assume that a project compression to duration θ' is performed and $\mathcal{C}_{\theta'}$ is the set of critical activities. Will $\mathcal{C}_{\theta'} \supseteq \mathcal{C}_\theta$? Discuss.

7-20 Consider the instance of the time–cost problem in Figure 7.19 where nonlinearities in activity time–cost functions are introduced. Create appropriate pseudoactivities and solve the corresponding formulation P_{TC} with duration times 20, 17, and 14.

7-21 Produce a solution to P_{CP} relative to the AOA instance in Figure 7.20. From this solution, produce a complementary (optimal) solution to the dual of P_{CP}.

7-22 Consider an *alternative* to P_{CP}:

$$\begin{array}{ll} \min & t \\ \text{s.t.} & t \geq L_p \text{ for } p \in \xi, \end{array}$$

where ξ is the set of directed paths (indexed by p) in the network and L_p is the length of path p. Will this alternative suffice? If it will, is it preferable to P_{CP}? Explain.

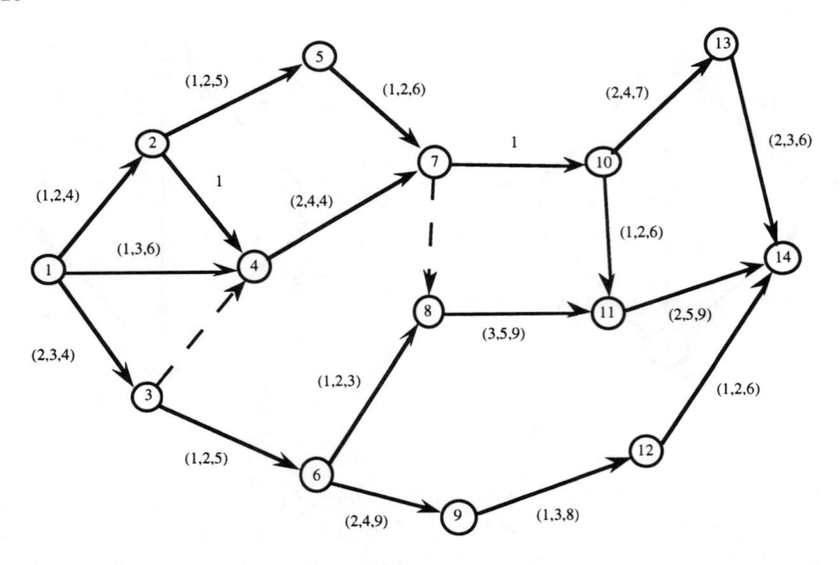

Figure 7.18. *Exercise 7-18*

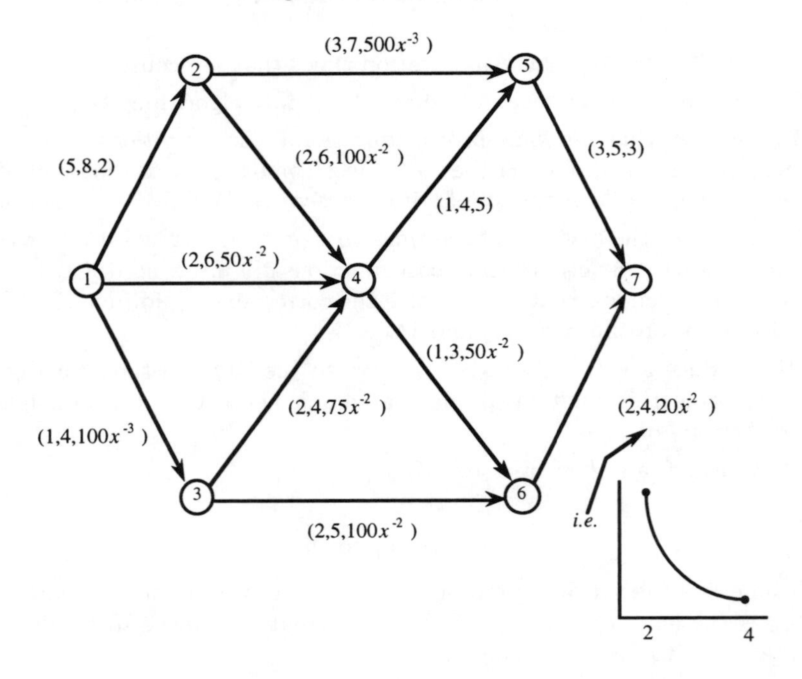

Figure 7.19. *Exercise 7-20*

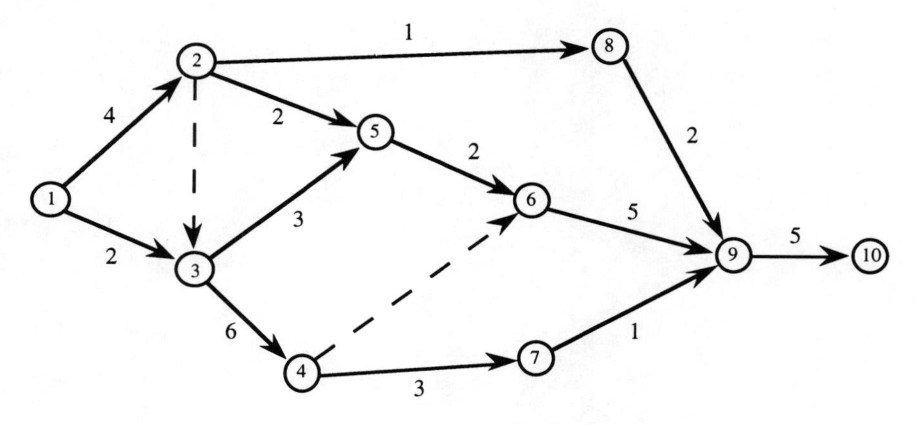

Figure 7.20. *Exercise 7-21*

CHAPTER 8

CHINESE POSTMEN AND TRAVELING SALESMEN

We present this chapter last since the corresponding subject matter is, by most honest judgments, somewhat peripheral to the material dealt with thus far. Until now, our settings have generally required little defense in order that the problems they generate qualify as "scheduling problems." With very few exceptions, the cases we have dealt with to this point fit quite easily into the subject description stated in the first sentence of Chapter 1.

Still, when we examine seriously what it is that provides the basis for our presentation of the subject of scheduling, namely the inherent *combinatorial* aspect of finding correct sequences or orderings, it is more than slightly disquieting to eliminate out-of-hand problems of the sort that make up the present chapter title. Indeed, we even suggested in Chapters 1 and 3 that the problem $1|seq.dep.|C_{\max}$ is really a traveling salesman problem in disguise.

So, we will take up in this chapter, two topics that often arise in scheduling contexts albeit perhaps in less than traditional ways. But these are problems that tend to appeal to the same sorts of intellectual interests as do the more traditional problems in combinatorial scheduling theory and for us, this is enough to warrant at least some space in this book.

8.1 Eulerian Traversals and the Chinese Postman Problem

Suppose we are given a graph, $G = (V, E)$, and let us suppose further that the edges are weighted as $w : E \rightarrow \mathbb{Z}^+$. Now, consider the following problem, posed initially by Kwan (1962): Find a walk or traversal in G which includes each edge *at least* once, begins and ends at the same vertex, and which is of least total (edge) weight. Taken out of this abstract setting, the problem arises in the delivery of mail, refuse collection, and other routing-like contexts. As a consequence, these sorts of walks are often referred to as *postman traversals* (defined similarly whether graphs are undirected, directed, or mixed) and finding the least cost such traversal has come to

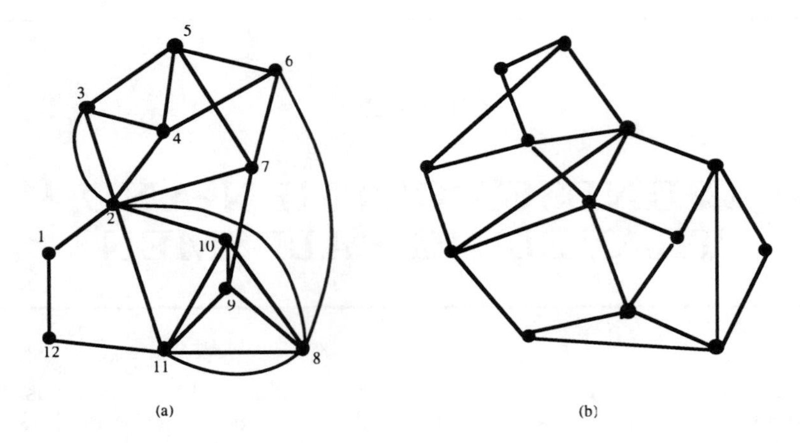

Figure 8.1. *Eulerian and non-Eulerian graphs*

be known as the *Chinese Postman's Problem* (Edmonds, 1965d).

8.1.1 Eulerian Graphs

Undirected graphs. Clearly, the postman problem sounds remarkably like a very old and well-known problem in graph theory: Given a graph $G = (V, E)$, does G possess a walk that begins and ends at the same vertex and includes each edge exactly once? Such a walk is called an *Eulerian* traversal and a graph admitting same is said to be Eulerian. The name follows from the resolution of the question on any undirected graph by a famous result of Euler (1736) (some contend that it was the first result in graph theory). We have:

Theorem 8.1 *A connected graph $G = (V, E)$ possesses a traversal that begins and ends at the same vertex and includes each edge exactly once if and only if the degree of every vertex in G is even.* □

Sometimes, graphs satisfying this degree condition are called "even." The graph in Figure 8.1, part a is Eulerian while the one in part b is not.

But just because a graph passes Euler's existence test it does not follow that we can exhibit the traversal easily (*i.e.*, polynomially). In this case, however, such exhibition is easily accomplished following a simple strategy:

A_F: Traversals in Eulerian Graphs

> Given a position in the traversal, select the next edge arbitrarily so long as its removal would not disconnect the graph *unless* this edge is the only choice. □

For example, starting with vertex 1 in Figure 8.1, we can apply $\boldsymbol{A_F}$ and produce the traversal given by the vertex sequence $\{1, 2, 3, 2, 4, 3, 5, 6, 4, 5, 7, 6, 8, 2, 7, 9, 8, 10, 2, 11, 10, 9, 11, 8, 11, 12, 1\}$. On the other hand, upon reaching vertex 11 (the first time), the edge leading to vertex 12 is not allowed, for its removal leaves us stranded in a component of a disconnected graph from which we cannot escape.

Directed graphs. Suppose our input is a digraph $G = (V, A)$ and we are interested in the same issues as before but now involving traversals using each arc exactly once. Clearly, in- and outdegree at each vertex must balance in an obvious way. When these values are equal in G, we say the graph is symmetric and the following analog of the previous theorem is apparent.

Theorem 8.2 *A connected and directed graph $G = (V, A)$ is Eulerian if and only if it is symmetric.* □

Now, producing an existing (directed) Eulerian traversal in a digraph G is a little more interesting than in the undirected case.

$\boldsymbol{A_{ED}}$: Traversals in Eulerian Digraphs

Step 1: In the input digraph G, find any spanning arborescence T having outdegree at most 1 at each vertex. Let the root of T be vertex i^* (*i.e.*, outdegree of $i^* = 0$).

Step 2: Label arcs in A as follows:

- **i)** For vertex $j \neq i^*$, having outdegree $(j) = t$, label the arc from j which is in T by t. Label the others arbitrarily as $1, 2, \ldots, t-1$.
- **ii)** Label arcs directed from i^* arbitrarily.

Step 3: Beginning at vertex i^*, traverse arcs in A in label order. □

Example 8.1 We can illustrate the algorithm on the digraph of Figure 8.2 (note its symmetry in the sense defined). The arcs have been labeled per the algorithm, and in particular with regard to the arborescence shown. □

Mixed graphs. Finally, let us suppose the input graph is given by $G = (V, E \cup A)$. Since G contains both directed as well as undirected edges, it is often referred to as a *mixed* graph. Unfortunately, our results (throughout) now start to become a little more tricky. The next theorem describes sufficient conditions only:

Theorem 8.3 *A connected, mixed graph $G = (V, E \cup A)$ is Eulerian if it is symmetric and the total degree at every vertex is even.* □

Note that when measuring "total" degree, we are relaxing the distinction between in- and outdegree. Also, it is trivial to see that the even degree requirement in this sense is certainly necessary while the requirement for

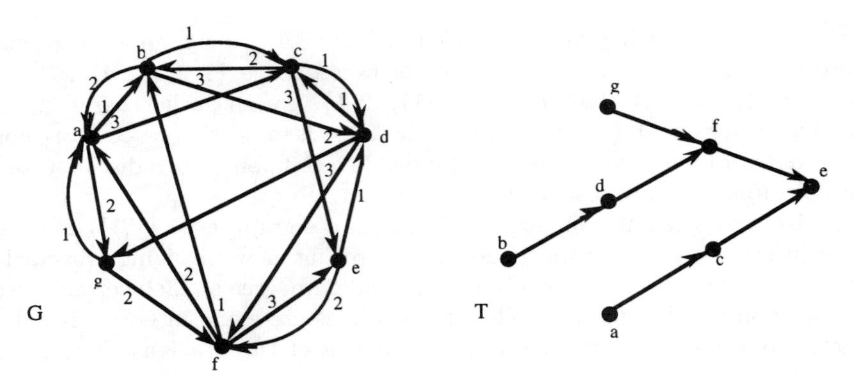

traversal: {e,d,c,d,g,a,b,c,b,a,g,f,b,d,f,a,c,e,f,e}

Figure 8.2. *Directed Eulerian traversal*

symmetry is not. The graph in Figure 8.3 illustrates. Note also that reversing the orientation of edge (3, 4) in the figure produces a graph that possesses no Eulerian traversal.

In any event, the issue in the mixed graph case is to create (if possible) a graph that satisfies the (sufficient) symmetry conditions of the theorem. That is, we need to orient some of the edges in such a way that symmetry is created. Fortunately, there is a very nice formulation that accomplishes this or (correctly) concludes that it is not possible.

Let us denote the difference between in- and outdegree at each vertex by b_k. In addition, let us create, for every edge in E, a pair of arcs, where each arc in the pair is oriented opposite to the other. Letting these arcs be given by a set U we have $U = \{(i,j),(j,i)|(i,j) \in E\}$. Consider now the following model:

$$P_{sym}: \quad min \quad \sum_{(i,j)\in U} x_{ij}$$

$$s.t. \quad \sum_{\{(i,j)\in U|i=k\}} x_{ij} - \sum_{\{(j,i)\in U|i=k\}} x_{ij} = b_k, \quad k \in V$$

$$0 \le x_{ij} \le 1 \text{ for } (i,j) \in U$$

If P_{sym} has a solution, then the input mixed graph is Eulerian. Accordingly, we deal with each edge (i,j) in E in the following way: Create (orient the edge as) arc (i,j) if $x_{ij} = 1$ in the solution to P_{sym}, create arc (j,i) if $x_{ji} = 1$ in the solution, and leave the edge undirected otherwise.

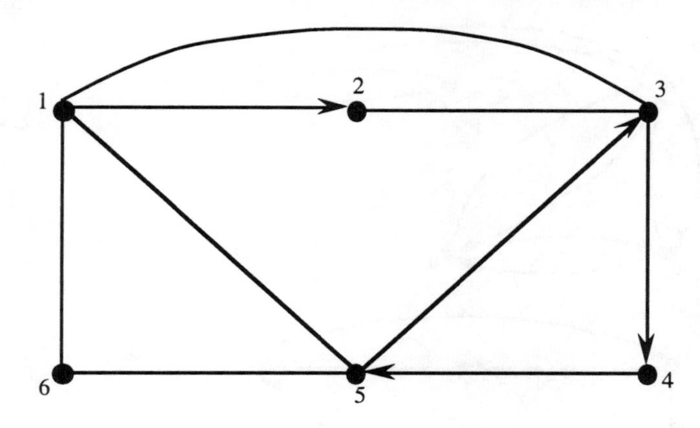

Figure 8.3. *Eulerian traversal in a mixed graph*

Example 8.2 Returning to the in Figure 8.3, if P_{sym} is formulated and solved, the solution results as indicated in Figure 8.4. Note that we have specified the b_k values next to each vertex and the arcs in the solution are indicated. The resultant interpretation for the original graph is then specified at the bottom where it can be seen that the symmetry conditions are satisfied. The generation of the existing traversal in the graph resulting from the solution of P_{sym} follows by employing the same notions as in the other two cases. We ask the reader to pursue this as an exercise (see Exercise 8-12). □

8.1.2 Postman Problems

If the input graph satisfies the appropriate conditions (or can be made to do so *vis-a-vis* Theorem 8.3 whereby traversals exist that use each edge/arc exactly once, then there is nothing more to do. That is, if the input graph G is Eulerian, the corresponding Chinese Postman Problem is not interesting; we simply produce the existing traversal which, by definition, involves no backtracking and thus exhibits minimum total edge and/or arc weight. On the other hand, when the input graph is not Eulerian or P_{sym} has no solution, the postman problem becomes meaningful. Required accordingly is a least weight edge/arc duplication (if possible), which in turn corresponds to a supergraph of G that does satisfy the relevant eulericity conditions. An Eulerian traversal is produced in this multigraph where its interpretation relative to a postman's route in G is obvious. Following, we begin with the

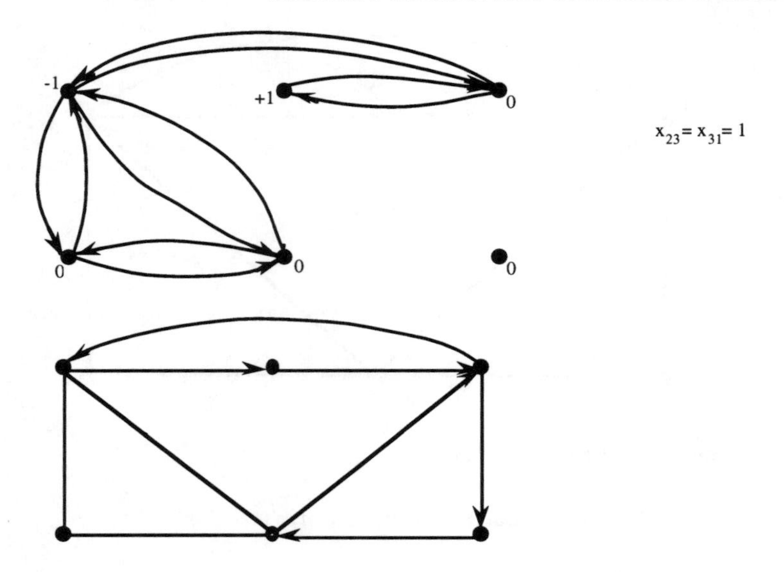

Figure 8.4. *P_{sym} solution for a mixed graph*

undirected case.

Postman problems on undirected graphs. The first algorithm is generally well known, following the seminal works on matchings by Edmonds (1965a,c). Input is taken to be a graph $G = (V, E)$ with at least two vertices having odd degree. Recall that edges are weighted as nonnegative integers.

$\mathcal{A}_{UCP}$: Edmonds' algorithm for the Undirected Chinese Postman's Problem

Step 0: Let G be the input graph and denote the set of odd-degree vertices in G by V_0.

Step 1: Determine a shortest path in G and the respective lengths between all pairs of vertices in V_0.

Step 2: Form a complete graph on vertex set V_0 with edges weighted by the respective shortest path lengths from Step 1. Find a least weight perfect matching in $K_{|V_0|}$ and duplicate the implied edges in G. Call this supergraph G'.

Step 3: Produce the existing Eulerian traversal in G'. □

This algorithm is polynomial with the matching effort required in Step 3 dominating. It should also be clear that the requirement for nonnegativity

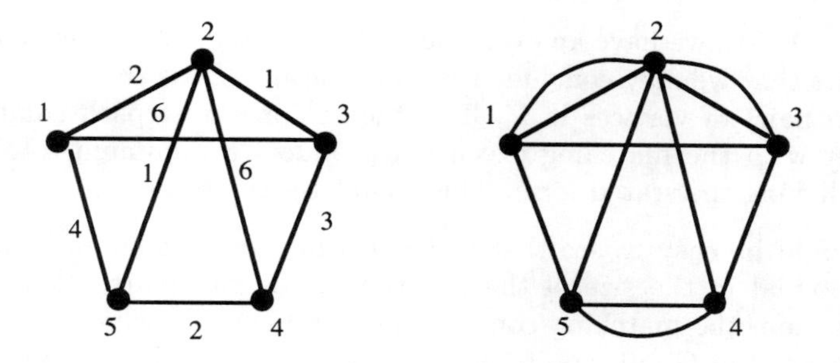

Figure 8.5. *Graphs of Example 8.3*

on edge-weights is crucial since shortest path problems are hard in the presence of so-called negative cycles.

Example 8.3 We can demonstrate the algorithm for the undirected case with the graph G in Figure 8.5 where weights are written next to each edge. The set V_0 results as $\{1,3,4,5\}$ and the six shortest paths along with their lengths are found, resulting in the following:

vertex pair	path	length
1, 3	$1-2-3$	3
1, 4	$1-2-5-4$	5
1, 5	$1-2-5$	3
3, 4	$3-4$	3
3, 5	$3-2-5$	2
4, 5	$4-5$	2

The matching step would produce the edge set $M = \{(1,3),(4,5)\}$ with weight 5 corresponding to total edge weight of the two paths 1–2–3 and 4–5. The respective edges of these paths are duplicated in G, producing the graph on the right in Figure 8.5 which clearly is Eulerian. A suitable traversal is given by the vertex sequence $\{1,2,3,2,1,3,4,2,5,4,5,1\}$, the total weight of which is 30 or $\sum_{(i,j)\in E} w_{ij} + \sum_{(i,j)\in M} w_{ij}$ as expected. □

The validity of Edmonds' postman algorithm is not difficult to establish. We have:

Theorem 8.4 *Edmonds' algorithm* $\boldsymbol{A_{UCP}}$ *will produce a minimum weight postman traversal in any connected, undirected graph with nonnegative edge weights.*

Proof. Let $G = (V, E)$ be the input graph and let E^* be the set of duplicated edges traversed by an optimal walk in G. Then $G^* = (V, E \cup E^*)$ is even and for each $i \in V_0$, E^* contains an odd number of edges incident to

i. For $i \in V \backslash V_0$, we have an even (possibly zero) number of incident edges. It follows that we may conclude that any suitable E^* is the union of paths between pairs of vertices in V_0. But then the shortest path computation together with the matching execution produces a minimum total weight set E^* having the stated form. This completes the proof. □

It should be easy to see that the algorithm requires effort bounded by a polynomial in the size of the input, *i.e.*, the "all pairs" shortest path computation, the matching computation, and the generation of the Eulerian traversal in G' all require polynomial effort. It should also be easy to see that the algorithm will never duplicate an edge more than once (see Exercise 8-3).

Postman problems on directed graphs. Conceptually, the resolution of the directed case is the same as for the undirected version. Clearly, there are some technical differences, however, in order to handle degree requirements which now involve more than a simple parity test. Whereas before, our interest was in building up to even degree everywhere, we now have to create/preserve symmetry at each vertex. But this is also easy to do. Consider the input digraph to be $G = (V, A)$.

$\mathcal{A}_{DCP}$: Algorithm for the Directed Postman Problem

Step 1: Determine shortest paths (and their lengths) from every vertex with excessive indegree to every vertex with excessive outdegree. We assume that some path exists between every pair of vertices.

Step 2: Solve a transportation problem with sources the excessive indegree vertices and demands, the excessive outdegree vertices. Per unit shipment costs are the relevant shortest path values from Step 1. Availability and demand values are given by the respective differences between indegree and outdegree at those vertices where symmetry is not satisfied.

Step 3: Duplicate the arcs in G relative to the transportation solution. Call this supergraph G' and produce the existing Eulerian traversal in G'. □

Example 8.4 Let us orient the edges of the graph used in the previous example to produce the digraph shown in Figure 8.6. The transportation problem is constructed with source vertices 3 and 4 each having availability of 1 and demands at vertices 1 and 5 with requirements of 1 in each position. Shortest path lengths result between the relevant pairs and are shown in the transportation network accordingly. We can solve the problem by inspection, "shipping" from 3 to 1 and from 4 to 5 at total cost of 11 (clearly there is another solution as well). The directed paths are duplicated on G

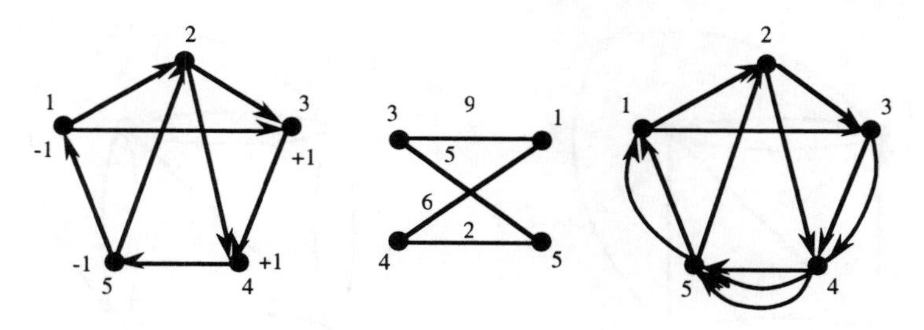

Figure 8.6. *Digraph of Example 8.4*

producing G' as shown and a traversal can then be found. One possiblity is $\{1,2,3,4,5,1,3,4,5,2,4,5,1\}$. □

The reader will observe from the illustration that in the directed postman case, it is possible to duplicate arcs more than once. In addition, it should also be clear that unlike the undirected version, it may be that no postman traversal exists at all in the directed case, *i.e.*, no amount of arc duplication will suffice. In order to guarantee the existence of a postman solution the input digraph must be *strongly connected.* That is, every vertex must be reachable from any other vertex. But strong connectivity is easy to test for, which justifies the assumption stated in Step 1 of $\boldsymbol{A_{DCP}}$.

Postman problems on mixed graphs. When the input is defined on a graph containing undirected as well as directed edges, our luck begins to run out insofar as the polynomial-time resolution of the postman problem is concerned. The next result makes this more formal.

Theorem 8.5 *The (recognition version) of the mixed postman problem is $\mathcal{NP}$-Complete.*

Proof. The reduction is from 3-SATISFIABILITY the details of which are given in Papadimitriou (1976). □

There is some small piece of good news for the mixed graph case however. If $G = (V, E \cup A)$ is even, then we can employ formulation P_{sym} to see if symmetry can be achieved. If not, then duplication of edges and/or arcs must be attempted without losing the even-degree property (recall that our interpretation of "even" in the mixed case discounts the distinction between in- and outdegree). Consider the graph in Figure 8.7, part a. It is easy to see that no orientation of edges exists that creates symmetry. But, upon resorting to edge and arc duplication, we can produce the structure in part b which is symmetric but not even. On the other hand, the multigraph in part c has the same weight (we are considering a cardinality version of the problem) *and* is symmetric and even; it clearly is the preferred outcome.

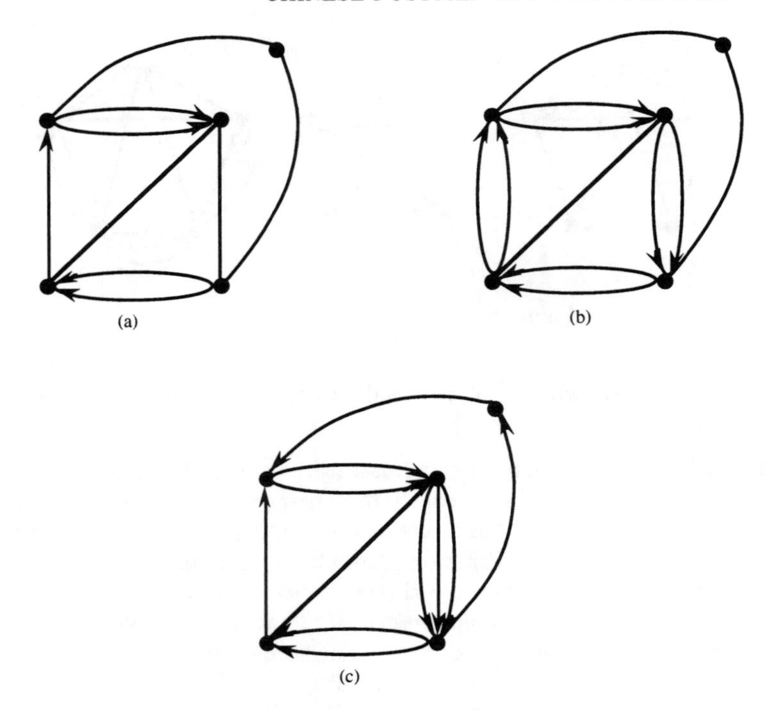

Figure 8.7. *Postman construction for mixed graphs*

While problematical, we can still employ a network flow formulation to deal with some of these issues. The work of Edmonds and Johnson (with some tricks by others) is key (Edmonds and Johnson, 1973). Regardless, the real culprit contributing to the intractability outcome is the apparent inability to negotiate effectively the even degree requirement. That is, do we first create even degree and then try to produce symmetry, or vice versa? As a consequence, we are thus led to the application of heuristics for the mixed case. To this end, consider the following, "even-symmetric" strategy suggested in the Edmonds and Johnson work:

A_{ES}: ES-Heuristic for Mixed Postman Problems

Step 1: From the input graph $G = (V, E \cup A)$ create an even degree graph as in the undirected postman case (disregard edge orientation).

Step 2: Apply the Edmonds–Johnson procedure to obtain symmetry while preserving even degree. Produce the existing Eulerian traversal.

□

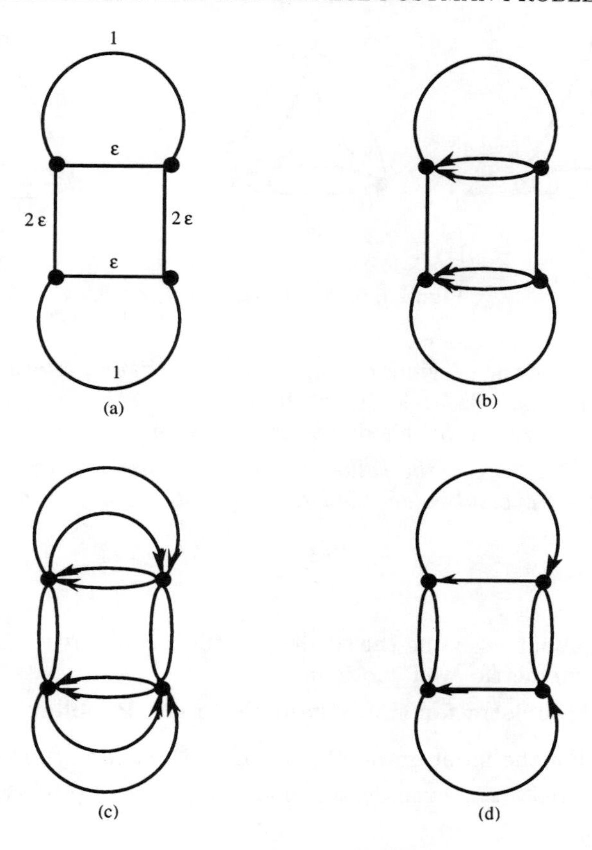

Figure 8.8. *Application of A_{ES}*

Example 8.5 Consider the instance in Figure 8.8, part a, where edge weights are specified as shown. The even-degree creation phase of the ES heuristic would duplicate the arcs (considered as undirected edges for Step 1 purposes) having weight ε yielding the multigraph in part b. Operating on this graph to produce symmetry while preserving the even degree condition yields the graph in part c with total weight $4 + 12\varepsilon$. We leave it as an exercise to produce the existing Eulerian traversal. More importantly, we note that the multigraph produced by the heuristic is not optimal. Indeed, had we been a little less greedy in our even-degree creation phase, duplicating instead the edges with weight 2ε, the graph in part d would have ultimately resulted yielding the optimal solution directly. □

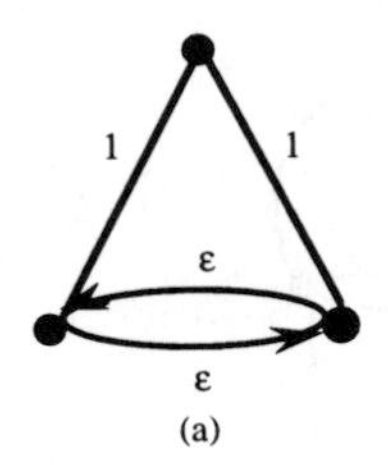

(a)

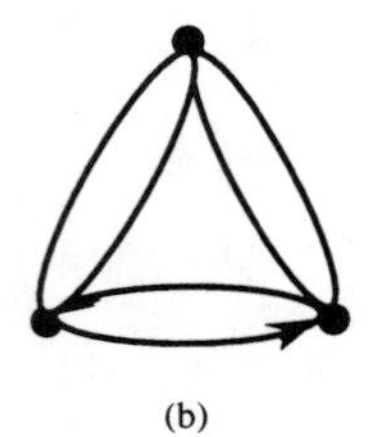
(b)

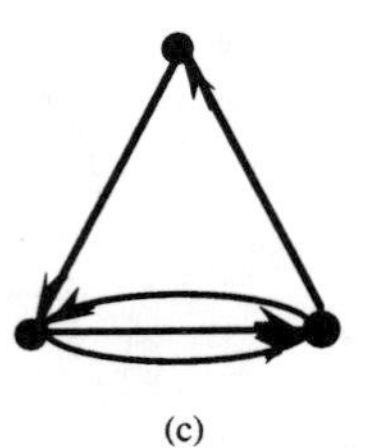
(c)

Figure 8.9. *Application of A_{SE}*

The instance of the previous example indicates that a performance bound for the ES heuristic is at least on the order of 2. In fact, we have the following formalization of this due to Frederickson (1979):

Theorem 8.6 *If ν_{ES} is the value of a traversal produced by $\mathbf{A_{ES}}$ and ν^* is the optimal traversal value, then for any instance of the mixed postman problem,*

$$\frac{\nu_{ES}}{\nu^*} \leq 2.$$

□

But what about reversing the strategy of the ES heuristic? Consider the following, "symmetric-even" notion:

A_{SE}: SE-Heuristic for the Mixed Postman Problem.

Step 1: For the input graph $G = (V, E \cup A)$, create symmetry.

Step 2: Create an "even-degree" multigraph while preserving symmetry. □

Note that Step 2 above is performed over the undirected, edge-induced subgraph of the output of Step 1.

Example 8.6 Consider the graph in Figure 8.9, part a. Applying A_{SE}, we see that the input is already symmetric and the even-degree creation phase would then duplicate the two edges, yielding the graph in part b which is Eulerian and has weight $4 + 2\varepsilon$. But this is clearly not optimal, as the graph in part c indicates. □

The analog of Theorem 8.6 can be given by the following result, also in Frederickson (1979):

Theorem 8.7 *If ν_{SE} is the value of a postman traversal produced by $\mathbf{A_{SE}}$ and ν^* is the optimal value, then for any instance of the mixed postman problem,*

$$\frac{\nu_{SE}}{\nu^*} \leq 2.$$

□

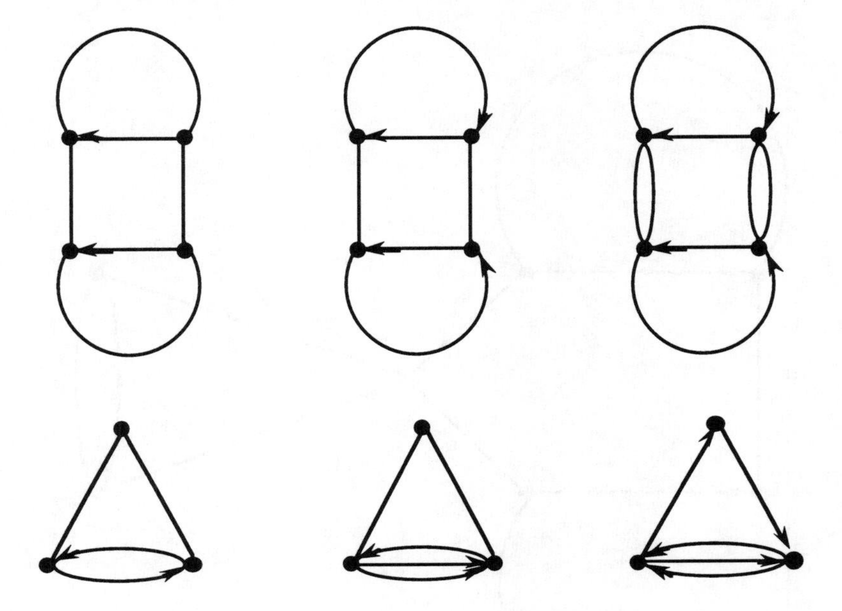

Figure 8.10. *Behavior of A_{ES} and A_{SE}*

So, in the worst case sense, both heuristics perform similarly. Suppose, however, that we apply each of these to the worst case instance of the other. The outcome is summarized in Figure 8.10. Interestingly, $\boldsymbol{A_{ES}}$ solves correctly the "bad" instance for $\boldsymbol{A_{SE}}$ while the latter does the same on the worst case instance of $\boldsymbol{A_{ES}}$. This outcome begs an obvious question; put loosely, perhaps $\boldsymbol{A_{ES}}$ and $\boldsymbol{A_{SE}}$ realize their respective worst case outcomes on different classes of graphs. If so, then it might be the case that they could be used in a composite manner, comparing the result of each and selecting the best, and in so doing, improve on the bound of 2 arising when each is employed singularly. In fact, this is more than a possibility. We have

Theorem 8.8 *[Frederickson (1979)] If the performance ratios for $\boldsymbol{A_{ES}}$ and $\boldsymbol{A_{SE}}$ are denoted by ν_{ES}/ν^* and ν_{SE}/ν^* respectively, then for any instance of the mixed postman problem,*

$$\min\{\nu_{ES}/\nu^*, \nu_{SE}/\nu^*\} \leq \frac{5}{3}.$$

□

Interestingly, the bound of the composite strategy just described is not presently known to be realizable. The instance shown in Figure 8.11 yields

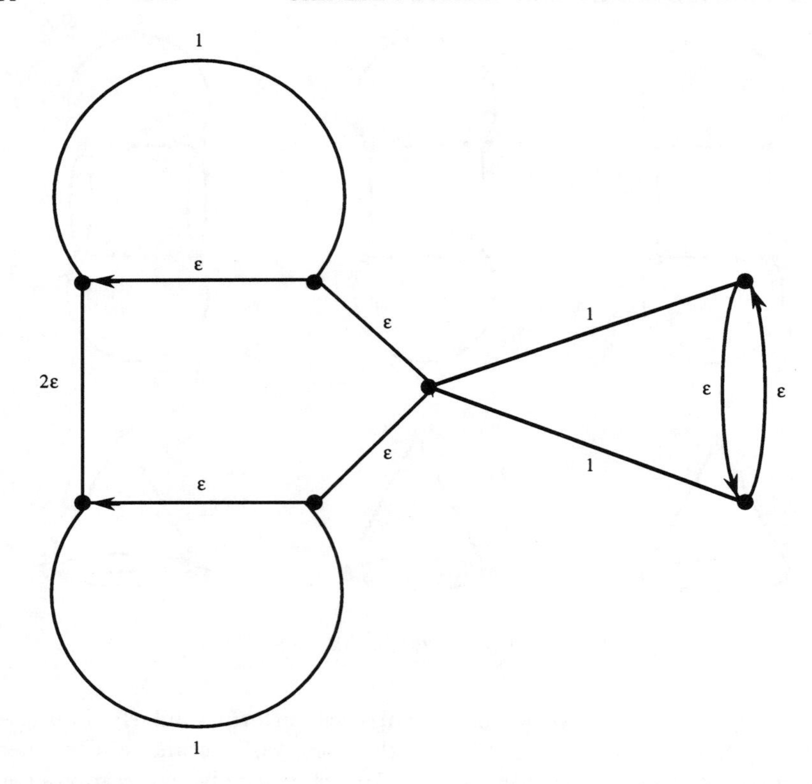

Figure 8.11. *Worst-known graph for the composite use of A_{ES} and A_{SE}*

the tightest value thus far obtained.

8.2 Hamiltonian Cycles and the Traveling Salesman Problem

To claim that the traveling salesman problem (TSP) is the most celebrated problem in combinatorial optimization might invite debate in some quarters. Here, of course, we have no interest in engaging in such an exercise; however, it is surely quite safe to anoint the TSP as one of the most "worked on" problems in the area. Lending testimony to this enormous concentration of effort is the vast literature that has evolved on the subject, including an excellent book of the same name (Lawler *et al.*, 1985).

Our coverage here of the TSP will be very limited, since the problem is so well studied with most major results readily accessible in the open literature. In addition, most of the successful (contemporary) approaches to the problem derive from inherently enumerative strategies, themselves involving machinery (*i.e.*, polyhedral approaches) well beyond the scope of this book. Rather, we will examine briefly the complexity of the problem after which we will examine some results pertaining to TSP heuristics.

8.2.1 Hamiltonian Cycles

The classic (and somewhat loosely stated) TSP asks for a tour through n "cities" that covers least total travel distance (equivalently, cost, time, etc.). The tour must begin and end at the same city with no city visited more than once. In graph-theoretic language, the problem is usually defined on a complete graph where edges are weighted as $w : E \rightarrow \mathbb{Z}$ and the aim is to find a spanning cycle of least total edge weight. These spanning cycles are referred to as *Hamiltonian cycles* after the work on same by the mathematician W.R. Hamilton (*e.g.*, Biggs *et al.*, 1976).

But what if our graph is not complete, and further, suppose that all edges have weight 1? Then, if our interest is in finding a so-called optimal traveling salesman tour, we would observe immediately that the only interest in the exercise would be to decide if the given graph admitted *any* tour. That is, every Hamiltonian cycle in the graph would have the same length and so we would need only find one such cycle, or, conversly, establish that no such cycles exist at all. To the truly uninitiated, this rather primitive problem of simply trying to find any Hamiltonian cycle in a graph is occasionally guessed to be much less difficult (if difficult at all) than the general TSP. To the slightly less uninitiated, however, this guess is well known to be wrong.

If a graph (digraph) G possesses a Hamiltonian cycle (circuit), it is said to be *Hamiltonian.* In this regard, a major question in graph theory asks: "Which graphs are Hamiltonian?" That is, given an arbitrary graph G, are there easily testable conditions that will reveal, unambiguously, whether or not G is Hamiltonian? Equivalently, are there *necessary and sufficient* conditions for a graph to be Hamiltonian?

No matter how posed, the answer to this question is *no.* This, in turn, leaves the task of deciding the hamiltonicity of an arbitrary graph or digraph problematical, requiring nothing less than a search of possibly exponential length in order to settle the issue. There is a formal way of stating this outcome:

Theorem 8.9 *Given a graph (digraph) $G = (V, E), ((V, A))$ the problem of deciding if G possesses a Hamiltonian cycle (circuit) is $\mathcal{NP}$-Complete.* □

The proof of this result is given in the original complexity paper by Karp (1972) and a more lucid exposition can be found in Garey and John-

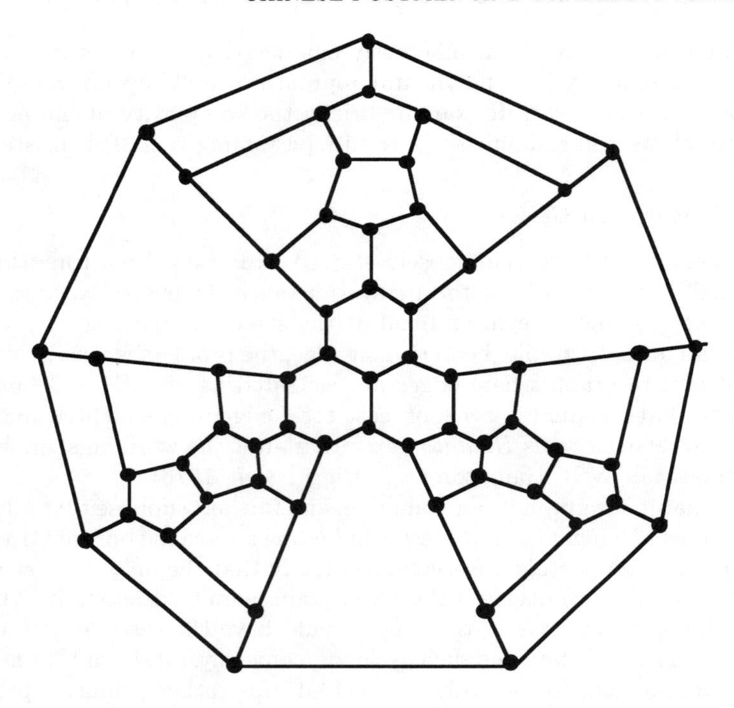

Figure 8.12. *A non-Hamiltonian graph*

son (1979). Moreover, the degree to which the problem remains difficult is exposed most starkly by the fact that intractability remains even if the input graphs are planar and regular in degree 3 (resp. digraphs with every vertex having total of in- and outdegree of 3). The graph in Figure 8.12 helps to make the point.

There are some easy *necessary* conditions for a graph to be Hamiltonian. For example, all Hamiltonian graphs must be biconnected. The more interesting exercise, however, is to examine *sufficiency* conditions, which if met guarantee hamiltonicity. Of these, the conditions of Chvátal (1972) are particularly notable because in a well-defined sense, they can be shown to be the best possible.

Let us assume our graph $G = (V, E)$ to be simple, and further, let us suppose the vertices to be labeled in *degree sequence*, *i.e.*, $d_1 \leq d_2 \leq \ldots \leq d_p$. Then Chvatal's conditions assert that for any such graph G with $p = |V| \geq 3$, we have that G is Hamiltonian if $d_k \leq k < p/2 \Rightarrow d_{p-k} \geq p - k$.

Example 8.7 The application of these conditions can be demonstrated on the graphs in Figure 8.13. Considering first the one on the left with degree

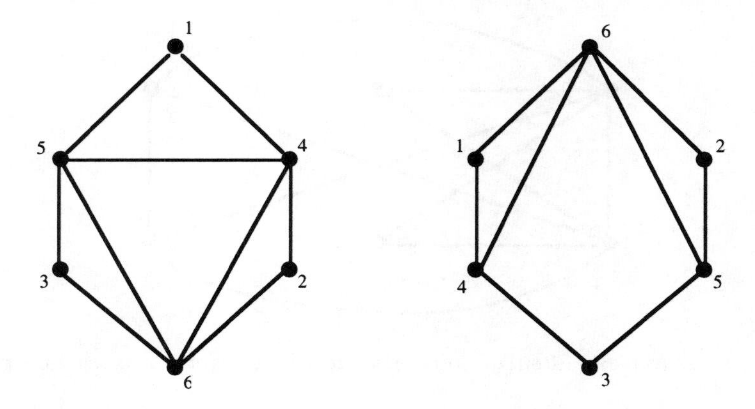

Figure 8.13. *Graphs of Example 8.7*

sequence (2,2,2,4,4,4) we have that $d_2 \leq 2 \Rightarrow d_4 \geq 4$ is true so we can expect the graph shown to be Hamiltonian which indeed is the case. On the other hand, the graph to the right is also clearly Hamiltonian but its degree sequence (2,2,2,3,3,4) fails the stated conditions, leaving us unable to make any claims accordingly, *i.e.*, $d_2 \leq 2 \not\Rightarrow d_4 \geq 4$ ($d_4 = 3$). □

Let us now examine the assertion that this hamiltonicity test of Chvatal's is unimprovable (among this sort of class of degree sequence-based conditions). Accordingly, we say that a sequence $d_1 \leq d_2 \leq \ldots \leq d_p$ is *graphic* if it is the degree sequence of a graph. If $D = (d_1, d_2, \ldots, d_p)$ and $D^* = (d_1^*, d_2^*, \ldots, d_p^*)$ are two graphic sequences and $d_i^* \geq d_i$ for $1 \leq i \leq p$, then we say that D^* *majorizes* D. Clearly, if a graphic sequence satisfies a given condition then so does every majorization of it. Now, Chvátal was able to show that *every* graphic sequence that fails the conditions stated above is majorized by the sequence $(d_1^*, d_2^*, \ldots, d_p^*)$ where $d_i^* = k$ for $1 \leq i \leq k$, $d_i^* = p - k - 1$ for $k + 1 \leq i \leq p - k$, and $d_i^* = p - 1$ for $p - k + 1 \leq i \leq p$ for some $k < p/2$. But the degree sequence formed in this way is graphic, and moreover, corresponds to the degree sequence of a non-Hamiltonian graph!

Chvátal (1972) also gives a general construction for these graphs while in Figure 8.14 we only demonstrate a specific case. Letting $p = 6$ and $k = 2$ Chvátal's model would produce the graph shown with degree sequence (2,2,3,3,5,5) and we have that $d_2 \leq 2 < p/2 \not\Rightarrow d_4 \geq 4$. It is also easy to see that this graph is not Hamiltonian.

Hence, if there existed a stronger condition than Chvátal's, there must be a graph that is Hamiltonian and has degree sequence satisfying the new condition but that fails Chvátal's. If such a sequence existed, say $\bar{D}$, then $\bar{D}$ is majorized by the sequence D^* (given above) which is the degree sequence

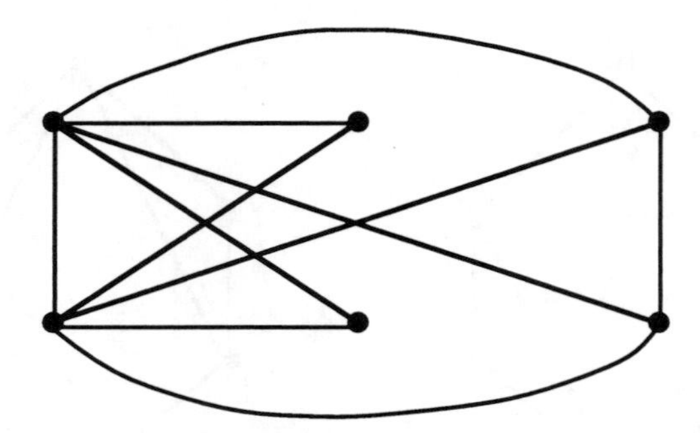

Figure 8.14. *Non-Hamiltonian graph generated by Chvátal's construction procedure*

of a non-Hamiltonian graph. Thus if $\bar{D}$ satisfies the new condition then so must every majorization of it (*i.e.*, D^*) but this would deny the validity of the new condition.

On the other hand, we can improve on the sufficiency condition just described if we are willing to do a little more work than testing the complexion of a graph's degree sequence. We define the *closure* of a graph G to be the graph obtained from G by recursively joining pairs of nonadjacent vertices whose degree sum is at least p until no such pair remains. We denote the closure of G by $C_p(G)$. Note also that a graph's closure is well defined.

Then Bondy and Chvátal (1976) proved that for any simple graph G of order $p \geq 3$, if $C_p(G) = K_p$, then G is Hamiltonian. Observe that satisfaction of the earlier condition of Chvátal relative to some G guarantees that $C_p(G)$ is complete. It is also clear that this new condition is strictly stronger than the previous one, *i.e.*, the closure of the second graph in the example above is complete. On the other hand, the limitation of this closure-based condition is easily (but by interesting means) established by recalling a result of Nash-Williams which holds that every r-regular graph on $2r + 1$ vertices is Hamiltonian (Nash-Williams, 1966), *i.e.*, such graphs are their own closures.

But let us suppose that the Bondy–Chvátal condition holds for a graph G. Then we know that G is Hamiltonian but this does not imply that we can produce the existing cycle quickly. Fortunately, however, this is not an issue in this case. Consider the algorithm below (Bondy and Chvátal, 1976):

A_{HC}: Hamiltonian Cycles in Graphs Satisfying the Bondy–Chvátal Closure Condition

Step 0: Let $G = (V, E)$ be the input graph upon which the closure

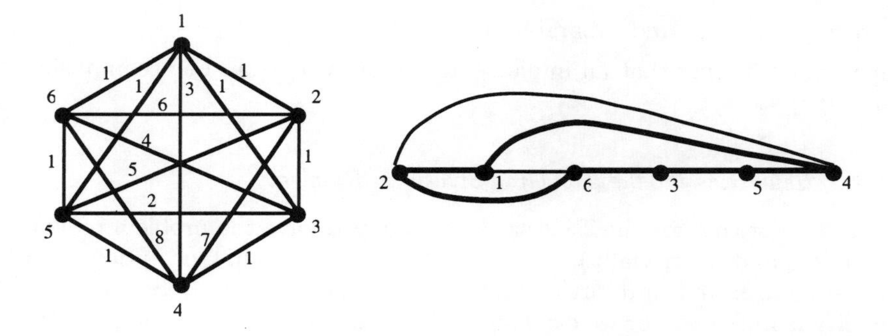

Figure 8.15. *Graph of Example 8.8*

construction is applied: Beginning with edges $(i, j) \in E$ labeled as $a_{ij} = 1$, proceed by adding edges, labeling each with consecutive integer values $2, 3, \ldots$. Select any Hamiltonian cycle in the final graph.

Step 1: Let the vertices in the present cycle be given by an ordering denoted as $(v_1, v_2, \ldots v_p, v_1)$ where $a_{v_p,v_1} \geq a_{v_i,v_{i+1}}$, $i = 1, 2, \ldots, p-1$. If $a_{v_p,v_1} = 1$ we may stop with a desired cycle in the original graph G. Otherwise, proceed to Step 2.

Step 2: Locate a vertex v_s in the present cycle (from Step 1) such that $a_{v_1,v_{s+1}} < a_{v_p,v_1}$ and $a_{v_p,v_s} < a_{v_p,v_1}$. Change the present cycle to take the form $(v_1, v_{s+1}, v_{s+2}, \ldots v_p, v_s, v_{s-1}, \ldots, v_1)$. Return to Step 1 with this new cycle. □

We will not take space to establish the validity of $\boldsymbol{A_{HC}}$, leaving this as an exercise (see Exercise 8-13). Following, we provide an illustration of how it works.

Example 8.8 Let us consider the earlier graph and in particular, its closure shown in Figure 8.15. The numbers on the edges indicate the corresponding labels per the closure construction. Suppose we select the initial cycle as (1,6,3,5,4,2,1). Here, $a_{42} = 7$ is maximum and so a_{v_p,v_1} is identified with edge (4,2) in the reoriented list (2,1,6,3,5,4,2). We have that $v_s = 1$ and $a_{v_1,v_{s+1}} = a_{26} = 6 < 7$, $a_{v_p,v_s} = a_{41} = 3 < 7$ and so the adjustment of Step 2 produces the new cycle (2,6,3,5,4,1,2). This update is depicted to the right in Figure 8.15.

We continue with $a_{26} = 6$ resulting as the greatest label value in the current cycle. Reorienting the same as (6,3,5,4,1,2,6), we have $v_s = 3$ whereupon we find $a_{65} = 1 < 6$ and $a_{23} = 1 < 6$ and we alter the cycle again to obtain (6,5,4,1,2,3,6). Finally, we have $a_{36} = 4$, v_s (per the reorientation) $=$ 4 and $a_{61} = 1 = a_{34} < 4$, and we can create the final cycle (6,1,2,3,4,5,6) which yields a desired subgraph in G. □

It is worth noting that an implementation of $\boldsymbol{A_{HC}}$ can be accomplished in $O(p^4)$ steps.

8.2.2 Heuristics for the Traveling Salesman Problem

A fast algorithm for the TSP would certainly resolve the problem of Theorem 8.9 and so (trivially) the TSP is $\mathcal{NP}$-Hard as well. Our options then are standard; we can resort to enumerative/exponential procedures if optimality is important or we can employ heuristic approaches if the premium on reduced computational effort is greater than that on finding exact solutions. As suggested previously, the former option is rich with elegant and well-documented results but that require a background beyond our present scope. As a consequence, we will devote our coverage in this section to some key results dealing with nonexact procedures, and even then, we will deal only with the problem defined on undirected graphs (the so-called symmetric TSP).

So, let an instance of our p-vertex problem be defined on a complete graph $G = (V, E) = K_p$, with edges weighted by $w : E \rightarrow \mathbb{Z}^{\geq 0}$, and further, where these edge weights satisfy the *triangle inequality*; that is $w_{ij} \leq w_{ik} + w_{kj}$ for all triples $\{i, j, k\} \in V$. We will be more specific later as to why these fairly stringent stipulations may not be easily relaxed. Regardless, consider the following heuristic due to Christofides (1976).

$\boldsymbol{A_C}$: Christofides' Heuristic for the TSP

Step 1: Find a minimum weight spanning tree in G. Let this tree be given by $T \subseteq E$.

Step 2: Let the odd-degree vertices in the tree of Step 1 be denoted by V_0 and find a minimum weight perfect matching in the subgraph induced by V_0. Let this matching be denoted by $M \subseteq E$.

Step 3: The graph formed as $M \cup T$ is Eulerian. Produce the corresponding (Eulerian) cycle and, interpreting it as a vertex sequence, form a TSP tour by beginning at the initial vertex, and proceeding in order, "shortcutting" past duplicated vertices until the starting vertex is reached again. □

It should be clear that this is a polynomial-time strategy. Following, we illustrate its application on an unweighted graph.

Example 8.9 Consider the graph in Figure 8.16. In part a, the tree is given. The matching of Step 2 is shown in part b, and in parts c and d, the Eulerian cycle and generated tour are shown respectively. □

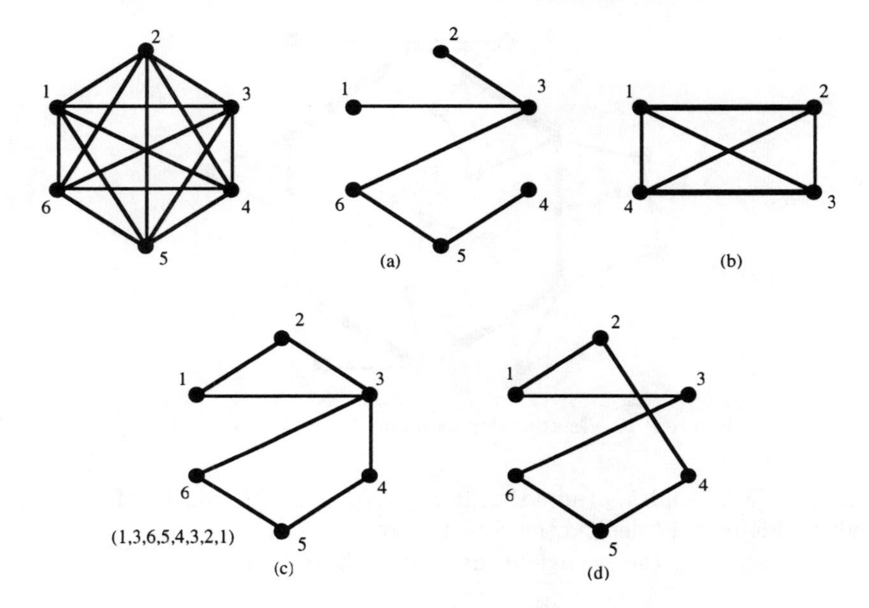

Figure 8.16. *Application of algorithm A_C*

The performance guarantee of $\boldsymbol{A_C}$ is relatively easy to establish. We can state the result in the following:

Theorem 8.10 *If the value of the solution produced by $\boldsymbol{A_C}$ is given by ν_c and the optimal value by ν^*, then*

$$\frac{\nu_c}{\nu^*} \leq \frac{3}{2}.$$

Proof. By the triangle inequality, it is clear that the value of a solution produced by the algorithm is bounded from above by the weight of the Eulerian graph resulting at Step 3. The latter is the sum of edge weights of the tree of Step 1 and that of the matching from Step 2. That is,

$$\nu_c \leq w(T) + w(M)$$

But we also know that the weight of the optimal tree provides a lower bound on the optimal TSP length, *i.e*,

$$w(T) \leq \nu^*$$

Now, consider the (nonempty) set V_0 and let us assume the $2k$ vertices that comprise it to be ordered in the same way that they appear in an optimal tour. Then the implied cycle through only these $2k$ vertices can be expressed as the disjoint union of two perfect matchings. Call these

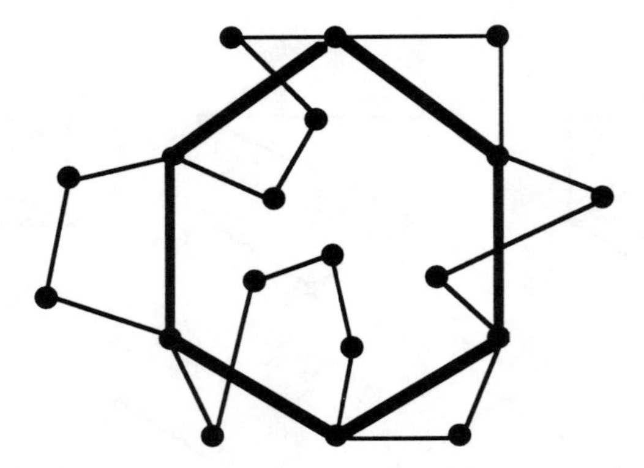

Figure 8.17. *Relationship between V_0 vertices and a tour*

matchings M_1 and M_2 (note that if $k = 1, M_1$, and M_2 consist of the same edge). Figure 8.17 depicts the construction.

Now, again by the triangle inequality, we have that

$$\nu^* \geq w(M_1) + w(M_2)$$

or equivalently,

$$\nu^* \geq 2w(M),$$

and upon combining, we are led to the desired result, *i.e.*,

$$\nu_c \leq \nu^* + \frac{\nu^*}{2}.$$

□

It is important to note that the bound of the theorem above has been sharpened in Cornuéjols and Nemhauser (1978) to $\frac{3m-1}{2m}$, where $m = \left\lfloor \frac{p}{2} \right\rfloor$ and $p \geq 3$ is the number of vertices in the instance. The following example establishes realizability.

Example 8.10 Consider the graph G in Figure 8.18 with edges weighted as shown. Note that for ease of depiction we have omitted many edges; those not shown are assumed to have arbitrary positive integer weights at least satisfying the triangle inequality. Observe also that we take m to have magnitude as indicated above.

Now, upon application of the heuristic, we could produce the minimum weight tree shown in Figure 8.19.

The matching over V_0 is particularly easy, resulting in the Eulerian graph in Figure 8.20 which, in turn, requires no application of Step 3, yielding a tour directly. The weight of the latter is $3m - 1$. Of course it is easy to

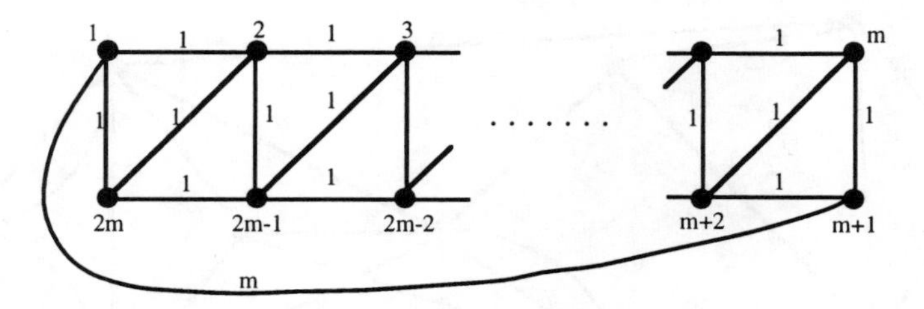

Figure 8.18. *Worst-case instance for A_C*

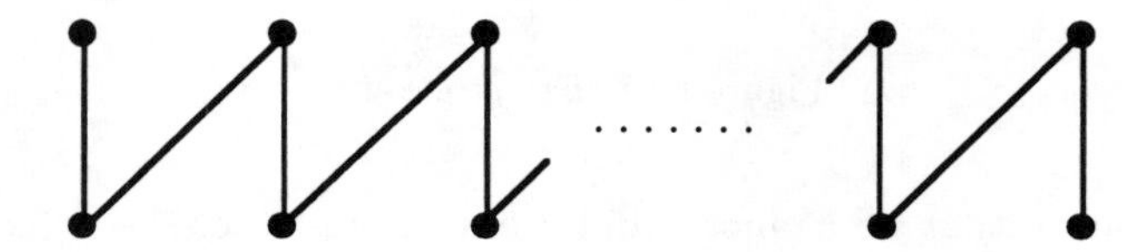

Figure 8.19. *Step 1 construction*

see that an optimal tour in G traces through the vertices as $1, 2, \ldots, m-1, m, m+1, \ldots, 2m-1, 2m, 1$ with total edge weight of $2m$. These two outcomes gives us the desired ratio. □

As one might suppose, numerous heuristics for the TSP have been proposed. Our choice here of Christofides' procedure is predicated upon its distinction as the approach that, to date, exhibits the best, constant performance, worst-case bound. In this regard, however, we note that there exists no result that suggests that an improvement on the 3/2 bound is not possible (unless $\mathcal{P} = \mathcal{NP}$).

But what if we relaxed the requirement that edge weights satisfy the triangle inequality? Clearly, many instances of the TSP, especially when abstracted from a single-processor scheduling context, are not likely to even be symmetric, let alone have edge weights that satisfy the stated metric. Unfortunately, that edge weights satisfy the triangle inequality appears to

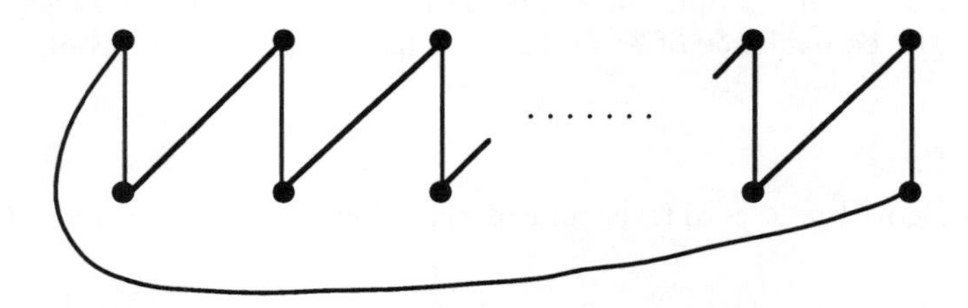

Figure 8.20. *Step 2 construction*

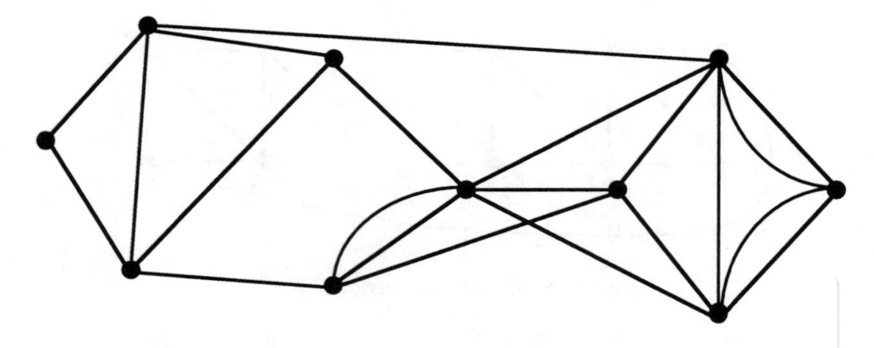

Figure 8.21. *Exercise 8-1*

be a stipulation that we cannot avoid if our interest includes a (constant) bound on the performance of a heuristic. And while the so-called worst-case perspective is often criticized in general, the following result from Sahni and Gonzalez (1976) (and others like it) should still serve as a stark reminder of our limitations.

Theorem 8.11 *If there exists any constant $\rho < +\infty$ and a polynomial time heuristic A for the arbitrary TSP such that the performance ratio $\nu_A/\nu^* \leq \rho$ is always satisfied, then $\mathcal{P} = \mathcal{NP}$.*

Proof. Assume such an algorithm exists with any constant performance bound. Now, consider an arbitrary graph $G = (V, E)$ and let us construct an instance of the TSP in the following way:

$$w_{ij} = \begin{cases} 1 & \text{if } (i,j) \in E \\ \rho|V| & \text{otherwise.} \end{cases}$$

Applying the hypothesized heuristic, we have that if G is Hamiltonian, then $\nu^* = |V|$ and thus $\nu_A \leq \rho|V|$ as claimed. Conversely, if G is not Hamiltonian, it must be that $\nu^* > \rho|V|$ from which it follows that $\nu_A > \rho|V|$. But this means that we have a polynomial time test for deciding the hamiltonicity of any graph and from the result of Theorem 8.9 this would establish the equivalence of $\mathcal{P}$ and $\mathcal{NP}$. This completes the proof. □

8.3 Exercises

8-1 Use algorithm $\boldsymbol{A_F}$ to trace an Eulerian traversal in the graph of Figure 8.21.

8-2 Use algorithm $\boldsymbol{A_{ED}}$ to find an Eulerian traversal in the digraph of Figure 8.22.

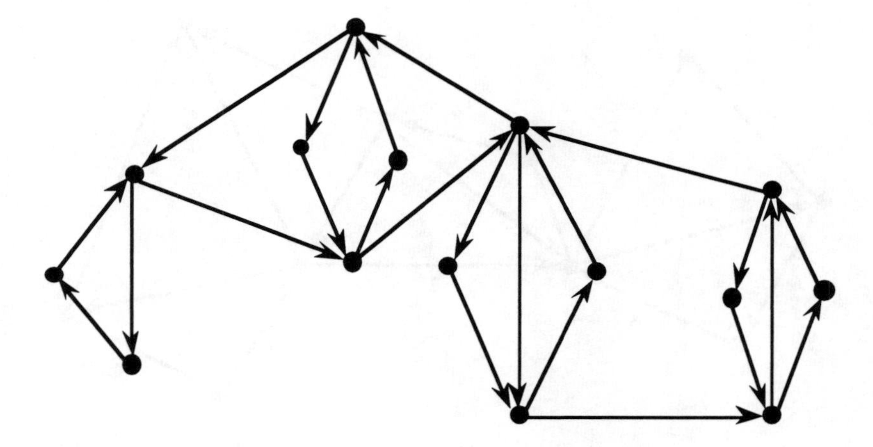

Figure 8.22. *Exercise 8-2*

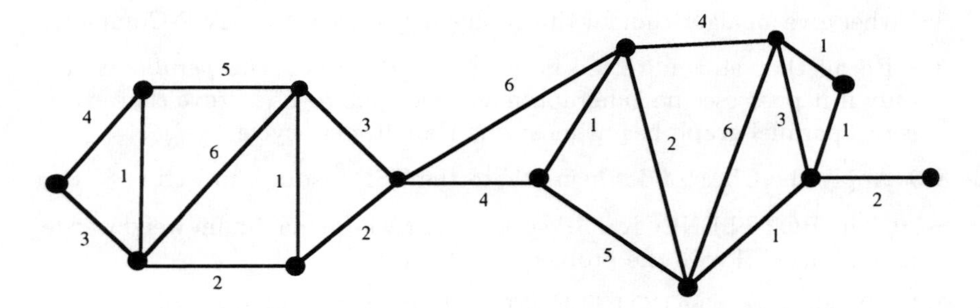

Figure 8.23. *Exercise 8-4*

8-3 Prove that Edmonds' algorithm $\boldsymbol{A_{UCP}}$ would never duplicate an edge more than once.

8-4 Apply $\boldsymbol{A_{UCP}}$ to the graph in Figure 8.23.

8-5 Apply $\boldsymbol{A_{DCP}}$ to the digraph in Figure 8.24.

8-6 Let us modify the conventional (undirected) postman problem to one where the traversal begins at a fixed vertex i and concludes at another vertex $j \neq i$. Either indicate a (fast) way to solve this new problem or prove it to be hard.

8-7 Consider the following EULERIAN SUBGRAPH problem:

Given a graph $G = (V, E)$, does G possess any spanning subgraph that is Eulerian?

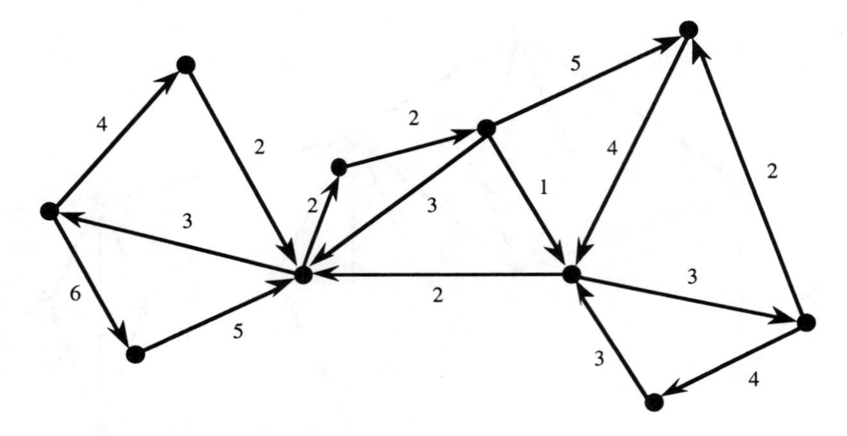

Figure 8.24. *Exercise 8-5*

Either give an algorithm for this problem or show it to be $\mathcal{NP}$-Complete.

8-8 Recall that an undirected graph $G = (V, E)$ is *series-parallel* if and only if it possesses no subgraph homeomorphic to K_4. Prove that every series-parallel graph has at most one Hamiltonian cycle.

8-9 Apply the Christofides heuristic to the TSP instance in Figure 8.25.

8-10 The BOTTLENECK TSP seeks a tour whose maximum weight edge is minimized. Prove the problem is $\mathcal{NP}$-Hard.

8-11 Relative to the BOTTLENECK TSP, state and prove a theorem analogous to Theorem 8.11.

8-12 Give a polynomial algorithm for producing the traversal in an Eulerian, mixed graph.

8-13 Prove that algorithm $\boldsymbol{A_{HC}}$ will stop with a valid Hamiltonian cycle in an input graph G satisfying the Bondy–Chvátal closure test.

8-14 Give a linear time algorithm for finding a maximum cardinaltiy Eulerian subgraph in the class of series-parallel graphs.

8-15 Repeat Exercise 8-7 given that the input graph is directed.

8-16 Repeat Exercise 8-14 for series-parallel digraphs (directed graphs with underlying undirected structure, series-parallel).

8-17 Relative to the (undirected) postman problem, suppose any subsequent traversal of an edge costs more than when it is used initially. Does this complicate the classic problem? Discuss.

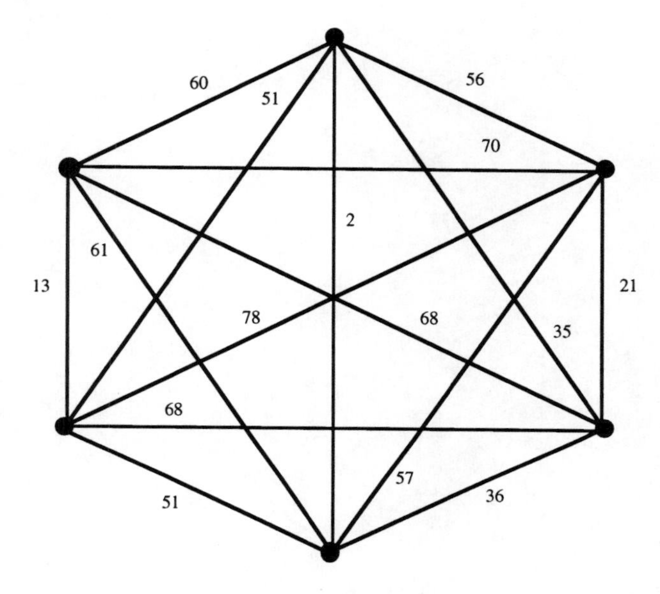

Figure 8.25. *Exercise 8-9*

8-18 In Nash-Williams (1966), it is shown that every r-regular graph in $2r+1$ vertices possesses a Hamiltonian cycle. So, for a graph G that satisfies this degree and order condition, give a fast algorithm for producing the existing Hamiltonian cycle.

8-19 A *Halin* graph is a planar graph whose edge set can be partitioned into a tree no vertex of which is degree 2 and a cycle through only and all degree-1 vertices of the tree. All Halin graphs are Hamiltonian. Repeat Exercise 8-18 for the class of Halin graphs.

8-20 Repeat Exercise 8-14 for the class of Halin graphs.

References

Abdul-Razaq, T.S. and Potts, C.N. (1988) Dynamic programming state-space relaxation for single-machine scheduling. *J. Oper. Res. Soc.*, **39**, 141-152.

Abernathy, W.J. and Demski, J.S. (1973) Simplification of activities in a network scheduling context. *Management Science*, **19**, 1052-1062.

Achugbue, J.O. and Chin, F.Y. (1981) Bounds on schedules for independent tasks with similar execution times. *J. Assoc. Comput. Mach.*, **28**, 81-99.

Achugbue, J.O. and Chin, F.Y. (1982a) Scheduling the open shop to minimize mean flow time. *SIAM J. Comput.*, **11**, 709-720.

Achugbue, J.O. and Chin, F.Y. (1982b) Complexity and solution of some three-stage flow shop scheduling problems. *Math. Oper. Res.*, **7**, 532-544.

Adams, J., Balas, E. and Zawack, D. (1988) The shifting bottleneck procedure for job shop scheduling. *Management Science*, **34**, 391-401.

Adiri, I. and Aizikowitz, N. (1986) *Openshop Scheduling Problems with Dominated Machines.* Operations Research, Statistics and Economics Mimeograph Series 383, Technion, Haifa.

Adiri, I. and Amit, N. (1983) Route-dependent open shop scheduling. *IIE Trans.*, **15**, 231-234.

Adiri, I. and Amit, N. (1984) Openshop and flow shop scheduling to minimize sum of completion times. *Comput. Opns. Res.*, **11**, 275-284.

Adolphson, D. and Hu, T.C. (1973) Optimal linear ordering. *SIAM J. Appl. Math.*, **25**, 403-423.

Agrawala, A.K., Coffman, Jr., E.G., Garey, M.R. and Tripathi, S.K. (1984) A stochastic optimization algorithm minimizing expected flow times on uniform processors. *IEEE Trans. Comput.*, **C-33**, 351-356.

Aho, A.V., Hopcroft, J.E. and Ullman, J.D. (1974) *The Design and Analysis of Computer Algorithms.* Addison-Wesley, Reading, Mass.

Akiyama, T., Nishizeki, T. and Saito, N. (1980) $\mathcal{NP}$-completeness of the Hamiltonian cycle problem for bipartite graphs. *J. Inform. Process.*, **3**, 73-76.

van den Akker, M., van Hoesel, S. and Savelsbergh, M. (1993) Facet inducing inequalities for single-machine scheduling problems. COC Report 93-03, School of Industrial and Systems Engineering, Georgia Institute of Technology.

Alasker, E.T. (1962) The basic technique: network analysis. J.W. Blood (ed.) *PERT, A New Management Planning and Control Technique*, American Management Assocation, 37-60.

Antill, J.M. and Woodhead, R.W. (1965) *Critical Path Methods in Construction Practice.* Wiley, New York.

Ashour, S. (1967) A decomposition approach for the machine scheduling problem. *Int. J. Prod. Res.*, **6**, 109-122.

Ashour, S. (1970a) An experimental investigation and comparative evaluation of flow-shop scheduling techniques. *Operations Research*, **12**, 541-549.

Ashour, S. (1970b) A branch-and-bound algorithm for the flow-shop scheduling problem. *AIIE Trans.*, **2**, 172-176.

Bagchi, U. and Ahmadi, R.H. (1987) An improved lower bound for minimizing weighted completion times with deadlines. *Operations Research*, **35**, 311-313.

Baker, K.R. (1974) *Introduction to Sequencing and Scheduling.* Wiley, New York.

Baker, K.R., Lawler, E.L., Lenstra, J.K. and Rinnooy Kan, A.H.G. (1983) Preemptive scheduling of a single machine to minimize maximum cost subject to release dates and precedence constraints. *Operations Research*, **31** 381-386.

Baker, K.R. and Schrage, L.E. (1978) Finding an optimal sequence by dynamic programming: an extension to precedence-related tasks. *Operations Research*, **26**, 111-120.

Baker, K.R. and Scudder, G.D. (1990) Sequencing with earliness and tardiness penalties: a review. *Operations Research*, **38**, 22-36.

Baker, K.R. and Su, Z.-S. (1974) Sequencing with due-dates and early start times to minimize maximum tardiness. *Naval Research Logistics Quarterly*, **21**, 171-176.

Balas, E. (1970) Project scheduling with resource constraints. E.M.L. Bale (ed.) *Applications of Mathematical Programming Techniques.* The English University Press, London, 187-200.

Balas, E. (1985) On the facial structure of scheduling polyhedra. *Math. Programming Stud.*, **24**, 179-218.

Balas, E. and Christofides, N. (1981) A restricted Lagrangean approach to the traveling salesman problem. *Mathematical Programming*, **21**, 19-46.

Banerjee, B.P. (1965) Single facility sequencing with random execution times. *Operations Research*, **13**, 358-364.

Barany, I. (1981) A vector-sum theorem and its application to improving flow shop guarantees. *Math. Oper. Res.*, **6**, 445-452.

Barker, J.R. and McMahon, G.B. (1985) Scheduling the general job-shop. *Management Science*, **31**, 594-598.

Barnes, J.W. and Brennan, J.J. (1977) An improved algorithm for scheduling jobs on identical machines. *AIIE Trans.*, **9**, 25-31.

Barr, J. and Howard, W.E. (1961) *Polaris.* Harcourt, Brace, Jovanovich, New York.

Bartholdi, J.J. (1981) A guaranteed-accuracy round-off algorithm for cyclic scheduling and set covering. *Operations Research*, **29**, 501-510.

Bartholdi, J.J. and McCroan, K.L. (1990) Scheduling interviews for a job fair. *Operations Research*, **38**, 951-960.

Bartholdi, J.J., Orlin, J.B. and Ratliff, H D. (1980) Cyclic scheduling via integer programs with circular ones. *Operations Research*, **28**, 1074-1085.

Bartholdi, J.J. and Ratliff, H.D. (1978) Unnetworks, with applications to idle time scheduling. *Management Science*, **24**, 850-858.

Bartusch, M. Möhring, R.H. and Radermacher, F.J. (1988a) M-machine unit time scheuling: a report of ongoing research. A. Kurzhanski, K. Neumann and D. Pallaschke (eds.) *Optimization, Parallel Processing, and Applications.* Lecture Notes in Economics and Mathematical Systems 304, Springer, Berlin, 165-212.

Bartusch, M., Möhring, R.H. and Radermacher, F.J. (1988b) Scheduling project networks with resource constraints and time windows. *Ann. Oper. Res.*, **16**, 201-240.

Battersby, A. (1967) *Network Analysis.* Macmillan, New York.

Battersby, A. (1970) *Network Analysis for Planning and Scheduling.* St. Martin's Press, New York, third edition.

Beardwood, J., Halton, J.H. and Hammersley, J.M. (1959) The shortest path through many points. *Proc. Cambridge Philos. Soc.*, **55**, 299-327.

Bellmore, M. and Nemhauser, G.L. (1968) The traveling salesman problem: a survey. *Operations Research*, **16**, 538-558.

Belouadah, H., Posner, M.E. and Potts, C.N. (1989) *A Branch and Bound Algorithm for Scheduling Jobs with Release Dates on a Single Machine to Minimize Total Weighted Completion Time.* Preprint OR14, Faculty of Mathematical Studies, University of Southampton.

Bennington, G.E. and McGinnis, L.F. (1973) A critique of project planning with constrained resources. S.E. Elmaghraby (ed.) *Symposium on Theory of Scheduling and Its Applications,* Springer-Verlag, New York, 1-28.

Berge, C. (1973) *Graphs and Hypergraphs.* North-Holland, Amsterdam.

Berman, E.B. (1964) Resource allocation in a PERT network under continuous time-cost functions. *Management Science*, **10**, 734-745.

Bermond, J.C. (1979) Hamiltonian graphs. L.W. Beineke and R.J. Wilson (eds.) *Selected Topics in Graph Theory*, Academic Press, New York, 127-167.

Bermond, J.C. and Thomassen, C. (1981) Cycles in digraphs—a survey. *J. Graph Theory*, **5**, 1-43.

Bianco, I. and Ricciardelli, S. (1982) Scheduling of a single machine to minimize total weighted completion time subject to release dates. *Naval Research Logistics Quarterly*, **29**, 151-167.

Bigelow, O.J. (1962) Bibliography on project planning and control by network analysis: 1959-1961. *Operations Research*, **10**, 728-731.

Biggs, N.L., Lloyd, E.K. and Wilson, R.J. (1976) *Graph Theory 1736-1936.* Clarendon Press, Oxford.

Bixby, R.E. and Wang, D. (1978) An algorithm for finding Hamiltonian circuits in certain graphs. *Math. Programming Stud.*, **8**, 35-49.

Blau, R.A. (1973) N-job, one machine sequencing problems under uncertainty. *Management Science*, **20**, 101-109.

Blazewicz, J. (1987) Selected topics in scheduling theory. *Ann. Discrete Math.*, **31**, 1-60.

Blazewicz, J., Finke, G., Haupt, R. and Schmidt, G. (1988) New trends in machine scheduling. *European J. Oper. Res.*, **37**, 303-317.

Blazewicz, J., Lenstra, J.K. and Rinnooy Kan, A.H.G. (1983) Scheduling subject to resource constraints: classification and complexity. *Discrete Applied Mathematics*, **5**, 11-24.

Bondy, J.A. (1969) Properties of graphs with constraints on degrees. *Studia Sci. Math. Hungar.*, **4**, 473-475.

Bondy, J.A. (1972) Variations on the Hamiltonian theme. *Canad. Math. Bull.*, **15**, 57-62.

Bondy, J.A. (1978) Hamilton cycles in graphs and digraphs. *Proceedings of the Ninth Southeastern Conference on Combinatorics, Graph Theory and Computing*, Utilitas Mathematica, Winnipeg, 3-28.

Bondy, J.A. and Chvátal, V. (1976) A method in graph theory. *Discrete Mathematics*, **15**, 111-135.

Bondy, J.A. and Murty, U.S.R. (1976) *Graph Theory with Applications.* MacMillan, London.

Borie, R., Parker, R.G. and Tovey, C.A. (1992) Automatic generation of linear-time algorithms from predicate calculus descriptions of problems on recursively constructable graph families. *Algorithmica*, **7**, 555-581.

Boxma, O.J. (1984) Probabilistic analysis of the LPT scheduling rule. E. Gelenbe (ed.) *Performance '84*. North-Holland, Amsterdam, 475-490.

Boxma, O.J. and Forst, F.G. (1986) Minimizing the expected weighted number of tardy jobs in stochastic flow shops. *Operations Research Letters*, **5**, 119-126.

Bräsel, H., Kluge, D. and Werner, F. (1994) A polynomial algorithm for the $n|m|O,\ t_{ij} = 1,\ tree|C_{\max}$ open shop problem. *European J. Oper. Res.*, **72**, 125-134.

Bratley, P., Florian, M. and Robillard, P. (1973) On sequencing with earliest starts and due dates with application to computing bounds for the $(n/m/G/F_{\max})$ problem. *Naval Research Logistics Quarterly*, **20** 57-67.

Bratley, P., Florian M. and Robillard, P. (1975) Scheduling with earliest start and due date constraints on multiple machines. *Naval Research Logistics Quarterly*, **22**, 165-173.

Brucker, P. (1981) Minimizing maximum lateness in a two-machine unit-time job shop. *Computing*, **27**, 367-370.

Brucker, P. (1982) A linear time algorithm to minimize maximum lateness for the two-machine, unit-time, job-shop scheduling problem. R.F. Drenick and F. Kozin (eds.) *System Modeling and Optimization.* Lecture Notes in Control and Information Sciences 38, Springer, Berlin, 566-571.

Brucker, P., Garey, M.R. and Johnson, D.S. (1977) Scheduling equal-length tasks under tree-like precedence constraints to minimize maximum lateness. *Math. Oper. Res.*, **2**, 275-284.

Brumelle, S.L. and Sidney, J.B. (1982) *The Two Machine Makespan Problem with Stochastic Flow Times.* Technical report, University of British Columbia, Vancouver.

Bruno, J.L., Coffman, Jr., E.G. and Sethi, R. (1974) Scheduling independent tasks to reduce mean finishing time. *Comm. ACM*, **17**, 382-387.

Bruno, J.L. and Downey, P.J. (1977) *Sequencing Tasks with Exponential Service*

Times on Two Machines. Technical report, Department of Electrical Engineering and Computer Science, University of California, Santa Barbara.

Bruno, J.L. and Downey, P.J. (1986) Probabilistic bounds on the performance of list scheduling. *SIAM J. Comput.*, **15**, 409-417.

Bruno, J.L., Downey, P.J. and Frederickson, G.N. (1981) Sequencing tasks with exponential service times to minimize the expected flowtime or makespan. *J. Assoc. Comput. Mach.*, **28**, 100-113.

Bruno, J.L. and Gonzalez T. (1976) *Scheduling Independent Tasks with Release Dates and Due Dates on Parallel Machines.* Technical Report 213, Computer Science Department, Pennsylvania State University.

Buer, H. and Möhring, R.H. (1983) A fast algorithm for the decomposition of graphs and posets. *Math. Oper. Res.*, **8**, 170-184.

Bulfin, R.L. and Liu, C.Y. (1988) Scheduling open shops with unit execution times to minimize functions of due dates. *Operations Research*, **4**, 553-559.

Campbell, H.G., Dudek, R.A. and Smith, M.L. (1970) A heuristic algorithm for the n job, m machine sequencing problem. *Management Science*, **16**, B630-637.

Carlier, J. (1982) The one-machine sequencing problem. *European J. Oper. Res.*, **11**, 42-47.

Carlier, J. (1987) Scheduling jobs with release dates and tails on identical machines to minimize makespan. *European J. Oper. Res.*, **29**, 298-306.

Carlier, J. and Pinson, E. (1989) A method for the job-shop problem. *Management Science*, **35**, 164-176.

Carpaneto, G. and Toth, P. (1980) Some new branching and bounding criteria for the asymmetric travelling salesman problem. *Management Science*, **26**, 736-743.

Carter, M.W. (1986) A survey of practical applications of examination timetabling algorithms. *Operations Research*, **34**, 193-202.

Carter, M.W. and Tovey, C.A. (1992) When is the classroom assignment problem hard? *Operations Research*, **40**, S28-S39.

Charlton, J.M. and Death, C.C. (1970) A generalized machine scheduling algorithm. *Oper Res. Quart.*, **21**, 127-134.

Chen, N.-F. (1975) *An Analysis of Scheduling Algorithms in Multiprocessing Computing Systems.* Technical Report UIUCDCS-R-75-724, Department of Computer Science, University of Illinois at Urbana-Champaign.

Chen, N.-F. and Liu, C.L. (1975) On a class of scheduling algorithms for multiprocessors computing systems. T.-Y. Feng (ed.) (1975) *Parallel Processing.* Lecture Notes in Computer Science 24, Springer, Berlin, 1-16.

Cheng, T.C.E. and Gupta, M.C. (1989) Survey of scheduling research involving due date determination decisions. *European J. Oper. Res.*, **38**, 156-166.

Chin, F.Y. and Tsai, L.-L. (1981) On J-maximal and J-minimal flow-shop schedules. *J. Assoc. Comput. Mach.*, **28**, 462-476.

Cho, Y. and Sahni S. (1980) Bounds for list schedules on uniform processors. *SIAM J. Comput.*, **9**, 91-103.

Cho, Y. and Sahni, S. (1981) Preemptive scheduling of independent jobs with release and due times on open, flow and job shops. *Operations Research*, **29**,

511-522.

Christofides, N. (1970) The shortest Hamiltonian chain of a graph. *SIAM J. Appl. Math.*, **19**, 689-696.

Christofides, N. (1976) *Worst-Case Analysis of a New Heuristic for the Travelling Salesman Problem.* Report 388, Graduate School of Industrial Administration, Carnegie-Mellon University, Pittsburgh, PA.

Christofides, N. (1979) The travelling salesman problem. N. Christofides, A. Mingozzi, P. Toth and C. Sandi (eds.) *Combinatorial Optimization*, Wiley, Chichester, 131-149.

Christofides, N., Alvarez-Valdes, R. and Tamarit, J.M. (1987) Project scheduling with resource constraints: a branch and bound approach. *European J. Oper. Res.*, **29**, 262-273.

Chvátal, V. (1972) On Hamilton's ideals. *J. Combin. Theory*, **12B**, 163-168.

Chvátal, V. (1973) Tough graphs and Hamiltonian circuits. *Discrete Math.*, **5**, 215-228.

Coffman, Jr., E.G. (ed.) (1976) *Computer & Job/Shop Scheduling Theory.* Wiley, New York.

Coffman, Jr., E.G., Flatto, L., Garey, M.R. and Weber, R.R. (1987) Minimizing expected makespans on uniform processor systems. *Adv. Appl. Probab.*, **19**, 177-201.

Coffman, Jr., E.G., Flatto, L. and Lueker, G.S. (1984) Expected makespans for largest-fit multiprocessor scheduling. E. Gelenbe (ed.) *Performance '84*, North-Holland, Amsterdam, 491-506.

Coffman, Jr., E.G., Garey, M.R. and Johnson, D.S. (1978) An application of bin-packing to multiprocessor scheduling. *SIAM J. Comput.*, **7**, 1-17.

Coffman, Jr., E.G. and Gilbert, E.N. (1985) On the expected relative performance of list scheduling. *Operations Research*, **33**, 548-561.

Coffman, Jr., E.G. and Graham, R.L. (1972) Optimal scheduling for two-processor systems. *Acta Informat.*, **1**, 200-213.

Coffman, Jr., E.G., Lueker, G.S. and Rinnooy Kan, A.H.G. (1988) Asymptotic methods in the probabilistic analysis of scheduling and packing heuristics. *Management Science*, **34**, 266-290.

Conway, R.W., Maxwell, W.L. and Miller, L.W. (1967) *Theory of Scheduling.* Addison-Wesley, Reading, MA.

Cook, S.A. (1971) The complexity of theorem-proving procedures. *Proc. 3rd Annual ACM Symp. Theory of Computing*, 151-158.

Cornuéjols, G., Naddef, D. and Pulleyblank, W.R. (1983) Halin graphs and the traveling salesman problem. *Math. Programming*, **26**, 287-294.

Cornuéjols, G. and Nemhauser, G.L. (1978) Tight bounds for Christofides' traveling salesman heuristic. *Mathematical Programming*, **14**, 116-121.

Crabill, T.B. and Maxwell, W.L. (1969) Single machine sequencing with random processing times and random due-dates. *Naval Research Logistics Quarterly*, **16**, 549-554.

Crowder, H., Johnson, E.L. and Padberg, M.W. (1983) Solving large-scale zero-one linear programming problems. *Operations Research*, **31**, 803-834.

Crowder, H. and Padberg, M.W. (1980) Solving large-scale symmetric travelling salesman problems to optimality. *Management Science*, **26**, 495-509.

Dannenbring, D.G. (1977) An evaluation of flow shop sequencing heuristics. *Management Science*, **23**, 1174-1182.

Davida, G.I. and Linton, D.J. (1976) A new algorithm for the scheduling of tree structured tasks. *Proc. Conf. Inform. Sci. Syst.*, Baltimore, Maryland, 543-548.

Davis, E. and Jaffe, J.M. (1981) Algorithms for scheduling tasks on unrelated processors. *J. Assoc. Comput. Mach.*, **28**, 721-736.

Davis, E.W. (1966) Resource allocation in project network models—a survey. *J. Ind. Eng.*, **17**, 177-188.

Davis, E.W. (1973) Project scheduling under resource constraints—historical review and categorization of procedures. *AIIE Trans.*, **5**, 297-313.

Day, J. and Hottenstein. M.P. (1970) Review of scheduling research. *Naval Research Logistics Quarterly*, **17**, 11-39.

Dempster, M.A.H., Lenstra, J.K. and Rinnooy Kan, A.H.G. (eds.) (1982) *Deterministic and Stochastic Scheduling*. Reidel, Dordrecht.

Dessouky, M.I. and Deogun, J.S. (1981) Sequencing jobs with unequal ready times to minimize mean flow time. *SIAM J. Comput.*, **10**, 192-202.

Dessouky, M.I., Lageweg, B.J. and van de Velde, S.L. (1989) *Scheduling Identical Jobs on Uniform Parallel Machines*. Report BS-89xx. Centre for Mathematics and Computer Science, Amsterdam.

Dijkstra, E.W. (1959) A note on two problems in connection with graphs. *Numer. Math.*, **1**, 269-271.

Dileepan, P. and Sen, T. (1988) Bicriterion static scheduling research for a single machine. *Omega*, **16**, 53-59.

Dirac, G.A. (1952) Some theorems on abstract graphs. *Proc. London Math. Soc.*, **2**, 69-81.

Dirac, G.A. (1960) In abstrakten Graphen vorhandene vollständige 4-Graphen und ihre Untereilungen. *Math. Nachr.*, **22**, 61-85.

Dobson, G. (1984) Scheduling independent tasks on uniform processors. *SIAM J. Comput.*, **13**, 705-716.

Dolev, D. and Warmuth, M.K. (1984) Scheduling precedence graphs of bounded height. *J. Algorithms*, **5**, 48-59.

Dolev, D. and Warmuth, M.K. (1985a) Scheduling flat graphs. *SIAM J. Comput.*, **14**, 638-657.

Dolev, D. and Warmuth, M.K. (1985b) Profile scheduling of opposing forests and level orders. *SIAM J. Algebraic Discrete Methods*, **6**, 665-687.

Du, J. and Leung, J.Y.-T. (1988a) Scheduling tree-structure tasks with restricted execution times. *Inform. Process. Lett.*, **28**, 183-188.

Du, J. and Leung, J.Y.-T. (1988b) *Minimizing Mean Flow Time with Release Time and Deadline Constraints*. Technical Report, Computer Science Program, University of Texas, Dallas.

Du, J. and Leung, J.Y.-T. (1989) Scheduling tree-structured tasks on two processors to minimize schedule length. *SIAM J. Discrete Math.*, **2**, 176-196.

Du, J. and Leung, J.Y.-T. (1990) Minimizing total tardiness on one machine is

$\mathcal{NP}$-hard. *Math. Oper. Res.*, **15**, 483-494.

Du, J., Leung, J.Y.-T. and Wong, C.S. (1989) *Minimizing the Number of Late Jobs with Release Time Constraints.* Technical Report, Computer Science Program, University of Texas, Dallas.

Du, J., Leung, J.Y.-T. and Young, G.H. (1988) *Minimizing Mean Flow Time with Release Time Constraint.* Technical Report, Computer Science Program, University of Texas, Dallas.

Du, J., Leung, J.Y.-T. and Young, G.H. (1991) Scheduling chain-structured tasks to minimize makespan and mean flow time. *Inform. Comput.*, **92**, 219-236.

Dudek, R.A., Panwalker, S.S. and Smith, M.L. (1992) The lessons of flow shop scheduling research. *Operations Research*, **40**, 7-13.

Eastman, W.L., Even, S. and Isaacs, I.M. (1964) Bounds for the optimal scheduling of n jobs on m processors. *Management Science*, **11**, 268-279.

Edmonds, J. (1965a) Paths, trees, and flowers. *Canad. J. Math.*, **17**, 449-467.

Edmonds, J. (1965b) Minimum partition of a matroid into independent subsets. *J. Res. Nat. Bur. Standards*, **69B**, 67-72.

Edmonds, J. (1965c) Maximum matching and a polyhedron with 0,1-vertices. *J. Res. Nat. Bur. Standards*, **69B**, 125-130.

Edmonds, J. (1965d) The Chinese postman's problem (abstract). *Operations Research*, **13**, Suppl. 1, B73.

Edmonds, J. (1971) Matroids and the greedy algorithm. *Math. Programming* , **1**, 127-136.

Edmonds, J. and Johnson, E.L. (1973) Matching, Euler tours, and the Chinese postman. *Math. Programming*, **5**, 88-124.

Elmaghraby, S.E. (1968) The one-machine sequencing problem with delay costs. *J. Ind. Eng.*, **19**, 105-108.

Elmaghraby, S.E. (1970) *Some Network Models in Management Science.* Springer-Verlag, New York. Lecture Notes in Operations Research and Mathematical Systems, **29**.

Elmaghraby, S.E. (1977) *Activity Networks.* John Wiley, New York.

Elmaghraby, S.E. and Park, S.H. (1974) Scheduling jobs on a number of identical machines. *AIIE Trans.*, **6**, 1-12.

Emmons, H. (1969) One machine sequencing to minimize certain functions of job tardiness. *Oper. Res.*, **17**, 701-715.

Erschler, J., Fontan, G., Merce, C. and Roubellat, F. (1982) Applying new dominance concepts to job schedule optimization. *European J. Oper. Res.*, **11**, 60-66.

Erschler, J., Fontan, G., Merce, C. and Roubellat, F. (1983) A new dominance concept in scheduling n jobs on a single machine with ready times and due dates. *Operations Research*, **31**, 114-127.

Euler, L. (1736) Solutio problematis ad geometriam situs pertinentis. *Commentarii Academiae Petropolitanae*, **8**, 128-140.

Euler, L. (1759) Solution d'une question curieuse qui ne paroit soumise à aucune analyse. *Mem. Acad. Sci. Berlin*, **15**, 310-337.

Evarts, H.F. (1964) *Introduction to PERT.* Allyn and Bacon, Boston, Mass.

Falk, J.E. and Horowitz, J.L. (1972) Critical path problems with concave cost-time curves. *Management Science*, **19**, 446-455.

Federgruen, A. and Groenevelt, H. (1986) Preemptive scheduling of uniform machines by ordinary network flow techniques. *Management Science*, **32**, 341-349.

Fiala, T. (1983) An algorithm for the open-shop problem. *Math. Oper. Res.*, **8**, 100-109.

Fischetti, M. and Martello, S. (1987) Worst-case analysis of the differencing method for the partition problem. *Math. Programming*, **37** 117-120.

Fisher, H. and Thompson, G.L. (1963) Probabilistic learning combinations of local job-shop scheduling rules. J.F. Muth and G.L. Thompson (eds.) *Industrial Scheduling*. Prentice-Hall, Englewood Cliffs, New Jersey, 225-251.

Fisher, M.L. (1976) A dual algorithm for the one-machine scheduling problem. *Math. Programming*, **11**, 229-251.

Fisher, M.L. and Krieger, A.M. (1984) Analysis of a linearization heuristic for single-machine scheduling to maximize profit. *Math. Programming*, **28**, 218-225.

Fisher, M.L., Lageweg, B.J., Lenstra, J.K. and Rinnooy Kan, A.H.G. (1983) Surrogate duality relaxation for job scheduling. *Discrete Applied Mathematics*, **5**, 65-75.

Fleischner, H. (1974) The square of every two-connected graph is Hamiltonian. *J. Combin. Theory, Ser. B*, **16**, 29-34.

Flood, M.M. (1956) The traveling-salesman problem. *Operations Research*, **4**, 61-75.

Forst, F.G. (1984) A review of the static, stochastic job sequencing literature. *Opsearch*, **21**, 127-144.

Frederickson, G.N. (1979) Approximation algorithms for some postman problems. *J. Assoc. Comput. Mach.*, **26**, 538-554.

Frederickson, G.N. (1983) Scheduling unit-time tasks with integer release times and deadlines. *Inform. Process. Lett.*, **16**, 171-173.

Frederickson, G.N., Hecht, M.S. and Kim, C.E. (1978) Approximation algorithms for some routing problems. *SIAM J. Comput.*, **7**, 178-193.

French, S. (1982) *Sequencing and Scheduling: An Introduction to the Mathematics of the Job-Shop*. Horwood, Chichester.

Frenk, J.B.G. (1988) , *A General Framework for Stochastic One-Machine Scheduling Problems with Zero Release Times and No Partial Ordering*. Economic Institute, Erasmus University, Rotterdam.

Frenk, J.B.G. and Rinnooy Kan, A.H.G. (1986) The rate of convergence to optimality of the LPT rule. *Discrete Applied Mathematics*, **14**, 187-197.

Frenk, J.B.G. and Rinnooy Kan, A.H.G. (1987) The asymptotic optimality of the LPT rule. *Math. Oper. Res.*, **12**, 241-254.

Friesen, D.K. (1984) Tighter bounds for the multifit processor scheduling algorithm. *SIAM J. Comput.*, **13**, 170-181.

Friesen, D.K. (1987) Tighter bounds for LPT scheduling on uniform processors. *SIAM J. Comput.*, **16**, 554-560.

Friesen, D.K. and Langston, M.A. (1983) Bounds for multifit scheduling on uniform processors. *SIAM J. Comput.*, **12**, 60-70.

Friesen, D.K. and Langston, M.A. (1986) Evaluation of a MULTIFIT- based scheduling algorithm. *J. Algorithms*, **7**, 35-59.

Frieze, A.M., Galbiati, G. and Maffioli, F. (1982), On the worst-case performance of some algorithms for the asymmetric traveling salesman problem. *Networks*, **12**, 23-39.

Frostig, E. (1988) A stochastic scheduling problem with intree precedence constraints. *Operations Research*, **36**, 937-943.

Fujii, M., Kasami, T. and Ninomiya, K. (1969, 1971) Optimal sequencing of two equivalent processors. *SIAM J. Appl. Math.*, **17**, 784-789; Erratum. *SIAM J. Appl. Math.*, **20**, 141.

Fulkerson, D.R. (1961) A network flow computation for project cost curve. *Management Science*, **7**, 167-178.

Fulkerson, D.R. (1964) *Scheduling in Project Networks*. RAND memo, RM–4137–PR, RAND Corporation, Santa Monica.

Gabow, H.N. (1982) An almost linear-time algorithm for two-processor scheduling. *J. Assoc. Comput. Mach.*, **29**, 766-780.

Gabow, H.N. (1988) Scheduling UET systems on two uniform processors and length two pipelines. *SIAM J. Comput.*, **17**, 810-829.

Gabow, H.N. and Stallmann, M. (1986) An augmenting path algorithm for linear matroid parity. *Combinatorica*, **6**, 123-150.

Gabow, H.N. and Tarjan, R.E. (1985) A linear-time algorithm for a special case of disjoint set union. *J. Comput. System Sci.*, **30**, 209-221.

Garey, M.R., Graham, R.L. and Johnson, D.S. (1978) Performance guarantees for scheduing algorithms. *Operations Research*, **26**, 3-21.

Garey, M.R. and Johnson, D.S. (1975) Complexity results for multiprocessor scheduling under resource constraints. *SIAM J. Comput.*, **4**, 397-411.

Garey, M.R. and Johnson, D.S. (1976) Scheduling tasks with nonuniform deadlines on two processors. *J. Assoc. Comput. Mach.*, **23**, 461-467.

Garey, M.R., Johnson, D.S. and Tarjan, R.E. (1976) The planar Hamiltonian circuit problem is $\mathcal{NP}$-complete. *SIAM J. Comput.*, **5**, 704-714.

Garey, M.R. and Johnson, D.S. (1977) Two-processor scheduling with start-times and deadlines. *SIAM J. Comput.*, **6**, 416-426.

Garey, M.R. and Johnson, D.S. (1978) Strong $\mathcal{NP}$-completeness results: motivation, examples and implications. *J. Assoc. Comput. Mach.*, **25**, 499-508.

Garey, M.R. and Johnson, D.S. (1979) *Computers and Intractability: A Guide to the Theory of $\mathcal{NP}$-Completeness*. Freeman, San Francisco.

Garey, M.R., Johnson, D.S. and Sethi, R. (1976) The complexity of flowshop and jobshop scheduling. *Math. Oper. Res.*, **1**, 117-129.

Garey, M.R., Johnson, D.S., Simons, B.B. and Tarjan, R.E. (1981) Scheduling unit-time tasks with arbitrary release times and deadlines. *SIAM J. Comput.*, **10**, 256-269.

Garey, M.R., Johnson, D.S., Tarjan, R.E. and Yannakakis, M. (1983) Scheduling opposing forests. *SIAM Journal of Algebraic Discrete Methods*, **4**, 72-93.

Garey, M.R., Tarjan, R.E. and Wilfong, G.T. (1988) One-processor scheduling with symmetric earliness and tardiness penalties. *Math. Oper. Res.*, **13**, 330-348.

Gazmuri, P.G. (1985) Probabilistic analysis of a machine scheduling problem. *Math. Oper. Res.*, **10**, 328-339.

Gelders, L. and Kleindorfer, P.R. (1974) Coordinating aggregate and detailed scheduling decisions in the one-machine job shop: part I. Theory. *Operations Research*, **22**, 46-60.

Gelders, L. and Kleindorfer, P.R. (1975) Coordinating aggregate and detailed scheduling in the one-machine job shop: II. Computation and structure. *Operations Research*, **23**, 312-324.

Gens, G.V. and Levner, E.V. (1978) Approximation algorithm for some scheduling problems. *Engrg. Cybernetics*, **6**, 38-46.

Gens, G.V. and Levner, E.V. (1981) Fast approximation algorithm for job sequencing with deadlines. *Discrete Appl. Math.*, **3**, 313-318.

Gere, W.S. (1966) Heuristics in job shop scheduling. *Management Science*, **13**, 167-190.

Giffler, B. and Thompson, G.L. (1960) Algorithms for solving production-scheduling problems. *Operations Research*, **8**, 487-503.

Gilmore, P.C. and Gomory, R.E. (1964) Sequencing a one-state variable machine: a solvable case of the traveling salesman problem. *Operations Research*, **12**, 655-679.

Golumbic, M.C. (1980) *Algorithmic Graph Theory and Perfect Graphs*, Academic Press, New York.

Gonzalez, T. (1977) *Optimal Mean Finish Time Preemptive Schedules.* Technical report 220, Computer Science Department, Pennsylvania State University.

Gonzalez, T. (1979) A note on open shop preemptive schedules. *IEEE Trans. Comput.*, **C-28**, 782-786.

Gonzalez, T. (1982) Unit execution time shop problems. *Math. Oper. Res.*, **7**, 57-66.

Gonzalez, T., Ibarra, O.H. and Sahni, S. (1977) Bounds for LPT schedules on uniform processors. *SIAM J. Comput.*, **6**, 155-166.

Gonzalez, T. and Johnson, D.B. (1980) A new algorithm for preemptive scheduling of trees. *J. Assoc. Comput. Mach.*, **27**, 287-312.

Gonzalez, T., Lawler, E.L. and Sahni, S. (1990) Optimal preemptive scheduling of two unrelated processors. *ORSA J. Comput.*, **2**, 219-224.

Gonzalez, T. and Sahni, S. (1976) Open shop scheduling to minimize finish time. *J. Assoc. Comput. Mach.*, **23**, 665-679.

Gonzalez, T. and Sahni, S. (1978a) Flowshop and jobshop schedules: complexity and approximation. *Operations Research*, **26**, 36-52.

Gonzalez, T. and Sahni, S. (1978b) Preemptive scheduling of uniform processor systems. *J. Assoc. Comput. Mach.*, **25**, 92-101.

Goyal, S.K. (1975) A note on a simple CPM time-cost tradeoff algorithm. *Management Science*, **21**, 718-722.

Goyal, D.K. (1977) *Non-Preemptive Scheduling of Unequal Execution Time Tasks on Two Identical Processors.* Technical report CS-77-039, Computer Science

Department, Washington State University, Pullman.

Goyal, S.K. and Sriskandarajah, C. (1988) No-wait shop scheduling: computational complexity and approximate algorithms. *Opsearch*, **25**, 220-244.

Goyou-Beauchamps, D. (1982) The Hamiltonian circuit problem is polynomial for 4-connected planar graphs. *SIAM J. Comput.*, **11**, 529-539.

Grabowski, J. (1980) On two-machine scheduling with release dates to minimize maximum lateness. *Opsearch*, **17**, 133-154.

Grabowski, J. (1982) A new algorithm of solving the flow-shop problem. G. Feichtinger and P. Kall (eds.) *Operations Research in Progress*, Reidel, Dordrecht, 57-75.

Grabowski, J., Skubalska, E. and Smutnicki, C. (1983) On flow shop scheduling with release and due dates to minimize maximum lateness. *J. Oper. Res. Soc.*, **34**, 615-620.

Graham, R.L. (1966) Bounds for certain multiprocessing anomalies. *Bell Systems Tech. J.*, **45**, 1563-1581.

Graham, R.L. (1969) Bounds on multiprocessing timing anomalies. *SIAM J. Appl. Math.*, **17**, 263-269.

Graham, R.L. (1978) Combinatorial Scheduling Theory. L.A. Steen (ed.), *Mathematics Today*. Springer-Verlag, New York, 183-211.

Graham, R.L., Lawler, E.L., Lenstra, J.K. and Rinnooy Kan, A.H.G. (1979) Optimization and approximation in deterministic sequencing and scheduling: a survey. *Ann. Discrete Math.*, **5**, 287-326.

Graves, S.C. (1981) A review of production scheduling. *Operations Research*, **29**, 646-675.

Grötschel, M. (1980b) On the symmetric travelling salesman problem: solution of a 120-city problem. *Math. Programming Stud.*, **12**, 61-77.

Grötschel, M. and Padberg, M.W. (1974) Zur oberflächenstruktur des travelling salesman polytopen. H.-J. Zimmermann, A. Schub, H. Späth, and J. Stoer (eds.) *Proc. Operations Research 4*, Physica, Würzburg, 207-211.

Grötschel, M. and Padberg, M.W. (1975a) *On the Symmetric Travelling Salesman Problem.* Working paper, Institut für Oekonometrie und Operations Research, Universität Bonn.

Grötschel, M. and Padberg, M.W. (1975b) Partial linear characterizations of the assymmetric travelling salesman polytope. *Math. Programming*, **8**, 378-381.

Grötschel, M. and Padberg, M.W. (1977) Lineare charakterisierungen von travelling salesman problemen. *Z. Oper. Res.*, **21**, 33-64.

Grötschel, M. and Padberg, M.W. (1978) On the symmetric travelling salesman problem: theory and computation. R. Henn, B. Korte, and W. Oettli (eds.) *Optimization and Operations Research*, Lecture notes in Economics and Mathematical Systems 157, Springer, Berlin, 105-115.

Grötschel, M. and Padberg, M.W. (1979a) On the symmetric travelling salesman problem I: inequalities. *Math. Programming*, **16**, 265-280.

Grötschel, M. and Padberg, M.W. (1979b) On the symmetric travelling salesman problem II: lifting theorem and facets. *Math. Programming*, **16**, 281-302.

Grötschel, M. and Pulleyblank, W.R. (1986) Clique tree inequalities and the symmetric travelling salesman problem. *Math. Oper. Res.*, **11**, 537-569.

Grötschel, M. and Wakabayashi, Y. (1981a) On the structure of the monotone asymmetric travelling salesman polytope I: Hypohamiltonian facets. *Discrete Math.*, **34**, 43-59.

Grötschel, M. and Wakabayashi, Y. (1981b) On the structure of the monotone asymmetric travelling salesman polytope II: hypotraceable facets. *Math. Programming Stud.*, **14**, 77-97.

Gupta, J.N.D. and Reddi, S.S. (1978) Improved dominance conditions for the three-machine flowshop scheduling problem. *Operations Research*, **26**, 200-203.

Gupta, R.P. (1966) The chromatic index and the degree of a graph. *Notices Amer. Math. Soc.*, **13**, abstract 66T-429.

Gupta, S.K. and Kyparisis, J. (1987) Single machine scheduling research. *Omega*, **15**, 207-227.

Gusfield, D. (1984) Bounds for naive multiple machine scheduling with release times and deadlines. *J. Algorithms*, **5**, 1-6.

Hall, L.A. and Shmoys, D.B. (1992) Jackson's rule for single-machine scheduling: making a good heuristic better. *Math. Oper. Res.*, **17**, 22-35.

Hariri, A.M.A. and Potts, C.M. (1983) An algorithm for single machine sequencing with release dates to minimize total weighted completion time. *Discrete Appl. Math.*, **5**, 99-109.

Hariri, A.M.A. and Potts, C.M. (1984) Algorithms for two-machine flow-shop sequencing with precedence constraints. *European J. Oper. Rers.*, **17**, 238-248.

Haupt, R. (1989) A survey of priority rule-based scheduling. *OR Spektrum*, **11**, 3-16.

Hefetz, N. and Adiri, I. (1982) An efficient optimal algorithm for the two-machines unit-time jobshop schedule-length problem. *Math. Oper. Res.*, **7**, 354-360.

Helbig Hansen, K. and Krarup, J. (1974) Improvements of the Held-Karp algorithm for the symmetric traveling-salesman problem. *Math. Programming*, **7**, 87-96.

Held, M. and Karp. R.M. (1962) A dynamic programming approach to sequencing problems. *SIAM J. Appl. Math.*, **10**, 196-210.

Held, M. and Karp. R.M. (1970) The traveling-salesman problem and minimum spanning trees. *Operations Research*, **18**, 1138-1162.

Held, M. and Karp, R.M. (1971) The traveling-salesman problem and minimum spanning trees: part II. *Math. Programming*, **1**, 6-25.

Hochbaum, D.S. and Shmoys, D.S. (1987) Using dual approximation algorithms for scheduling problems: theoretical and practical results. *J. Assoc. Comput. Mach.*, **34**, 144-162.

Hochbaum, D.S. and Shmoys, D.B. (1988) A polynomial approximation scheme for machine scheduling on uniform processors: using the dual approximation approach. *SIAM J. Comput.*, **17**, 539-551.

Hodgson, S.M. (1977) A note on single machine sequencing with random processing times. *Management Science*, **23**, 1144-1146.

Holyer, I. (1981) The $\mathcal{NP}$-completeness of edge-coloring. *SIAM J. Comput.*, **10**, 718-720.

Horn, W.A. (1972) Single-machine job sequencing with treelike precedence ordering and linear delay penalties. *SIAM J. Appl. Math.*, **23**, 189-202.

Horn, W.A. (1973) Minimizing average flow time with parallel machines. *Operations Research*, **21**, 846-847.

Horn, W.A. (1974) Some simple scheduling algorithms. *Naval Research Logistics Quarterly*, **21**, 177-185.

Horowitz, E. and Sahni, S. (1976) Exact and approximate algorithms for scheduling nonidentical processors. *J. Assoc. Comput. Mach.*, **23**, 317-327.

Horvath, E.C., Lam, S. and Sethi, R. (1977) A level algorithm for preemptive scheduling. *J. Assoc. Comput. Mach.*, **24**, 32-43.

Hsu, N.C. (1966) Elementary proof of Hu's theorem on isotone mappings. *Proc. Amer. Math. Soc.*, **17**, 111-114.

Hu, T.C. (1961) Parallel sequencing and assembly line problems. *Operations Research*, **9**, 841-848.

Ibarra, O.H. and Kim, C.E. (1976) On two-processor scheduling of one- or two-unit time tasks with precedence constraints. *J. Cybernet.*, **5**, 87-109.

Ibarra, O.H. and Kim, C.E. (1977) Heuristic algorithms for scheduling independent tasks on nonidentical processors. *J. Assoc. Comput. Mach.*, **24**, 280-289.

Ibarra, O.H. and Kim, C.E. (1978) Approximation algorithms for certain scheduling problems. *Math. Oper. Res.*, **3**, 197-204.

Ignall, E. and Schrage, L. (1965) Application of the branch and bound technique to some flow-shop scheduling problems. *Operations Research*, **13**, 400-412.

Itai, A., Papadimitriou, C.H. and Szwarcfiter, J. (1982) Hamiltonian paths in grid graphs. *SIAM J. Comput.*, **11**, 676-686.

Jackson, J.R. (1955) *Scheduling a Production Line to Minimize Maximum Tardiness.* Research Report 43, Management Science Research Project, University of California, Los Angeles.

Jackson, J.R. (1956) An extension of Johnson's results on job lot scheduling. *Naval Research Logistics Quarterly*, **3**, 201-203.

Jaffe, J.M. (1980a) Efficient scheduling of tasks without full use of processor resources. *Theoret. Comput. Sci.*, **12**, 1-17.

Jaffe, J.M. (1980b) An analysis of preemptive multiprocessor job scheduling. *Math. Oper. Res.*, **5**, 415-421.

Johnson, D.S. (1983) The $\mathcal{NP}$-completeness column: an ongoing guide. *J. Algorithms*, **4**, 189-203.

Johnson, S.M. (1954) Optimal two- and three-stage production schedules with setup times included. *Naval Research Logistics Quarterly*, **1**, 61-68.

Johnson, S.M. (1958) Discussion: Sequencing n jobs on two machines with arbitrary time lags. *Management Science*, **5**, 299-303.

Kao, E.P.C. and Queyranne, M. (1982) On dynamic programming methods for assembly line balancing. *Operations Research*, **30**, 375-390.

Karmarkar, N. (1984) A new polynomial time algorithm for linear programming. *Combinatorica*, **4**, 375-395.

Karmarkar, N. and Karp, R.M. (1982) *The Differencing Method of Set Partitioning.* Report UCB/CSD 82/113, Computer Science Division, University of

California, Berkeley.

Karp, R.M. (1972) Reducibility among combinatorial problems. R.E. Miller and J.W. Thatcher (eds.) *Complexity of Computer Computations*, Plenum Press, New York, 85-103.

Karp, R.M. (1975) On the computational complexity of combinatorial problems. *Networks*, **5** 45-68.

Karp, R.M. (1977) Probabilistic analysis of partitioning algorithms for the traveling-salesman problem in the plane. *Math. Oper. Res.*, **2**, 209-224.

Karp, R.M. (1979) A patching algorithm for the nonsymmetric traveling-salesman problem. *SIAM J. Comput.*, **8**, 561-573.

Kaufman, M.T. (1974) An almost-optimal algorithm for the assembly line scheduling problem. *IEEE Trans. Comput.*, **C-23**, 1169-1174.

Kawaguchi, T. and Kyan, S. (1986) Worst case bound of an LRF schedule for the mean weighted flow-time problem. *SIAM J. Comput.*, **15**, 1119-1129.

Kawaguchi, T. and Kyan, S. (1988) Deterministic scheduling in computer systems: a survey. *J. Oper. Res. Soc. Japan*, **31**, 190-217.

Khachian, L.G. (1979) A polynomial algorithm in linear programming. *Soviet Mathematics Doklady*, **20**, 191-194.

Kise, H., Ibaraki, T. and Mine, H. (1978) A solvable case of the one-machine scheduling problem with ready and due times. *Operations Research*, **26**, 121-126.

Kise, H., Ibaraki, T. and Mine, H. (1979) Performance analysis of six approximation algorithms for the one-machine maximum lateness scheduling problem with ready times. *J. Oper. Res. Soc. Japan*, **22**, 205-224.

Knuth, D.E. (1976) The travelling salesman problem. W. Sullivan. Frontiers of Science, From Microcosm to Macrocosm. *The New York Times*, February 24, 1976.

Kohler, W.H. and Steiglitz, K. (1976) Exact approximate and guaranteed accuracy algorithms for the flow-shop problem $n/2/F/\bar{F}$. *J. Assoc. Comput. Mach.*, **22**, 106-114.

Krishnamoorthy, M.S. and Deo, N. (1979) Complexity of the minimum-dummy-activities problem in a pert network. *Networks*, **9**, 189-194.

Kruskal, Jr., J.B. (1956) On the shortest spanning subtree of a graph and the traveling salesman problem. *Proc. Amer. Math. Soc.*, **7**, 48-50.

Kumar, P.R. and Walrand, J. (1985) Individually optimal routing in parallel systems. *J. Appl. Probab.*, **22**, 989-995.

Kunde, M. (1981) Nonpreemptive LP-scheduling on homogeneous multiprocessor systems. *SIAM J. Comput.*, **10**, 151-173.

Kunde, M. and Steppat, H. (1985) First fit decreasing scheduling on uniform multiprocessors. *Discrete Appl. Math.*, **10**, 165-177.

Kwan, M.-K. (1962) Graphic programming using odd and even points. *Chinese Math.*, **1**, 273-277.

Labetoulle, J., Lawler, E.L, Lenstra, J.K. and Rinnooy Kan, A.H.G. (1984) Preemptive scheduling of uniform machines subject to release dates. Pulleyblank [1984] 245-261.

Lageweg, B.J., Lawler, E.L., Lenstra, J.K. and Rinnooy Kan, A.H.G. (1981) *Computer Aided Complexity Classification of Deterministic Scheduling Problems.* Report BW 138, Centre for Mathematics and Computer Science, Amsterdam.

Lageweg, B.J., Lenstra, J.K., Lawler, E.L. and Rinnooy Kan, A.H.G. (1982) Computer-aided complexity classification of combinatorial problems. *Comm. ACM*, **25**, 817-822.

Lageweg, B.J., Lenstra, J.K. and Rinnooy Kan, A.H.G. (1976) Minimizing maximum lateness on one machine: computational experience and some applications. *Statist. Neerlandica*, **30**, 25-41.

Lageweg, B.J., Lenstra, J.K. and Rinnooy Kan, A.H.G. (1977) Job-shop scheduling by implicit enumeration. *Management Science*, **24**, 441-450.

Lageweg, B.J., Lenstra, J.K. and Rinnooy Kan, A.H.G. (1978) A general bounding scheme for the permutation flow-shop problem. *Operations Research*, **26**, 53-67.

Lam, S. and Sethi, R. (1977) Worst case analysis of two scheduling algorithms. *SIAM J. Comput.*, **6**, 518-536.

Lamberson, L.R. and Hocking, R.R. (1970) Optimum time compression in project scheduling. *Management Science*, **16**, B597-B606.

Larson, R.E., Dessouky, M.I. and Devor, R.E. (1985) A forward-backward procedure for the single machine problem to minimize lateness. *IIE Trans.*, **17**, 252-260.

Lawler, E.L. (1971) A solvable case of the traveling salesman problem. *Math. Programming*, **1**, 267-269.

Lawler, E.L. (1973) Optimal sequencing of a single machine subject to precedence constraints. *Management Science*, **19**, 544-546.

Lawler, E.L. (1976a) Sequencing to minimize the weighted number of tardy jobs. *RAIRO Rech. Opér.*, **10**, 27-33.

Lawler, E.L. (1976b) *Combinatorial Optimization: Networks and Matroids.* Hold, Rinehart and Winston, New York.

Lawler, E.L. (1977) A pseudopolynomial algorithm for sequencing jobs to minimize total tardiness. *Ann. Discrete Math.*, **1**, 331-342.

Lawler, E.L. (1978) Sequencing jobs to minimize total weighted completion time subject to precedence constraints. *Ann. Discrete Math.*, **2**, 75-90.

Lawler, E.L. (1979a) *Premptive Scheduling of Uniform Parallel Machines to Minimize the Weighted Number of Late Jobs.* Report BW 105, Centre for Mathematics and Computer Science. Amsterdam.

Lawler, E.L. (1979b) *Efficient Implementation of Dynamic Programming Algorithms for Sequencing Problems.* Report BW 106, Centre for Mathematics and Computer Science, Amsterdam.

Lawler, E.L. (1982a) Preemptive scheduling of precedence-constrained jobs on parallel machines. Dempster, Lenstra & Rinnooy Kan [1982], 101-123.

Lawler, E.L. (1982b) *Scheduling a Single Machine to Minimize the Number of late Jobs.* Preprint, Computer Science Division, University of California, Berkeley.

Lawler, E.L. (1982c) A fully polynomial approximation scheme for the total tardiness problem. *Operations Research Letters*, **1**, 207-208.

Lawler, E.L. (1983) Recent results in the theory of machine scheduling. A. Bachem, M. Grötschel and B. Korte (eds.) *Mathematical Programming: the State of the Art—Bonn 1982*. Springer, Berlin, 202-234.

Lawler, E.L. and Labetoulle, J. (1978) On preemptive scheduling of unrelated parallel processors by linear programming. *J. Assoc. Comput. Mach.*, **25**, 612-619.

Lawler, E.L. and Lenstra, J.K. (1982) Machine scheduling with precedence constraints. I. Rival (ed.) *Ordered Sets*. Reidel, Dordrecht, 655-675.

Lawler, E.L., Lenstra, J.K. and Rinnooy Kan, A.H.G. (1981, 1982) Minimizing maximum lateness in a two-machine open shop. *Math. Oper. Res.*, **6**, 153-158; Erratum. *Math. Oper. Res.*, **7**, 635.

Lawler, E.L., Lenstra, J.K. and Rinnooy Kan, A.H.G. (1982) Recent developments in deterministic sequencing and scheduling: a survey. Dempster, Lenstra & Rinnooy Kan [1982] 35-73.

Lawler, E.L., Lenstra, J.K., Rinnooy Kan, A.H.G. and Shmoys, D.B. (eds.) (1985) *The Traveling Salesman Problem: a Guided Tour of Combinatorial Optimization*. Wiley, Chichester.

Lawler, E.L., Lenstra, J.K., Rinnooy Kan, A.H.G. and Shmoys, D.B. (1989) *Sequencing and Scheduling: Algorithms and Complexity*. Report BS-R8909, Centre for Mathematics and Computer Science, Amsterdam.

Lawler, E.L. and Martel, C.U. (1982) Computing maximal polymatroidal network flows. *Math. Oper. Res.*, **7**, 334-347.

Lawler, E.L. and Martel, C.U. (1989) Preemptive scheduling of two uniform machines to minimize the number of late jobs. *Operations Research*, **37**, 314-318.

Lawler, E.L. and Moore, J.M. (1969) A functional equation and its application to resource allocation and sequencing problems. *Management Science*, **16**, 77-84.

Lenstra, J.K. (1977) *Sequencing by Enumerative Methods*. Mathematical Centre Tracts 69, Centre for Mathematics and Computer Science, Amsterdam.

Lenstra, J.K. and Rinnooy Kan, A.H.G. (1975) Some simple applications of the travelling salesman problem. *Oper. Res. Quart.*, **26**, 717-733.

Lenstra, J.K. and Rinnooy Kan, A.H.G. (1978) Complexity of scheduling under precedence constraints. *Operations Research*, **26**, 22-35.

Lenstra, J.K. and Rinnooy Kan, A.H.G. (1979) Computational complexity of discrete problems. *Ann. Discrete Math.*, **4** 121-140.

Lenstra, J.K. and Rinnooy Kan, A.H.G. (1979) Computational complexity of discrete optimization problems. *Ann. Discrete Math.*, **4**, 121-140.

Lenstra, J.K. and Rinnooy Kan, A.H.G. (1980) Complexity results for scheduling chains on a single machine. *European J. Oper. Res.*, **4**, 270-275.

Lenstra, J.K. and Rinnooy Kan, A.H.G. (1984) New directions in scheduling theory. *Operations Research Letters*, **2**, 255-259.

Lenstra, J.K. and Rinnooy Kan, A.H.G. (1985) Sequencing and scheduling. O'hEigeartaigh, Lenstra & Rinnooy Kan [1985], 164-189.

Lenstra, J.K., Rinnooy Kan, A.H.G. and Brucker, P. (1977) Complexity of machine scheduling problems. *Ann. Discrete Math.*, **1**, 343-362.

Lenstra, J.K., Shmoys, D.B. and Tardos, E. (1990) Approximation algorithms for scheduling related parallel machines. *Math. Programming*, **46**, 259-272.

Lesniak-Foster, L. (1977) Some recent results in Hamiltonian graphs. *J. Graph Theory*, **1**, 27-36.

Leung, J.Y.-T. (1989) Bin packing with restricted piece sizes. *Inform. Process. Lett.*, **31**, 145-149.

Leung, J.Y.-T. and Young, G.H. (1989) *Minimizing Total Tardiness on a Single Machine with Precedence Constraints.* Technical Report, Computer Science Program, University of Texas, Dallas.

Lin, S. and Kernighan, B.W. (1973) An effective heuristic algorithm for the traveling-salesman problem. *Operations Research*, **21**, 498-516.

Little, J.D.C., Murty, K.G., Sweeney, D.W. and Karel, C. (1963) An algorithm for the traveling salesman problem. *Operations Research*, **11**, 972-989.

Liu, C.Y. and Bulfin, R.L. (1985) On the complexity of preemptive open-shop scheduling problems. *Operations Research Letters*, **4**, 71-74.

Liu, C.Y. and Bulfin, R.L. (1988) Scheduling open shops to minimize due-date functions. *Operations Research*, **36**, 553-559.

Liu, J.W.S. and Liu, C.L. (1974a) Bounds on scheduling algorithms for heterogeneous computing systems. J.L. Rosenfeld (ed.) *Information Processing 74*, Noth-Holland, Amsterdam, 349-353.

Liu, J.W.S. and Liu, C.L. (1974b) *Bounds on Scheduling Algorithms for Heterogeneous Computing Systems.* Technical Report UIUCDCS-R-74-632, Department of Computer Science, University of Illinois at Urbana-Champaign, 68pp.

Liu, J.W.S. and Liu, C.L. (1974c) Performance analysis of heterogeneous multiprocessor computing systems. E. Gelenbe and R. Mahl (eds.) *Computer Architectures and Networks.* North-Holland, Amsterdam, 331-343.

Lovász, L. (1979) *Combinatorial Problems and Exercises.* Akadémiai Kiadó, Budapest.

Lovász, L. (1981) The matroid matching problem. *Algebraic Methods in Graph Theory, Vol. II. Colloquia Mathematica Societatis Janos Bolyai*, **25**, North-Holland, Amsterdam, 495-518.

Marcotte, O. and Trotter, Jr., L.E. (1984) An application of matroid polyhedral theory to unit-execution time, tree-precedence constrained job scheduling. Pulleyblank [1984], 263-271.

Martel, C.U. (1982) Preemptive scheduling with release times, deadlines and due times. *J. Assoc. Comput. Mach.*, **29**, 812-829.

Matsuo, H., Suh, C.J. and Sullivan, R.S. (1988) *A Controlled Search Simulated Annealing Method for the General Jobshop Scheduling Problem.* Working paper 03-44-88, Graduate School of Business, University of Texas, Austin.

Maxwell, W.L. (1970) On sequencing n jobs on one machine to minimize the number of late jobs. *Management Science*, **16**, 295-297.

McCormick, S.T. and Pinedo, M.L. (1989) *Scheduling n Independent Jobs on m Uniform Machines with Both Flow Time and Makespan Objectives: a Parametric Analysis.* Department of Industrial Engineering and Operations Research, Columbia University, New York.

McHugh, J.A.M. (1984) Hu's precedence tree scheduling algorithm: a simple proof. *Naval Research Logistics Quarterly*, **31**, 409-411.

McMahon, G.B. (1969) Optimal production schedules for flow shops. *Canad. Oper. Res. Soc.*, **7**, 141-151.

McMahon, G.B. (1971) *A Study of Algorithms for Industrial Scheduling Problems.* Ph.D. thesis, University of New South Wales, Kensington.

McMahon, G.B. and Florian, M. (1975) On scheduling with ready times and due dates to minimize maximum lateness. *Operations Research*, **23**, 475-482.

McNaughton, R. (1959) Scheduling with deadlines and loss functions. *Management Science*, **6**, 1-12.

Meilijson, I. and Tamir, A. (1984) Minimizing flow time on parallel identical processors with variable unit processing time. *Operations Research*, **32**, 440-446.

Miliotis, P. (1976) Integer programming approaches to the travelling salesman problem. *Math. Programming*, **10**, 367-378.

Miliotis, P. (1978) Using cutting planes to solve the symmetric travelling salesman problem. *Math. Programming*, **15**, 177-188.

Mitten, L.G. (1958) Sequencing n jobs on two machines with arbitrary time lags. *Management Science*, **5**, 293-298.

Moder, J.J., Phillips, C.R. and Davis, E.W. (1983) *Project Management with CPM, PERT and Precedence Diagramming*, (3rd ed.). Van Nostrand Reinhold, New York.

Möhring, R.H. (1983) Scheduling problems with a singular solution. *Ann. Discrete Math.*, **16**, 225-239.

Möhring, R.H. (1984) Minimizing costs of resource requirements in project networks subject to a fixed completion time. *Operations Research*, **32**, 89-120.

Möhring, R.H. (1989) Computationally tractable classes of ordered sets. I. Rival (ed.) *Algorithms and Order.* Kluwer Academic, Dordrecht, 105-193.

Möhring, R.H. and Radermacher, F.J. (1985) Generalized results on the polynomiality of certain weighted sum scheduling problems. *Methods of Operations Research*, **49**, 405-417.

Monma, C.L. (1979) The two-machine maximum flow-time problem with series-parallel precedence constraints: an algorithm and extensions. *Operations Research*, **27**, 792-798.

Monma, C.L. (1980) Sequencing to minimize the maximum job cost. *Operations Research*, **28**, 942-951.

Monma, C.L. (1981) Sequencing with general precedence constraints. *Discrete Appl. Math.*, **3**, 137-150.

Monma, C.L. (1982) Linear-time algorithms for scheduling on parallel processors. *Operations Research*, **30**, 116-124.

Monma, C.L. and Rinnooy Kan, A.H.G. (1983) A concise survey of efficiently solvable special cases of the permutation flow-shop problem. *RAIRO Rech. Opér.*, **17**, 105-119.

Monma, C.L. and Sidney, J.B. (1979) Sequencing with series-parallel precedence constraints. *Math. Oper. Res.*, **4**, 215-224.

Monma, C.L. and Sidney, J.B. (1987) Optimal sequencing via modular decomposition: characterizations of sequencing functions. *Math. Oper. Res.*, **12**, 22-31.

Moore, J.M. (1968) An n job, one machine sequencing algorithm for minimizing the number of late jobs. *Management Science*, **15**, 102-109.

Morrison, J.F. (1988) A note on LPT scheduling. *Operations Research Letters*, **7**, 77-79.

Muntz, R.R. and Coffman, Jr., E.G. (1969) Optimal preemptive scheduling on two-processor systems. *IEEE Trans. Comput.*, **C-18**, 1014-1020.

Muntz, R.R. and Coffman, Jr., E.G. (1970) Preemptive scheduling of real time tasks on multiprocessor systems. *J. Assoc. Comput. Mach.*, **17**, 324-338.

Muth, J.F. and Thompson, G.L. (1963) *Industrial Scheduling*. Prentice-Hall, Englewood Cliffs, N.J.

Nabeshima, I. (1963) Sequencing on two machines with start lag and stop lag. *J. Oper. Res. Soc. Japan*, **5**, 97-101.

Nakajima, K., Leung, J.Y.-T. and Hakimi, S.L. (1981) Optimal two processor scheduling of tree precedence constrained tasks with two execution times. *Performance Evaluation*, **1**, 320-330.

Nash-Williams, C.St. J.A. (1966) On Hamiltonian circuits in finite graphs. *Proceedings of the American Mathematical Society*, **17**, 466-467.

Nawaz, M., Enscore, Jr., E.E. and Ham, I. (1983) A heuristic algorithm for the m-machine, n-job flowshop sequencing problem. *Omega*, **11**, 91-95.

Nemhauser, G. and Wolsey, L. (1988) *Integer and Combinatorial Optimization*. John Wiley, New York.

Nowicki, E. (1994) An approximation algorithm for a single-machine scheduling problem with release times, delivery times and controllable processing times. *European J. Oper. Res.*, **72**, 74-81.

Nowicki, E. and Smutnicki, C. (1987) On lower bounds on the minimum maximum lateness on one machine subject to release date. *Opsearch*, **24**, 106-110.

Nowicki, E. and Zdrzalka, S. (1986) A note on minimizing maximum lateness in a one-machine sequencing problem with release dates. *European J. Oper. Res.*, **23**, 266-267.

O'hEigeartaigh, M., Lenstra, J.K. and Rinnooy Kan, A.H.G. (eds.) (1985) *Combinatorial Optimization: Annotated Bibliographies*. Wiley, Chichester.

Ore, O. (1960) Note on Hamiltonian circuits. *Amer. Math. Monthly*, **67**, 55.

Orlin, J.B. and Vande Vate, J.H. (1990) Solving the linear matroid parity problem as a sequence of matroid intersection problems. *Math. Programming*, **47**, 81-106.

Osman, J.H. and Potts, C.N. (1989) *Simulated Annealing for Permutation Flow-Shop Scheduling*. Preprint OR17, Faculty of Mathematical Studies, University of Southampton.

Padberg, M.W. and Hong, S. (1980) On the symmetric travelling salesman problem: a computational study. *Math. Programming Stud.*, **12**, 78-107.

Padberg, M.W. and Rao, M.R. (1974) The travelling salesman problem and a class of polyhedra of diameter two. *Math. Programming*, **7**, 32-45.

Palmer, D.S. (1965) Sequencing jobs through a multistage process in the minimum total time - a quick method of obtaining a near optimum. *Oper. Res. Quart.*, **16**, 101-107.

Panwalkar, S.S. and Iskander, W. (1977) A survey of scheduling rules. *Operations Research*, **25**, 45-61.

Papadimitriou, C.H. (1976) On the complexity of edge traversing. *J. Assoc. Comput. Mach.*, **23**, 544-554.

Papadimitriou, C.H. and Kannelakis, P.C. (1980) Flowshop scheduling with limited temporary storage. *J. Assoc. Comput. Mach.*, **27**, 533-549.

Papadimitriou, C.H. and Steiglitz, K. (1977) On the complexity of local search for the traveling salesman problem. *SIAM J. Comput.*, **6**, 76-83.

Papadimitriou, C.H. and Steiglitz, K. (1978) Some examples of difficult traveling salesman problems. *Operations Research*, **26**, 434-443.

Papadimitriou, C.H. and Steiglitz, K. (1982) *Combinatorial Optimization: Algorithms and Complexity*. Prentice-Hall, Englewood Cliffs, N.J.

Papadimitriou, C.H. and Yannakakis, M. (1979) Scheduling interval-order- ed tasks. *SIAM J. Comput.*, **8**, 405-409.

Parker, R.G. and Rardin, R.L. (1982a) An overview of complexity theory in discrete optimization: Part I. Concepts. *IIE Transactions*, **14**, 3-10.

Parker, R.G. and Rardin, R.L. (1982b) An overview of complexity theory in discrete optimization: Part II. Results and implications. *IIE Transactions*, **14**, 83-89.

Parker, R.G. and Rardin, R.L. (1983) The traveling salesman problem: an update of research. *Naval Research Logistics Quarterly*, **30**, 69-96.

Parker, R.G. and Rardin, R.L. (1988) *Discrete Optimization*. Academic Press, New York.

Pósa, L. (1962) A theorem concerning Hamilton lines. *Magyar Tud. Akad. Mat. Kutató Int. Közl.*, **7**, 225-226.

Pósa, L. (1976) Hamiltonian circuits in random graphs. *Discrete Math.*, **14**, 359-364.

Posner, M.E. (1985) Minimizing weighted completion times with deadlines. *Operations Research*, **33**, 562-574.

Potts, C.N. (1980a) An adaptive branching rule for the permutation flow-shop problem. *European J. Oper. Res.*, **5**, 19-25.

Potts, C.N. (1980b) Analysis of a heuristic for one machine sequencing with release dates and delivery times. *Operations Research*, **28**, 1436-1441.

Potts, C.N. (1980c) An algorithm for the single machine sequencing problem with precedence constraints. *Math. Programming Study*, **13**, 78-87.

Potts, C.N. (1985a) Analysis of a linear programming heuristic for scheduling unrelated parallel machines. *Discrete Appl. Math.*, **10**, 155-164.

Potts, C.N. (1985b) Analysis of heuristics for two-machine flow-shop sequencing subject to release date. *Math. Oper. Res.*, **10**, 576-584.

Potts, C.M. (1985c) A Lagrangean based branch and bound algorithm for single machine sequencing with precedence constraints to minimize total weighted completion time. *Management Science*, **31**, 1300-1311.

Potts, C.M. and van Wassenhove, L.M. (1982) A decomposition algorithm for the single machine total tardiness problem. *Operations Research Letters*, **1**, 177-181.

Potts, C.N. and van Wassenhove, L.M. (1983) An algorithm for single machine sequencing with deadlines to minimize total weighted completion time. *European Journal of Operations Research*, **12**, 379-387.

Potts, C.N. and van Wassenhove, L.N. (1985) A branch and bound algorithm for the total weighted tardiness problem. *Operations Research*, **33**, 363-377.

Potts, C.N. and van Wassenhove, L.N. (1987) Dynamic programming and decomposition approaches for the single machine total tardiness problem. *European Journal of Operations Research*, **32**, 405-414.

Potts, C.N. and van Wasenhove, L.N. (1988) Algorithms for scheduling a single machine to minimize the weighted number of late jobs. *Management Science*, **34**, 843-858.

Pulleyblank, W.R. (ed.) (1984) *Progress in Combinatorial Optimization*. Academic Press, New York.

Rachamadugu, R.M.V. (1987) A note on the weighted tardiness problem. *Operations Research*, **35**, 450-452.

Radermacher, F.J. (1985/6) Scheduling of project networks. *Ann. Oper. Res.*, **4**, 227-252.

Raghavachari, M. (1988) Scheduling problems with non-regular penalty functions: a review. *Opsearch*, **25**, 144-164.

Ratliff, H.D. and Rosenthal, A.S. (1983) Order picking in a rectangular warehouse: a solvable case of the traveling salesman problem. *Operations Researsch*, **31**, 507-521.

Rayward-Smith, V.J. (1987a) UET scheduling with unit interprocessor communication delays. *Discrete Appl. Math.*, **18**, 55-71.

Rayward-Smith, V.J. (1987b) The complexity of preemptive scheduling given interprocessor communication delays. *Inform. Process. Lett.*, **25**, 123-125.

Reddi, S.S. and Ramamoorthy, C.V. (1972) On the flowshop sequencing problem with no wait in process. *Operations Research Quarterly*, **23**, 323-331.

Righter, R. (1988) Job scheduling to minimize expected weighted flowtime on uniform processors. *Syst. Control Lett.*, **10**, 211-216.

Rinnooy Kan, A.H.G. (1976) *Machine Scheduling Problems: Classification Complexity and Computations*. Nijhoff, The Hague.

Rinnooy Kan, A.H.G., Lageweg, B.J. and Lenstra, J.K. (1975) Minimizing total costs in one-machine scheduling. *Operations Research*, **23**, 908-927.

Rock, H. (1984a) The three-machine no-wait flowshop problem is $\mathcal{NP}$-complete. *J. Assoc. Comput. Mach.*, **31**, 336-345.

Rock, H. (1984b) Some new results in flow shop scheduling. *Z. Oper. Res.*, **28**, 1-16.

Rock, H. and Schmidt, G. (1982) *Machine Aggregation Heuristics in Shop Scheduling*. Bericht 82-11, Fachbereich 20 Mathematik, Technische Universitat Berlin.

Rothkopf, M.H. (1966) Scheduling independent tasks on parallel processors. *Management Science*, **12**, 437-447.

Sahni, S. (1976) Algorithms for scheduling independent tasks. *J. Assoc. Comput. Mach.*, **23**, 116-127.

Sahni, S. and Cho, Y. (1979a) Complexity of scheduling jobs with no wait in process. *Math. Oper. Res.*, **4**, 448-457.

Sahni, S. and Cho, Y. (1979b) Nearly on line scheduling of a uniform processor system with release times. *SIAM J. Comput.*, **8**, 275-285.

Sahni, S. and Cho, Y. (1980) Scheduling independent tasks with due times on a uniform processor system. *J. Assoc. Comput. Mach.*, **27**, 550-563.

Sahni, S. and Gonzalez, T. (1976) $\mathcal{P}$-complete approximation problems. *J. Assoc. Comput. Mach.*, **23**, 555-565.

Sarin, S.C., Ahn, S. and Bishop, A.B. (1988) An improved branching scheme for the branch and bound procedure of scheduling n jobs on m machines to minimize total weighted flowtime. *Internat. J. Production Res.*, **26**, 1183-1191.

Schmidt, G. (1983) *Preemptive Scheduling on Identical Processors with Time Dependent Availabilities.* Bericht 83-4. Fachbereich 20 Informatik, Technische Universität Berlin.

Schrage, L. (1970) Solving resource-constrained network problems by implicit enumeration-nonpreemptive case. *Operations Research*, **18**, 263-278.

Schrage, L. and Baker, K.R. (1978) Dynamic programming solution of sequencing problems with precedence constraints. *Operations Research*, **26**, 444-449.

Sethi, R. (1976a) Algorithms for minimal-length schedules. Coffman [1976], 51-99.

Sethi, R. (1976b) Scheduling graphs on two processors. *SIAM J. Comput.*, **5**, 73-82.

Sethi, R. (1977) On the complexity of mean flow time scheduling. *Mathematics of Operations Research*, **2**, 320-330.

Shmoys, D.B. and Tardos, E. (1989) Computational complexity of combinatorial problems. R.L. Graham, M. Grötschel and L. Lovász (eds.) *Handbook in Combinatorics.* North-Holland, Amsterdam.

Shmoys, D.B. and Tardos, E. (1993) An approximation algorithm for the generalized assignment problem. *Mathematical Programming*, **3**, 461-474.

Shwimer, J. (1972) On the N-jobs, one-machine, sequence-independent scheduling problem with tardiness penalties: a branch-and-bound solution. *Management Science*, **18B**, 301-313.

Sidney, J.B. (1973) An extension of Moore's due date algorithm. S.E. Elmaghraby (ed.) *Symposium on the Theory of Scheduling and its Applications.* Lecture Notes in Economics and Mathematical Systems 86, Springer, Berlin, 393-398.

Sidney, J.B. (1975) Decomposition algorithms for single-machine sequencing with precedence relations and deferral costs. *Operations Research*, **23**, 283-298.

Sidney, J.B. (1979) The two-machine maximum flow time problem with series parallel precedence relations. *Operations Research*, **27**, 782-791.

Sidney, J.B. (1981) A decomposition algorithm for sequencing with general precedence constraints. *Mathematics of Operations Research*, **6**, 190-204.

Sidney, J.B. and Steiner, G. (1986) Optimal sequencing by modular decomposition: polynomial algorithms. *Operations Research*, **34**, 606-612.

Simons, B. (1978) A fast algorithm for single processor scheduling. *Proc. 19th*

Annual Symp. Foundations of Computer Science, 246-252.

Simons, B. (1983) Multiprocessor scheduling of unit-time jobs with arbitrary release times and deadlines. *SIAM J. Comput.*, **12**, 249-299.

Simons, B. and Warmuth, M. (1989) A fast algorithm for multiprocessor scheduling of unit-length jobs. *SIAM J. Comput.*, **18**, 690-710.

Smith, M.L., Panwalkar, S.S. and Dudek, R.A. (1975) Flow shop sequencing with ordered processing time matrices. *Management Science*, **21**, 544-549.

Smith, M.L., Panwalkar, S.S. and Dudek, R.A. (1976) Flow shop sequencing problem with ordered processing time matrices: a general case. *Naval Research Logistics Quarterly*, **23**, 481-486.

Smith, W.E. (1956) Various optimizers for single-stage production. *Naval Research Logistics Quarterly*, **3**, 59-66.

Stockmeyer, L.J. (1990) Complexity theory. E.G. Coffman, Jr., J.K. Lenstra, and A.H.G. Rinnooy Kan (eds.) *Handbooks in Operations Research and Management Science; Volume 3: Computation*. North-Holland, Amsterdam, Chapter 8.

Sturm, L.B.J.M. (1970) A simple optimality proof of Moore's sequencing algorithm. *Management Science*, **17**, B116-B118.

Szwarc, W. (1968) On some sequencing problems. *Naval Research Logistics Quarterly*, **15**, 127-155.

Szwarc, W. (1971) Elimination methods in the $m \times n$ sequencing problem. *Naval Research Logistics Quarterly*, **18**, 295-305.

Szwarc. W. (1983) Optimal elimination methods in the $m \times n$ sequencing problem. *Operations Research*, **21**, 1250-1259.

Szwarc, W. (1983) Dominance conditions for the three-machine flow-shop problem. *Operations Research*, **26** 203-206.

Talbot, F.B. and Patterson, J.H. (1978) An efficient integer programming algorithm with network cuts for solving resource-constrained scheduling problems. *Management Science*, **24**, 1163-1174.

Tovey, C.A. (1990) A simplified anomaly and reduction for precedence constrained multiprocessor scheduling. *SIAM J. Disc. Math.*, **3**, 582-584.

Tucker, A. (1971) Matrix characterization of circular arc graphs. *Pacific J. Math.*, **39**, 535-545.

Turner, S. and Booth, D. (1987) Comparison of heuristics for flow shop sequencing. *Omega*, **15**, 75-78.

Tutte, W.T. (1946) On Hamiltonian circuits. *J. London Math. Soc.*, **21**, 98-101.

Ullman, J.D. (1975) $\mathcal{NP}$-Complete scheduling problems. *J. Comput. System Sci.*, **10**, 384-393.

Ullman, J.D. (1976) Complexity of sequencing problems. Coffman [1976], 139-164.

Valdes, J., Tarjan, R.E., and Lawler, E.L. (1982) The recognition of series-parallel digraphs. *SIAM J. Comput.*, **11**, 298-313.

Vandermonde, A.T. (1771) Remarques sur les problèms de situation. *Histoire de l'Académie des Sciences (Paris)*, 566-574.

van de Velde, S.L. (1988) *Minimizing Total Completion Time in the Two-Machine*

Flow Shop by Lagrangian Relaxation. Report OS-R8808. Centre for Mathematics and Computer Science, Amsterdam.

van Laarhoven, P.J.M., Aarts, E.H.L. and Lenstra, J.K. (1988) *Job Shop Scheduling by Simulated Annealing.* Report OS-R8809. Centre for Mathematics and Computer Science, Amsterdam.

Villarreal, F.J. and Bulfin, R.L. (1983) Scheduling a single machine to minimize the weighted number of tardy jobs. *AIIE Trans.*, **15**, 337-343.

Vizing, V.G. (1964) On an estimate of the chromatic class of a $\mathcal{P}$-graph. *Diskretnyi Analiz*, **3**, 25-30 (in Russian).

Whitehouse, G.E. (1973) Project management techniques. *Industrial Engineering*, **5**, 24-29.

Wiest, J.D. and Levy, F.K. (1977) *A management guide to PERT/CPM.* Prentice-Hall, Englewood Cliffs, New Jersey.

Wismer, D.A. (1972) Solution of the flowshop scheduling problem with no intermediate queues. *Operations Research*, **20**, 689-697.

Yue, M. (1990) On the exact upper bound for the multifit processor scheduling algoriothm. *Ann. Oper. Res.*, **24**, 233-259.

Zaloom, V. (1971) On the resource constrained project scheduling problem. *AIIE Trans.*, **3**, 302-305.

Zdrzalka, S. and Grabowski, J. (1989) An algorithm for single machine sequencing with release dates to minimize maximum cost. *Discrete Appl. Math.*, **23**, 73-89.

Index

1-factors 31, 170
3-PARTITION 45, 80, 138
3-SAT 183

Achugboe, J.O. 171
active schedules 158
  algorithm for generation, A_{AS} 159
Adiri, I. 171
agreeable job weights 49
Amit, N. 171
AOA (activity-on-arc) 204
AON (activity-on-node) 204
arborescence 31, 58

Bartholdi, J.J. 186, 197
Biggs, N.L. 245
BIN-PACKING 91
  first-fit decreasing weight heuristic,
    A_{FFD} 91—92
bipartite graph 30
Bondy, J.A. 195, 248
Borie, R. 33
BOTTLENECK TSP 256
branch-and-bound 36ff
branchings 31
Bruno, J.L. 99,132
Bulfin, R.L. 171

candidate problem 39
canonical schedules 48
Carter, M.W. 181, 183
chain 29
Chinese Postman 231
  see postman problems
Chin, F.Y. 171
Cho, Y. 102, 171
Christofides, N. 250
chromatic number 34, 179
chromatic index 36, 193
Chvátal, V. 246—248
circuit 29
class-1 graphs 193
class-2 graphs 193
Classroom Assignment Problem 5,
  177ff
clique 33
closure 248
Coffman, Jr., E.G. 91—92, 113
column circularity 186
  see also row circularity
complete graph 31
completion time 12
  see makespan
complexity theory 16ff
computational effort 17
connected graph 29
co-$\mathcal{NP}$ 24
Cook, S. 25
co-$\mathcal{P}$ 24
Cornuéjols, G. 252
covering 33
  see vertex cover
critical paths 208
cycle 29

decision problems 22
  see language recognition problems

decomposition tree 63
Deo, N. 206
dependent jobs 58
  see precedence constraints
digraphs (directed graphs) 28
Dobson, G. 102
Du, J. 48
Dudek, R.A. 153
due-date 11
dummy activities 205
  complexity of minimizing 206
dynamic programming 51ff

EDD (earliest due-date) 45, 54
Edmonds, J. 23, 33, 236, 240
edge colorings 36, 192
  see chromatic index
edge connectivity 30
Elmaghraby, S.E. 203
encoding 19
Euler, L. 232
Eulerian graph 232
  traversal 232
EULERIAN SUBGRAPH 255
EVEN–ODD PARTITION 46
expected performance 28

fathom 39
Fiala 171
flow shop 11, 135
  $F\|C_{\max}$ 135
  $F3\|C_{\max}$ 138
    special cases 146
  $F2\|C_{\max}$ 141
    2-processor algorithm, A_{J2} 143
  $F2\|\Sigma c_i$ 140
  lower bound for general case 149ff
forbidden subgraph 65
forest 31
  in-forest 58
  out-forest 58
  antiforest 109
Frederickson, G.N. 242—243
Friesen, D.K. 102
Fujii, M. 109
Fulkerson, D.R. 215
fully polynomial-time approximation
  scheme 49

Gabow, H.N. 23
Garey, M.R. 17—18, 26, 116—117,
  138, 140, 185
Golumbic, M.S. 36
Gonzalez, T. 102, 166, 171, 254
good algorithm 23
Graham, R.L. 2, 85, 119, 124
graph 28
graphic degree sequence 247
Gupta, R.P. 36

Halin graph 257
Hamiltonian cycle 244
  complexity of finding 245
  necessary conditions 246
  sufficient conditions 246ff
Hamilton, W.R. 245
  see Biggs *et al.*
Holyer, I. 36
homeomorphism 31
Hu, T.C. 107

incomparability graph 110
incumbent solution 39
independent jobs 43
independent set 31
  see stable set
instance 11, 17
interval graph 181
isomorphic graphs 31

job characteristics 10—11
job fair 197—198
job shop 11, 153
  $J2\|C_{\max}$ with at most 3 operations
    per job 154
  $J2\|C_{\max}$ with at most 2 operations
    per job; algorithm A_{Ja} 156
  lower bounds in the general case
    162

Jackson, J.R. 54, 155
Johnson, S.M. 141, 148
  Johnson's rule 142
Johnson, D.S. 17—18, 26, 116—117, 185
Johnson, E. 240

Karmarker, N. 23
Karp, R.M. 245
Kasami, T. 110
k-clique 33, 71
  see clique
Kleitman, D. 85
Knuth, D.E. 85
Khachian, L.G. 23
KNAPSACK 21
Krishnamoorthy, M.S. 206
Kwan, M. 231

Lam, S. 131
Langston, M.A. 102
language recognition problem 22
lateness 12
Lawler, E.L. 10, 49—53, 56, 59, 63, 68—69, 75, 107, 171, 208
least bound selection strategy 41
least expansion factor 95
Lenstra, J.K. 45, 75, 80, 102, 107, 110
lexicographic ordering 112
Leung, J.Y.-T. 48
Levy, F. 207
Liu, C.L. 102, 171
Liu, J.W.S. 102
line graph 67
LINEAR ARRANGEMENT 68
list processing 84
Lovász, L. 23
LPT (longest processing time) 88

machine environment 11
makespan 12
Marriage Theorem 201
matching 31
mathematical preliminaries 15

McCroan, K. 197
McHugh, J.A.M. 107
MIN-SET UNION 106
mixed graph 233
Moore, J.M. 53
Morrison, J.F. 102
MULTIFIT 91
multigraph 28
Murty, U.S.R. 195

Nash-Williams, C. St. J.A. 248, 257
Nemhauser, G.L. 39, 252
Ninomiya, K. 110
nonactive schedule 158
nondeterministic polynomial 24
$\mathcal{NP}$ 24
$\mathcal{NP}$-Complete 25
$\mathcal{NP}$-Hard 25

open shop 11, 165
  $O||C_{\max}$ 165
  $O3||C_{\max}$ 167
  $O2||C_{\max}$ 167
    algorithm, A_{GS} 167
  $O||pmtn|C_{\max}$ 170
    algorithm 169
opposing forest 58
optimality criteria 12
order (of a graph) 28
Orlin, J.B. 23, 186

$\mathcal{P}$ 23
Papadimitriou, C.H. 39, 239
parallel-processor problems 11, 83ff
  $P||C_{\max}$ 83
    Knuth-Kleitman heuristic, A_{KK} 85
    MULTIFIT heuristic, A_{MF} 92
    LPT heuristic, A_{LPT} 88
  $P|prec|C_{\max}$ 119
    longest path heuristic A_{LP} 119
    arborescence structure 122
  $P|prec, \tau_i = 1|C_{\max}$ 102, 105—106

tree structure; Hu's algorithm, A_H 107
$P2||C_{\max}$ 84
$P2||\Sigma w_i c_i$ 101
$P2||prec, \tau_i = 1|C_{\max}$ 109
algorithm of Coffman-Graham, A_{CG} 112, 115
algorithm of Garey and Johnson, A_{GJ} 118
algorithm if Fujii *et al.*, A_{FKN} 110
$P2|prec, \tau_i \in \{1,2\}|C_{\max}$ 112
$R||\Sigma c_i$ 99
$Q|\tau_i = 1|f_{\max}$ 102
$Q|\tau_i = 1|\Sigma w_i u_i$ 102
$Q|\tau_i = 1|\Sigma T_i$ 102
$Q|\tau_i = 1|\Sigma w_i c_i$ 102
list processing, algorithm, A_L 84
Parker, R.G. 17, 39
partial enumeration 36ff
see branch-and-bound
PARTITION 83, 166
PARTITION SCHEDULING 75
path 29
perfect graph 36, 181
perfect matching 31
see 1-factors
performance guarantee 28
permutation schedules 136
polynomial reduction 20
postman problem 231, 235
undirected graphs 232
Eulerian traversals; algorithm A_F 232
postman's algorithm, A_{UCP} 236
directed graphs 233
Eulerian traversals; algorithm A_{ED} 233
postman's algorithm, A_{DCP} 238
mixed graphs 239
complexity 239
heuristics for mixed graphs, 240
A_{ES} and A_{SE}; combined use 240ff
problem classes 22
precedence constraints 2, 11, 58
special structures 58
chains 58
forests 58
series-parallel 59
reversal (inverse) 144
preemption 11
processing time 10
project scheduling 203
project network construction 204
basic scheduling computations 208
slack time 211
free slack 211
pseudopolynomial algorithm 20

r-regular graph 31
Rardin, R.L. 17, 39
Ratliff, H.D. 186
regular measure 12
release time 11
RESTRICTED EVEN-ODD PARTITION 46
Rinnooy Kan, A.H.G. 71, 75, 102, 110, 155
robust schedule 122
row circularity 186

Sahni, S. 102, 166, 171, 254
SAT (satisfiability problem) 25
scheduling
definition 1
classification 9ff
schedule length 12
see makespan
semi-active schedule 12
series-parallel graph 59
series-parallel digraph 59
edge series-parallel 59, 65
2-terminal graph 65
vertex series-parallel 59
vertex transitive series-parallel 59
series composition 59, 69
parallel composition 60, 68
Sethi, R. 131
Shmoys, D.B. 75

Sidney, J.B. 69
single-processor problems 43ff
  $1||\Sigma w_i c_i$ 44
  $1||\Sigma c_i$ 44
  $1||\Sigma w_i L_i$ 45
  $1||\Sigma L_i$ 45
  $1||L_{\max}$ 45
  $1||T_{\max}$ 46
  $1||\Sigma T_i$ 46
    complexity 46
    Lawler's approach 49
  $1||\Sigma w_i T_i$ 48
  $1||\Sigma w_i u_i$ 52
  $1||\Sigma u_i$ 53
    Moore's algorithm, A_M 53
  $1||f_{\max}$ 56
    Lawler's algorithm, A_{LF_m} 56, 164
  $1|r_i \geq 0|\Sigma c_i$ 45
  $1|r_i \geq 0|L_{\max}$ 57
  $1|r_i \geq 0, \tau_i = 1|\Sigma w_i T_i$ 58
  $1|r_i \geq 0, \tau_i = 1|\Sigma w_i u_i$ 58
  $1|prec, \tau_i = 1|\Sigma w_i c_i$ 68
  $1|prec, w_i = 1|\Sigma c_i$ 68
  $1|prec, \tau_i = 1|\Sigma T_i$ 71
  $1|prec|T_{\max}$ 72
  $1|prec|L_{\max}$ 72
  $1|seq.dep.|C_{\max}$ 74, 231
STP (shortest processing time) 44
stable set 31, 179
staffing problem 184
  algorithm for cyclic case, A_{SC} 189
Stallmann, M. 23
Steiglitz, K. 39
strong connectivity 239
Sturm., L.B.J.M. 54
subgraphs 30
  induced 30
  edge induced 30
  spanning 30

tardiness 12, 46
Tarjan, R.B. 63
terminal vertex 65
time–cost problem 212
  formulation 215
  optimization; algorithm A_{TC} 215
  linear time–cost data 212ff
  convex time–cost data 220
  nonconvex time–cost data 220
  pseudo-activities 220
time-complexity function 18
threshold value 23
timetabling problem 190
Tovey, C.A. 105, 125, 129, 181, 183
transitive reduction 113
transitive closure 60, 116
transversal matroid 53
trees 31
  in-tree 58
  out-tree 58
traveling salesman problem (TSP) 10
  Hamiltonian cycles in graphs
    satisfying Bondy and Chvatal's
    condition; algorithm A_{HC} 248
  heuristics 250
    Christofides' heuristic, A_C 250
triangle inequality 250
Tucker, A. 190

unary encoding 20
unit duration times 12, 48
  (unit execution time, UET)
unrelated parallel-processors 11
uniform parallel-processors 11, 100

Valdes, J. 63
Vande Vate, J.H. 23
vertex
  adjacency 28
  degree 28
  in-degree 29
  out-degree 29
vertex coloring 34, 178
  see chromatic number
vertex connectivity 30
vertex cover 33, 206
Vizing, V.G. 36

walk 29